AF522320

# Nutrition of Fruit Trees

# Nutrition of Fruit Trees

Pankaj Goyal

RANDOM PUBLICATIONS
NEW DELHI (INDIA)

**Nutrition of Fruit Trees**

---

ISBN 978-93-5111-360-7
© Reserved

All Rights Reserved. No Part of this book may be reproduced in any manner without written permission.

Published in 2014 in India by

**RANDOM PUBLICATIONS**

4376-A/4B, Gali Murari Lal, Ansari Road
New Delhi-110 002
Phone : +91-11-43580356, +91-11-23289044
e-mail: randomexports@gmail.com, sales@randompublications.com, info@randompublications.com

Reprinted 2018

*Type Setting by* : Keystoneprintads, Delhi-110051
*Digitally Printed at:* Replika Press Pvt. Ltd.

# Preface

Fruit trees, ranging from orange to peach to apple and beyond, are lovely additions to your yard. They are not only attractive, but they earn their keep by supplying you with fresh, nutritional fruit. Each type of tree has its own particular needs, but there are some tips that apply to all fruit trees.

Fruit trees are fertilized to ensure continued growth and fruit production. In the backyard orchard, proper pruning in addition to the application of nitrogen in the spring prior to or at bud break helps maintain this productive status.

Fruit trees need well-drained soil that is at least 3 feet deep. If you find hardpan or thick clay too close to the surface, break it up and amend it with good-quality compost and sand. Fruit trees' roots will not grow in very compacted soil.

To produce good-quality fruit, your tree must receive at least 6 hours of full sun each day. Fruit trees will grow with less sunlight, but they will not produce well, if at all.

Nitrogen is of special importance because plants need it in large amounts. It is also easily lost from soil and fairly expensive to supply. A major factor in successful farming is the grower's ability to manage N efficiently.

Nitrogen has numerous functions in plants, and essentially all life processes depend on it. Nitrogen occurs chiefly in amino acids, proteins, and sugars. The most active nitrogenous compounds occur largely in the protoplasm and nuclei of plant cells. Among them are the enzymes that speed up biological processes.

These are some of the reasons which have promoted us to write this book and we hope it will help in furthering the knowledge about fruit tree physiology.

I thank all members of my team who have helped in the preparation of the book. My special thanks go to "Random Publications" who have published the book.

*– Pankaj Goyal*

# Contents

# 1

# Fruit and Orchard Nutrition

## IMPORTANT FACTORS IN FRUIT PRODUCTION

### LOCATION AND ORCHARD SITE SELECTION

When choosing a potential site for an orchard, issues relating to the land, such as soil type, topography, aspect, slope and irrigation supply need to be considered carefully. Climatic factors for the chosen site including temperature, rainfall, wind, hail, chill units and frost potential also need to be taken into consideration. These factors are discussed in more detail under the following headings.

#### Climate

Site selection for an orchard must take into account local and regional weather patterns. Many deciduous fruit and nut trees are adapted to a Mediterranean climate of cool, wet winters and hot, dry summers. Frosts in spring can injure flowers and affect fruit set and shoot growth. Rain during spring and summer can increase the risk of fungal and bacterial diseases while hail and strong winds can cause physical damage to fruit and limbs. To avoid the risk of sunburn on fruit and limbs, covering trees with shade cloth, sunburn protection sprays and painting of tree limbs may be necessary.

Deciduous fruit and nut trees develop their vegetative and fruiting buds in the summer. As winter approaches, the already developed buds go dormant in response to both shorter day lengths and low temperatures. This dormancy or sleeping stage protects buds from the effects of cold weather. Once buds have started dormancy, they will be tolerant to temperatures much below freezing and will not grow in response to mid-winter warm spells. Buds remain dormant until they have accumulated sufficient chilling units (CU) of cold weather. A chill unit is allocated when temperatures occur within certain parameters (chill accumulation models). When enough chilling accumulates, the buds are ready to grow in response to high temperatures. As long as enough CU have been accumulated the flower and leaf buds develop normally. If the buds do not receive sufficient CU during winter to completely release dormancy, trees may have uneven flowering, poor fruit set and shoot dieback. Fruit and nut species and cultivars have different requirements for CU and the selection needs to be matched to the climate of the area to be planted. When selecting cultivars consideration of the future effects of climate change and likely reductions in CU for the production area.

**Soil Management**

Tree crops tend to be planted on light-textured soils such as sandy loams or loams. Soil types are classified by reference to the proportions of silt, sand and clay, referred to as texture. Because texture may change with depth, the thickness of each soil layer (horizon) should also be considered. Field assessments of soil include texture, colour, aggregation, rooting depth of vegetation, presence of lime or gravel, hardpans and water-tables. Soil analysis in the laboratory may include pH, strength, porosity, water-holding capacity, organic matter, nutrient status, salt content, cation exchange capacity (nutrient retention) slaking and dispersion. A soil test is essential before planting to allow application of soil treatments and amendments which cannot effectively be carried out after planting.

Many tree crops are sensitive to poor drainage (waterlogging) of the soil profile. The perennial nature of tree roots increases the risk of root disease from heavy rainfall, flooding or poor irrigation practice. Shallow topsoils (15-20cm deep) when formed into a treeline bank can provide good drainage, deeper surface soil and when managed carefully will support a healthy root system. Shallow roots can take advantage of the high levels of aeration, temperature, biological activity, and low levels of soil strength as well as being closer to irrigation and nutrients applied at the soil surface. By growing cover crops and adding organic mulch, porosity and stability of the soil is increased allowing frequent irrigation with a reduced risk of waterlogging and disease of tree roots. Shallow soils can also have an important part to play in controlling excessive vegetative growth as roots find it harder to explore deeper due to an impeding layer or an abrupt change in soil texture.

In the DPI Tatura System, to improve drainage and to optimise land use, the topsoil is hilled into a treeline bank approximately 0.5 m high. From soil tests, the specified amount of lime is incorporated and gypsum spread on the surface in a one metre wide strip on the treeline. Ryegrass is sown over the entire orchard. Where compacted (hardpan) subsoil is present, a ripper with a winged-tine is used to till the soil to a depth of 60 cm to create aggregates in the subsoil. The trees are planted and the bare soil, created by the tillage operation, is covered with straw mulch 1 metre wide.

Steps recommended as a guide to setting up a new orchard:

- In late summer/autumn, peg out the orchard treelines accurately and install the irrigation mains.

- Use a road grader to move the topsoil from the centre of the traffic line to the treeline to create a bank approximately 0.5 m high.
- For acid soils (pH <6.0), apply lime (amount determined by a soil test) in a 1 metre wide strip along the treeline and incorporate with a rotary-hoe.
- Install irrigation laterals and microjet sprinklers (output 5-10 mm/hr) and irrigate for 4-5 hours.
- When the soil has drained to around field capacity (2-3 days), cultivate the entire orchard with a tined implement, power harrow or a rotary hoe and smooth the soil surface.
- For dispersive soils, apply gypsum (amount determined by a soil test) in a 1 metre wide strip along the treeline.
- Sow the orchard to ryegrass or a ryegrass and clover mix and irrigate for 2-3 hours.
- In late winter, mow the grass/clover sward close to the ground.
- In compacted subsoils, use a winged-tine ripper to a depth of 60 cm in 3 passes in increments of 20 cm.
- Cultivate the 1 metre wide strip with a tined implement, power harrow or a rotary hoe and smooth the soil surface.
- Plant the trees without compacting the soil and water-in lightly to prevent slumping of the soil.
- Apply a surface mulch of straw in a 1 metre wide strip on the treeline.
- In spring/summer, use herbicides to control weeds in a 1 metre wide strip on the treeline.
- Slash the orchard and deliver the clippings onto the treeline to supplement the straw mulch.

## Salinity and Sodicity

Fruit and nut crops generally have a low tolerance to salt (Table 2). Soil salinity can be measured by electrical conductivity (EC) of a soil solution present mainly as sodium chloride (NaCl2) and high levels can result in soil structure damage due to sodicity. A sodic soil (soils high in sodium) will disperse (run together) upon wetting restricting moisture penetrating the soil and set hard upon drying limiting the movement of water, oxygen and the growth of tree roots. Over years of crop production and irrigation, many of our horticultural soils have accumulated salt in the soil profile. In addition, saline groundwater levels can rise to the soil surface or to the root zone during wet weather and flooding of orchards.

The chloride (Cl) component present in saline soils can be leached out of the root zone of the tree by irrigation and rainfall. However, the sodium (Na) component can be left behind firmly attached to clay particles. Affected soils have poor infiltration and drainage resulting in waterlogging, increased runoff, poor water storage and surface crusting. Gypsum or calcium sulphate is used to manage soil salinity and sodicity.

**Table. Salt tolerance of Horticultural Crops (Source: Salt Action Salinity Calculator for Horticulture, 1999).**

| Crop | Soil salinity EC (dS/m) of the saturation extract | |
|---|---|---|
| | Threshold | 25% yield loss |
| Almond | 1.5 | 2.8 |
| Apple & pear | 1.7 | 3.3 |
| Apricot | 1.6 | 2.6 |
| Olive | 2.7 | 5.5 |
| Orange | 1.7 | 3.2 |
| Peach | 1.7 | 2.9 |

## PLANTING STOCK

Most orchard trees have the desired cultivar (variety) grafted onto a rootstock or are grown as cuttings on their own roots to ensure cloning of selected traits. Nursery trees should be selected from high-performing cultivars, genetically uniform, free of pathogens, trained to a nominated style and managed to come into production at an early age. The root system on these trees should be fibrous and dense and handled in such a way as to ensure there is no check in tree growth after transplanting to the orchard. The roots may be treated with beneficial fungi and bacteria to resist root infections and to improve the uptake of water and nutrients by the tree.

The use of vigorous stocks are useful for early growth in the life of an orchard but vigorous rootstocks increase the cost of pruning and canopy management. Vegetative growth must be controlled and the resources of the tree directed into fruit and nut yield and quality. This can be achieved using pruning and training methods as well as irrigation management. Dwarfing rootstocks are available in some tree species and may prove useful e.g. apple rootstocks from the East Malling series, when grafted to commercial apple cultivars, produce compact trees with high yields of fruit.

### Variety Selection

Variety selection is critical and needs to be based on market demand. The choice also depends on region and climate e.g. low chill, susceptibility to frost or rain (some fruit varieties are prone to splitting of the stone and/or the flesh). Some stone fruit quickly become unfashionable and so there is a need to review and renew the varieties every few years by replanting.

### Irrigation Management

Water availability is essential to consider as commercial orchards must have reliability of supply and cannot produce optimum yields without irrigation. Water management is one of the largest and most important inputs into an orchard. Dry summers combined with shallow, fragile soils means that in mid-summer following irrigation, there may only be around a two or three day water supply at optimum levels held in the soil. Orchard trees are dependent on a regular irrigation supplied on demand, and dictated by the prevailing weather conditions.

Careful scheduling of irrigation based on a calculation of tree water use from meteorological data, monitoring of soil water status and measurement of the internal water potential of the tree are the tools to decide the amount and frequency of the irrigation applied. Water Budget Models are available to calculate tree water use. Advances in irrigation technology have improved water use efficiency through the use of microjet or drip irrigation. Use of microjets produces uniform distribution because water is spread through the air, wets the soil slowly and reduces structural breakdown but there is potential for more water loss due to evaporation. Drip irrigation has less even distribution over the root zone because the soil is saturated under the dripper and lateral movement of water is dictated by soil properties as well as by gravity.

Irrigation can be managed in strategic ways to reduce vigour without penalising orchard quality and yield. Regulated Deficit Irrigation (RDI) is a technique where water is withheld at strategic times to reduce excessive shoot growth in the tree. In some crops, RDI can give water savings without a yield penalty whilst fruit size and quality are unaffected. The objective is to produce enough vegetative vigour to ensure future bud development and to channel the remaining energy of the tree into orchard yield.

## TREE TRAINING

Tree training may involve pruning, pinching, tying, cincturing, notching, bending, twisting and spreading of shoots plus the use of growth regulators. The size, angle, placement and type of branching must be manipulated to create a specific tree architecture that is efficient and uniform in capturing sunlight throughout the orchard. The amount of intercepted sunlight in the orchard has a profound effect on shoot growth, flower formation, fruit set, fruit development, fruit yield and quality. For optimum light interception the maximum tree height in summer should not exceed 80% of the row width.

Trees may be trained as a vase, bush, central leader, Tatura Trellis or espalier or to the many other styles available. Trees may be free-standing or alternatively in a hedgerow where trees touch each other within but not across the row. Trees trained into a central leader (pyramid) shape, e.g. pome fruit and walnuts or Tatura Trellis e.g. stone fruit, will intercept more sunlight than umbrella-shaped or vase trees through the increased surface area of the canopy and by less shading of the lower parts of the tree.

## NUTRITION

For function and growth, plants obtain most of the mineral nutrients from the soil-water solution. The major nutrients needed are nitrogen, phosphorus and potassium. Secondary nutrients are calcium, magnesium and sulphur whilst micronutrients are iron, manganese, zinc, copper, boron, molybdenum and chlorine.

Soil texture, minerals and organic matter that make up the soil mass have a strong bearing on the buffering capacity to hold nutrients and resist the loss of nutrients due to leaching. Organic matter plays an important part in controlling the availability of nutrients, especially nitrogen, phosphorous and sulphur. Sandy soils have less buffering capacity and are readily leached of nutrients. Clay soils have a strong buffering capacity and because they tend to drain slowly, nutrients are not leached readily but de-nitrification (loss of nitrogen) can occur under waterlogged conditions.

Soil pH affects both the availability and absorption of mineral nutrients. Measurements of soil pH can be a good guide to the diagnosis of nutrient deficiencies and should be corrected before nutrients are applied. Low pH (<5.5) may result in deficiencies of calcium, magnesium, phosphorus, or molybdenum and perhaps excesses of manganese, iron, or aluminium. High pH (>7.5) may result in deficiencies in manganese, iron, zinc or copper. Some nitrogen fertilizers will acidify the soil over time so the choice of product must be matched to soil pH.

A soil analysis will indicate the level of nutrients present, but not necessarily available to the tree. It is essential that an annual program of leaf or sap analysis is carried out to provide the best guide of plant nutrient levels to tailor fertilizer application to meet the plant needs. Fertigation where soluble nutrients are injected into the irrigation system provides a cheap and efficient means to apply nutrients to the orchard. As a general rule a mature, high-yielding fruit or nut orchard may remove between 200 – 300 kg/ha of pure nitrogen each year. To maintain this level of production the nitrogen needs to be replaced by application of an appropriate fertilizer.

### Pollination

Fruit and nut trees may have male and female parts as one flower or as separate flowers. Pollen released by the male flower may not coincide with a receptive female flower on the same tree. For effective pollination, many fruit and nut trees require a different cultivar (variety) or, in some cases, a different species from the same plant family. Tables of suitable pollinators can be accessed through nurseries, texts or from the internet. Most fruit trees are pollinated by bees and other insects and to ensure good yields bee hives may need to be introduced into the orchard. Most nut trees are pollinated by wind so do not need bees. The exception in nut trees is the almond which is also pollinated by bees. The number of pollinator trees versus the rest of the orchard may be from <5% to 50% depending on the species and variety. Some varieties are self-fertile and do not need separate pollinator trees.

## INTEGRATED PEST AND DISEASE MANAGEMENT

All fruit and nut crops will have some potential pest and disease problems. It is essential to monitor crops for pests and diseases as well as the incidence of beneficial insects. Integrated Pest Management (IPM) is an effective combination of chemical, cultural (such as, farm management practices) and biological methods to keep, weeds, insect pest numbers, disease pressure and other crop production problems low enough to prevent significant economic loss. Crop monitoring is carried out by visual inspection of the crop for the presence of pests & diseases, crop damage, and beneficial insects. Monitoring can also be done by setting up traps for specific pests and these will need to be inspected regularly. It is important to consider that if only low levels of pests or diseases are present it may not be necessary to apply any control practices and existing pests may be controlled by beneficials (e.g. beneficial insects) accepting that the pest or disease may not cause economic damage.

If control is warranted then it is important to consider what beneficials are present, are they controlling other pests and are there targeted, soft chemicals (that protect the beneficial insects) available. For some pests such as aphids or caterpillars there are targeted pesticides that will not kill a number of the beneficials but will control the targeted pest, e.g. specific aphicides. It is important to remember that if a broad spectrum chemical is used it will not only kill the targeted pest but also a range of insects that control pests. Adoption of IPM helps protect the environment for chemical use may be reduced, or replaced by less toxic, targeted chemicals and/or, the introduction of natural predators. Fruit fly is an important consideration depending on location and if present, there will be a need to meet control and management requirements to supply different markets. It also likely that due to climate change and warmer winters that there will be an increased risk of infestations. Other key pests include Light Brown Apple Moth, Codling Moth, Two-Spotted Mite and Carpophilus Beetle.

### Chemical Control

The spray program needed for fruit and nut production will be dependant on a range of factors such pest levels, weather (temperature and rainfall), and quality requirements for the crop i.e. what is the tolerance level for damage to the marketable crop. Some organisms have been shown to be resistant to certain chemicals or chemical groups. It is important to minimise the development of resistance using a number of resistance management strategies that have been developed. Chemical labels also give information on how to minimize resistance that may include rotating chemical groups or limiting the number of sprays per crop.

There are a range of legal requirements that must be complied with when using chemicals. These include having the appropriate chemical user certification, adhering to withholding periods (length of time after spraying that a crop can be harvested) and strict rules on record keeping. More information about chemical use can be found on the DEPI website or with the regional Chemical Standards Officers (contact the DEPI call centre on 136 186, or see the DEPI website). It is also advisable to undertake a Farm Chemical User Course which is regularly run by training institutions such as TAFE. This is also a precursor to obtaining an Agricultural Chemical User's Permit, which is essential for the legal use of some chemicals.

Advice on chemical control of pests and diseases can be obtained from authorised dealers in agricultural chemicals or registrations from Australian Pesticides and Veterinary Medicines Authority f**www.apvma.gov.au.

## ORCHARD INFRASTRUCTURE AND EQUIPMENT

There is a significant long term investment in infrastructure in orchard development both in trees and equipment. A list of basic farm infrastructure and operating equipment may include the following:

- Trees
- Irrigation infrastructure and controller
- Fertigation tank and injector
- Machinery shed
- Grading and packing machinery
- Coolrooms, CA, forced air or hydro cooling
- Tractor
- Orchard sprayer
- Herbicide sprayer
- Slasher
- Forklift
- Bulk bins & bin trailers
- Trellising
- Ladders
- Pruning equipment
- Dryer
- Netting (shade, hail and bird)

### Quality Assurance

Quality Assurance (QA) allows a business to document in a formal management plan all of the actions that are needed to deal with risks identified in producing a product for market. This allows the business to define the real risks, manage them appropriately and be able to prove through verifiable records and external auditing that risks and management processes have been identified and followed. There is an expectation from major retail chains that suppliers will have a QA system in place and some retailers have their own QA systems. QA adoption will help establish the grower as a reliable supplier of quality produce services in national and international markets.

**Markets**

Produce can be sold at the farm gate, at farmer's markets, to processors, direct to supermarkets, through wholesale markets or export markets. Growers may choose to sell their produce themselves, through cooperatives, agents or growers' organisations. Where the product is sold will influence a range of other factors, such as, which quality assurance system is needed, how product is packed, transport costs, quality specifications and payment options. Selling direct to the consumer such at a roadside stall or farmer's market has the lowest costs and lowest number of constraints. Another option is selling to restaurants and other food service providers. Conversely the volume of product able to be sold to these markets is lower, demand may fluctuate and time needs to be spent on marketing and distribution not just production.

Selling direct to a retailer cuts out the middle-man and may be based on a contracted price so there is some surety of income. However it will also mean having to meet exacting standards, packing the product the way the buyer dictates and probably having to comply with an audited QA system. Retailers will also want volume and continuity of supply to service all their outlets. This may lead to penalties if obligations of volume and quantity cannot be met, rejected product having to be repacked for other markets, and leaves little option for taking advantage of higher-priced markets elsewhere.

There are wholesale markets in all capital cities and they are perhaps the easiest option for selling product. This can be done directly by taking out a market stall in the growers section of the market or by supplying an agent or merchant who will market the product for the grower. There are different requirements for agents and merchants and these can be obtained from market authorities. The wholesale markets can supply a list of agents or merchants. Some agents specialise in certain crops or regions and it is worth talking to range of them. It is important to have an understanding of payment policy, issues relating to unsold product or rejected product, QA requirements, packaging, and freight arrangements.

**Labour Requirements**

Fruit orchards have a high requirement for manual labour to carry out detailed management of the trees including pruning, training, thinning and harvesting. Nut trees require less manual maintenance and crops are harvested mechanically. The labour requirement varies significantly for different crops depending on picking frequency, pruning and training requirements. In choosing what crop to grow the level and frequency of labour needed to manage the crop should be taken into account. It is essential to have enough labour available to carry out key operations at critical production times. Fruit and nut orchards have a range of machinery operations for slashing, weed spraying as well as spraying for pests and diseases. Nut orchards have fewer pests and diseases to manage and therefore require fewer machinery operations.

### Post harvest/Cooling Requirements

Post harvest cooling and refrigerated storage of fruit is important to maintain shelf life and preserve quality and it is essential to have access to cool rooms to pre-cool fruit and hold fruit prior to shipment to market. Controlled Atmosphere Storage (CA) can be used to extend the shelf life of many fruits even further. A combination of adjustment of temperature (T), oxygen (O2), carbon dioxide (CO2) and relative humidity (RH) levels can extend the shelf life of fruit for up to 12 months. Typical figures for storing peaches require -0.5 – 00C, 2% O2, 4-5% CO2 and 95% RH.

### Climate Change/Variability

Climate change and associated global warming will require management of the higher potential risk of sunburn, lack of chilling and water stress of orchard trees. Climate change may also increase climate variability and the likelihood of extreme events such as heat, hail and late frost. The impact of climate change on rainfall and the security of irrigation water supply is one of the key factors to consider for sustainable production of tree crops into the future.

## MANAGING TREES THAT SHOW SIGNS OF LIFE

If trees show signs of life (bark not shrivelling and new shoots starting to emerge), the irrigation system needs to be re-established so that trees can be irrigated and enable regrowth to mature before autumn leaf fall. Trees may need to be protected from sunburn, particularly where defoliated. Trunks and large limbs can be painted with white wash or diluted water-based white paint to minimise sunburn problems. Remaining fruit should be removed to prevent pest and disease build-up and unwanted stress on the trees. Pruning should be delayed until regrowth has been established. Fertiliser should also be with held until there is sufficient growth to utilise it. It is essential not to force growth with extra fertiliser.

## LONG TERM VIABILITY OF FIRE DAMAGED TREES

Experience from fire damage to South Australian orchards in the Ash Wednesday fires indicates that whilst trees may survive the fire damage they will take at least 2-3 seasons to return to cropping and may have seriously reduced potential. Trees may also continue to collapse from secondary problems after producing new shoots. This means that even though trees may survive fire damage, it may not be economically viable to try to continue production in a block.

Decisions about long term block viability and possible replacement strategies should be made in consultation with technical specialists and your insurance company. Decisions should take into account the severity of the damage through the block, tree age, variety, rootstock and planting densities. Younger trees, if not killed outright, may be easier to re-shape into productive trees and worth saving whereas older trees may not.

**Key Messages for Fire Affected Fruit Trees**

- Assess whether the block is worth saving as soon as possible, based on the extent of damage and signs of regrowth.
- Assess if damaged trees still have a healthy cambium (layer of tissue beneath outer bark)
- If cambium is still alive and trees show signs of life, sunburn protection of trunk and limbs is essential
- If block is judged viable, repair or replace the irrigation system and irrigate as soon as possible.
- Remove any remaining fruit to prevent pest and disease build-up and unwanted stress on the trees
- With-hold fertilizer until there is sufficient growth to utilise it. Do not force growth with extra fertilizer
- Do not prune out damaged limbs until regrowth has been established.

## ORCHARD NUTRITION IN SOIL AND LEAF ANALYSIS

Chemical analyses of soil and leaf samples give an objective guide that helps us to choose the most appropriate orchard fertiliser. If a fertiliser program is based on analytical results rather than on routine applications of NPK fertiliser every year irrespectively of the trees' needs, then it is likely that money can be saved and the quality of fruit will improve. For example, a survey of 20 apple growers from southern Victoria showed that growers often used too much nitrogen, to the detriment of fruit quality.

Data from analysis of soil is not used for the same purposes as that from leaf analysis. Soil analysis is best used before planting. The soil pH, its salt content, and the phosphorus and potash content are more easily manipulated before planting than they are once the block is planted. In planted blocks, soil analysis still gives useful background information, especially if salt problems or problems associated with low or high pH are suspected. Once the trees are established, leaf analysis gives a better guide to the tree's nutrient status than soil analysis does because it shows what is happening in the tree itself. Soil analysis only shows what is happening to the reservoir of nutrients the tree has to draw on.

## SAMPLING THE SOIL

Soil samples for chemical analysis should be taken from at least 30-40 different spots in an orchard block and mixed together in one container. This should be done because soil fertility in an orchard block can vary as a result of previous lime and fertiliser applications. It is useful to sample two depths, for example, 0-150 mm and 150-300 mm, to indicate how far nutrients have penetrated and also to find how the fertility of the soil changes with depth. The State Chemistry Laboratory makes six determinations on soil samples. These are generally all that are needed.

### Description of Soil

This provides a visual identification of soil colour and texture (sand, loam, clay loam, clay). These factors can affect lime and fertiliser recommendations.

### Soil Acidity (pH)

This will indicate how much lime is needed. Fruit trees can grow well even on rather acid soils, but if acidity drops below pH 5.5 valuable soil nutrients become less available and may be lost. Lime will help the growth of grasses and clover and hence preserve the content of organic matter in your soil. Sufficient lime should be added to bring the pH to above 6.0. Various types of lime are available. Dolomitic lime (which contains magnesium) should not be applied unless it has been found that the soil is deficient in magnesium.

### Electrical Conductivity (EC)

This measures the total amount of soluble salts in the soil, both naturally occurring salts such as common salt and gypsum, and those derived from fertiliser. A conductivity of 0.22 deci-Siemens (ds) per metre is equal to 0.033% of Total Soluble Salts. Values above 0.3 ds/m may retard the growth of roots but other factors, such as soil type and the kind of salt present, must also be considered.

### Sodium Chloride (NaCI - Common Salt)

This is determined if EC values are high. More than 0.05% NaCl in sandy soils and more then 0.1% in clay soils could harm fruit trees; such values indicate that better drainage would improve the growth of trees and prevent the accumulation of salt in the soil.

### Available Potassium (K)

Potash fertiliser is unlikely to be needed if values are above 100 ppm in sandy soils or above 150 ppm in loams and clay loams, provided these levels are present throughout the top 300 mm of soil.

### Available Phosphorus (P)

Values above 20 ppm (30 ppm in clay soils) in the surface sample suggest that superphosphate is unlikely to be needed. But unless phosphorus values are also high in the subsurface soil, it would be useful to add superphosphate before planting young trees to supply them with a reserve of phosphate for the future.

### Soil Structural Tests

Other tests, such as slaking, dispersion and extractable cations (calcium, potassium, sodium and magnesium) may be useful in particular situations. Cation analysis indicates existing or potential soil structure problems and can be used in determining how much gypsum should be added to aid root and water penetration.

## SAMPLING THE LEAVES

### Time of Sampling

Leaf samples should be collected during the most favourable period for the particular crop; this is usually mid-summer.

**Table. Standardised tissue sampled and the best period to sample different crops in Victoria**

| Crop | Tissue sampled | Best sampling period |
|---|---|---|
| Apple, peach, pear, plum, sweet cherry | Mid-shoot*leaves | January-February |
| Apricot | Mid-shoot leaves from first main growth flush of current season | January-February |
| Orange | Spring-flush leaves | January-March |

*Mid-shoot leaf: a fully expanded leaf from the mid-portion of the current season's terminal growth. Shoots selected should be of average length and vigour (not water shoots) and free from disease, insect or mechanical damage.

### Sample Collection

Obtain a bulk sample of 200 leaves from the block by sampling 20 trees in a systematic pattern (for example, by following two diagonals in an "X" pattern across the block). Take 10 leaves from each tree, from around the periphery of the tree at about shoulder height. Record those observations or information on orchard history that may aid the accurate interpretation of the analytical data - such as visual symptoms, crop load (heavy, light), tree vigour, soil management and fertiliser practices, irrigation, soil drainage and possible foliar spray contaminants. If a deficiency/toxicity is suspected, two blocks from nearby may be sampled; the suspect block and a "control" block of healthy trees of the same crop, cultivar and soil type. Comparison of the analytical results may confirm or deny the deficiency/ toxicity. Transfer samples to the laboratory in polyethylene bags packed in crushed ice in a portable ice-chest, then store under refrigeration (0°C-4°C).

**Interpretation of leaf Analysis**

Results of the chemical analyses are compared with standard values, so that a rating of the sampled orchard's nutritional status is obtained. Tables 2.1 - 2.6 give standard values for a range of fruit trees. The terms deficient, low, normal, high, and excess are defined as follows:

- Deficient: Deficiency symptoms are present; level is too low for best performance.
- Low: No visual symptoms; level is below normal and may be insufficient for best performance.
- Normal: No visual symptoms; level is normal and should be adequate for best performance.
- High: No visual symptoms; level is above normal and may be causing a nutrient imbalance.
- Excess: Toxicity symptoms may or may not be present; level is too high for best performance.

**Orchard factors affecting the results of leaf analysis**

Seasonal conditions and orchard management have important effects on the content of nutrient in leaves. These effects may be just as large as those caused by the addition of fertiliser. The following factors affect the content of nutrients in leaves:

Dry seasons- In dry seasons the concentration in leaves of almost all elements (except manganese, sodium and chloride) will be lower because of decreased uptake. In dry years, manganese, sodium and chloride levels may be higher than normal. Weeds- Competition from weeds and grasses will lead to lower levels of nitrogen in leaves. Conversely, the use of herbicides around trees will increase the uptake of nitrogen and its concentration in leaves often more effectively than would nitrogen fertiliser. Irrigation also tends to increase nitrogen levels in leaves but decrease calcium. Waterlogging- Waterlogging caused by poor drainage and/or excessive irrigation can cause many problems including loss of nitrogen because of denitrification. Waterlogging also reduces the uptake of calcium and potassium, but may increase the uptake of sodium.

Crop load- Heavy crops are associated with higher concentrations of nitrogen, calcium, and magnesium in leaves. Conversely, light crops are associated with lower concentrations of nitrogen, calcium and magnesium in leaves. Variety/rootstock- Red Delicious apples have high levels of nitrogen and low levels of calcium compared with other varieties of apples. Dwarfing rootstock's, such as M9 and MM106, result in the scion leaves having higher levels of calcium than trees on seedling rootstock. Fungicides/foliar fertilisers- When fungicides or foliar fertilisers containing zinc and/ or manganese are used, the levels of these two elements will be high even after the leaves are washed. When copper fungicides are used during the season, they will raise the levels of copper in leaves. Other elements- High levels of one element may result in low levels of another. A good example is the relationship between potassium, calcium and magnesium. Applying too much potash fertiliser may result in magnesium deficiency. Applying magnesium may result in bitter pit or poor performance in storage caused by low levels of calcium in the leaves and fruit.

## SOIL AND LEAF SAMPLING FOR NUTRIENT ANALYSIS

Measuring the nutrient needs of orchards should be done by using both soil and leaf (also known as foliar) nutrient analyses. Soil analysis is best done before planting a new orchard so that lime and fertilizers can be incorporated deeper into the soil profile. Adjusting soil pH and getting less mobile nutrients into the rootzone is difficult once the orchard has been established, though there are times when these practices can and should be done. Soil sampling should continue throughout the life of an orchard every five or so years to monitor soil pH. In peach blocks where sulfur has been repeatedly applied for disease control, soil pH should be checked more frequently to ensure that the pH hasn't fallen out of the acceptable range.

Since soil pH and nutrient content changes through the soil profile, soil samples should be taken from a consistent depth. Many labs will recommend a soil depth to sample, and then use that depth to calculate fertilizer and lime recommendations. The Virginia Tech Soil Lab recommends samples be taken from the top 6-8 inches of soil. When soil sampling prior to planting, it is recommended to take separate top- and sub-soil samples. Additional information on soil sampling can be found in the Virginia Tech Agronomy Handbook.

Soil nutrient analysis is a measure of the "plant available" minerals in the soil. However, due to the complex nature of the soil-root interface, minerals that are "plant available" are not necessarily taken up by plants and therefore soil testing does not always provide an accurate measure of the tree's nutrient status in the orchard. Additionally, most soil analyses do not directly measure nitrogen content, but instead base nitrogen fertilizer recommendations on soil type and organic matter content. Additional information on soil nutrient analyses for orchards can be found in Virginia Tech's

- Soil Test Note 1: Explanation of Soil Tests
- Soil Test Note 10: Commercial Apple Production
- Soil Test Note 11: Commercial Peach Production

For commercial growers within Virginia, the Virginia Tech Soil Testing Lab conducts soil analyses at no charge. However, there is an additional charge for measurement of soluble salts ($2) and organic matter ($4). For the quickest processing time, be sure to use the official soil sampling boxes. Sample boxes are available from your local Virginia Cooperative Extension office. Also, new this year the Virginia Tech Soil Testing Lab will send you results via email. Please use the most recent submission form when submitting samples.

For mature fruit trees, leaf mineral content is a more accurate measure of the nutrient status than soil analyses. Leaf analysis can help determine if additional fertilizers are needed before there are any visual signs of deficiency in the tree. Leaf samples should be taken when most vegetative growth has subsided (terminal buds are set), which for most bearing apple trees happens in late July or early August. For the most part, by the time leaf samples are taken and analyzed there is little that can be done to impact the current season's crop and next year's flower buds. Therefore, to maintain optimal performance of your orchard, it is recommended to perform a leaf analysis for each block once every three years to ensure that you are consistently maintaining adequate nutrient levels. Below are some additional recommendations for taking accurate leaf samples.

- Collect 60-100 leaves to get a representative sample. Leaves should be taken at about the midpoint of the new growth and from all actively growing parts of smaller trees and between waist and head high on larger trees.
- Choose between 8 and 10 separate trees that are representative of the block.
- Don't mix leaves from different varieties, rootstocks, ages, soil types, blocks, or farms.
- Sample poorly growing or otherwise problematic trees separately from healthy trees, so that a comparison can be made between them.

# 2

# Plant Nutrients for Citrus Trees

## INTRODUCTION

Seventeen elements are considered necessary for the growth of green plants: carbon (C), hydrogen (H), oxygen (O), nitrogen (N), phosphorus (P), potassium (K), calcium (Ca), magnesium (Mg), sulfur (S), iron (Fe), zinc (Zn), manganese (Mn), boron (B), copper (Cu), molybdenum (Mo), chlorine (Cl), and nickel (Ni). These elements are commonly referred to as essential plant nutrients. An element is considered essential if a plant cannot complete its life cycle without it, and if the problem that develops in its absence is curable only by its addition. Plants obtain C, H, and O from carbon dioxide and water. The remaining elements, called the "mineral nutrients," are obtained from the soil.

Mineral nutrients are classified as macronutrients and micronutrients. The term "macronutrients" refers to those elements that plants require in large amounts (N, P, K, Ca, Mg, S). The term "micronutrients" applies to plant nutrients that are essential to plants but are needed only in small amounts (Fe, Zn, Mn, B, Cu, Mo, Ni, Cl). The use of the terms "minor element" or "trace element" for some of the nutrients can be misleading. For example, the role of Fe in plant metabolism should not be considered less important than the role of K. Iron deficiency can result in total crop loss, so its role is not a "minor" one, and it is not of minor importance. The difference between Fe and K is in the amount required by plants, so the use of the terms "micronutrients" and "macronutrients" is more appropriate.

## HOW CITRUS TREES ARE GROWN

Plant a citrus seed from a Valencia orange and what do you get? A Valencia orange tree just like the one your seed came from? Nope -- you can never be sure exactly what you'll get because plants grown from seed are all slightly different. (There are, however, some exceptions. Read on.) Trees vary in individual size and in the size and sweetness of their fruit. And, some trees may be more resistant to diseases. Naturally, citrus growers, to ensure the best possible crop, want to make sure all their trees are of the same high quality. To do this, each new tree they plant is grown, or propagated, not from a seed but by grafting or budding. (More on this later.) In most citrus trees, the scion, or top of the tree, is a different variety from the roots or rootstock of the tree. Citrus growers plant trees whose tops will grow Washington navel oranges or Eureka lemons on a rootstock that has special characteristics like disease resistance, quick to bear fruit or restricting tree size for easier harvesting.

Nurserymen begin the process of propagating a new citrus tree by planting a seed for the rootstock. Most citrus have an unusual characteristic -- they can produce trees that are genetically identical to the parent through a natural process in seed development called nucellar embryony. While many rootstocks are produced by seed, because of nucellar embryony, they are all the same. After the rootstock is a year old, a single bud is taken from a branch of the desired scion variety and inserted into the bark of the young seedling. This bud grows into the top of the tree that produces the fruit.

## FRUIT DEVELOPMENT

Citrus flower buds begin to form in early winter and develop through late winter and spring. Most flowers don't result in the formation of fruit because more than 99% of them usually fall off. Because the number that do become fruit depends largely on temperature and moisture, adequate water is very important during and immediately after flowering. Most pollination is done by insects, but due to a rather interesting process natural to some citrus called parthenocarpy, fruit can develop without pollination. Some varieties, such as the Clementine Mandarin, require cross pollination with another citrus variety. Why do some citrus, like Valencia oranges, have seeds, and others, like Washington navels, generally have none? Parthenocarpy is the main reason. If parthenocarpic flowers are not pollinated, they generally don't produce seeds. Some, like the Washington navel, don't produce viable pollen. After bloom, fruit develops from five to 18 months, depending on the variety and growing area. Unlike many other types of fruit, most citrus can be left on the tree without becoming overripe.

### Soil Type

Citrus will grow in most soils from sandy to adobe clay, provided it drains well. Sandy soils must be watered and fertilized more frequently than soils with a higher clay content and growers can add organic matter such as manure or compost to improve water and nutrient holding abilities.

### Irrigating and Fertilizing

Water quality is very important. Water high in salt content, common in some desert regions, can cause injury to leaves, burning leaf tips and margins.

Lower levels of salts can cause the tree to grow poorly or to produce fewer or smaller fruit. Nutrients that citrus needs in relatively large amounts are nitrogen, phosphorous, potassium, sulfur and calcium. In lesser amounts, citrus requires iron, magnesium, copper, zinc, manganese, molybdenum, boron and chlorine. In California and Arizona soils, usually only nitrogen must be added, most often in late winter to meet the demands of spring bloom.

## Pruning

Unlike most other fruit trees, citrus trees don't require regular pruning. Commercial growers trim tree tops to keep them smaller so they are easier, safer and less expensive to pick, and "hedge" the sides to let more sunlight into the trees to improve yields. Hand pruning opens up the trees to allow more sunlight into the center.

## Frost Control

Protection from frost is critical. Most citrus will freeze when fruit temperature drops to 27-28°F. The main methods of frost protection in California and Arizona are by wind and water. Wind machines -- large fans on poles about fifty feet above the grove -- are turned on when the temperatures near freezing. The fans mix the slightly warmer air above the grove with the colder air near the ground, which warms the air around the tree. By applying water, the heat built up in the soil during the day is lost more slowly, and air temperatures around the fruit stay warm a little longer. A few growers still use oil-burning orchard heaters, but this once common method is seldom used now because of the cost.

## Packing Citrus

Getting fruit from the tree to the table is complicated. A citrus fruit is a living organism which must be protected from injury, decay and water loss on its journey across the country or across the ocean. Fruit are carefully picked by hand and put into bins which hold about 900 pounds. The bins are hauled by truck to a packinghouse, where the fruit is pre-graded to eliminate the obvious culls (bad fruit) and washed to clean away field dirt and dust. The fruit's natural wax, removed during the washing phase, is replaced by a food grade, non-animal source wax to restrict moisture loss and extend shelf life. The fruit is then graded, most often by electronic camera systems. After grading, the fruit is divided by size and the first grade fruit is stamped or labeled "Sunkist." An unmarked second, or choice, grade is also packed. The remaining fruit is sent to the juice plant. Most oranges and grapefruit are packed with robotic carton packing machines in 40 pound cartons. Lemons and most tangerine cartons are volume-filled with a set number of fruit. The cartons are stacked on pallets and placed in a pre-cooler before being loaded on trucks or rail cars for transportation to markets.

## GROWING CITRUS TREES

A lot of people worry that citrus trees will not be easy to grow or will take forever to bear fruit. Put that thought behind you because neither of those assumptions is necessarily true. Even if you live where winter can toss a frost or two your way you can still enjoy fresh, organic citrus grown at home. For those who live in frigid climates, you too can succeed at growing citrus quite easily.

One of the biggest problems anyone has with homegrown citrus, such as lemons, oranges and clementine trees, is that they don't purchase super vigorous, healthy young trees to begin with. No matter what type of plants you try to grow, it will be much harder to do so easily if you purchase weak, immature or sickly plants in the first place. Start with a plant that has a well developed root system and not a twig in a dixie cup. The next step to success is to decide if you live in a location that the citrus trees can be planted outside in your garden soil or is best grown in a pot. If frost and below freezing temperatures is a possibility, you will want to container grow your chemical free citrus so you can winter it indoors. No matter where you will be growing your new lifetime supply of delicious Vitamin C, your young tree will need very good drainage and lots of sunshine. So, if container growing for wintering indoors, it is necessary to plant these types of fruit trees only in pots that have drainage holes in the bottom.

### Selecting Your Outdoor Planting Site

Obviously, most homeowners don't have lots of choices as to where you will plant your citrus trees. Except for some very southern regions, if you live north of zone 8, you should be prepared to container grow your citrus. In zone 8, it is best to plant citrus trees on the south or southeast side of your home. This provides added protection from northwestern cold fronts. Planting your new citrus tree close to your home's walls is also advisable as the heat the structure gives off will work to give it added warmth during such bouts of chillier than normal temperatures. Do be sure and place the trunk of the citrus plants a minimum of 6-8 feet away from walls, fences and paved surfaces like drives and sidewalks. Also, space your citrus trees 6-8 feet apart for ease of picking and maintenance. Naturally, should you have a septic system, choose your planting location away from it and your drain field. In the city, you won't want to place such a deep rooted plant near or over sewer lines as they can cause you a nightmare of clogged sewer lines as your new trees mature. You should never plant citrus trees beneath other trees, as they need full sunlight for ultimate growth and fruiting.

## The Right Soil Requirements for Citrus Trees

All citrus trees require deeply draining soil. This means runoff water leaves the area and the drainage beneath the surface is excellent and will not saturate the root system. Finding out what kind of soil you have to a depth that your new citrus will grow can be as simple as taking stock of what the other trees in your yard look like. Healthy shade and ornamental trees in your landscaping is a good sign that your soil should be great for growing citrus trees. You can also plant your new citrus orchard in raised beds to create better drainage below the surface. Your soil pH should be in the range of 6-8 and not be high in salt, as citrus won't be happy about dwelling in the wrong conditions. You can get your soil pH tested easily with pH test kits from your local garden center.

## Planting Your New Citrus Trees in the Yard

Unlike many types of nursery container grown plants, citrus trees will take a long time to root past the potting medium and into your ground soil. To help them spread their roots faster and improve initial new growth rates, remove the exterior of the potting soil just as you are ready to slip the tree in its installation hole. Don't take all of it off, that would be harmful – just the outer half inch or so to allow the roots to be surrounded with your ground soil. This is very simple to do by washing it off with the hose immediately before sinking your young citrus tree in its new real soil home. You want to cut a 4 ft. circle out of your lawn as the start of your planting hole. Discard the grassy top layer and dig your hole half again wider than each citrus tree's root ball. When planting citrus you will want to 'plant it high', meaning you want the top of the finished planting surface to be crowned 1 inch higher than the lawn. This creates your surface run-off of water, and will be a great assistance to your tree in wet weather.

You won't need to do a lot of amending with good soil, so don't worry about adding compost and peat unless your soil is not good quality for drainage. With your tree set in the planting hole, fill around it half way to the top and thoroughly saturate the back fill soil. This helps it to settle around the roots of your new citrus tree's ball. Finish filling the hole and tamp your loose soil into place lightly. Cover the roots with no more than 1 inch of ground soil to seal the potting soil from direct air contact. Soilless medium used in production nurseries is design to air dry too rapidly for home growing, blocking the air with a layer of soil stops you from having rapid drying problems. To let your citrus tree be able to have deep root watering, build a water retention ring around the perimeter of your planting circle that is 5 – 6 inches tall and 6 – 8 inches thick. When completed, your watering ring should be slightly wider than the removed circle of lawn for your planting hole. Don't skimp, should you be short on soil left from the planting, buy some from your local garden center or borrow it from your garden. Fill your new citrus tree's watering basin with water, touching up any sinking spots in the planting hole that may settle after this first deep root watering.

## ESTABLISHING CITRUS TREES IN YOUR YARD

### Mulching

Do not mulch over 3 inches deep and never put mulch any closer to the trunk of citrus trees than 12-24". Nature didn't create mulch, man did. Mulch can cause fungus issues with citrus plants.

### Watering

When freshly planted, citrus trees need a deep, thorough watering two to three times the first week and one to two times per week for the next few weeks, depending upon soil type, rainfall and the time of year. Once this adjustment and establishment period is over, water them deeply whenever the soil begins to get dry an inch or so below the surface. Simply fill the water ring each time. Within about four, perhaps six months, your watering reservoir will erode away. Don't worry about rebuilding it, by this time your new citrus trees will be established enough that you can use a soaker hose to supply that deep root moisture as the water in your soil level drops.

### Nutrition

Don't worry about fertilizing your new citrus trees when planting. Once your plants begin putting on new growth you will want to begin your fertilization schedule of once a month from February through October. Citrus trees are hungry plants and excellent performance will come from applying citrus fertilizer on a timely basis.

| Citrus Tree Fertilizing: | February –October |
|---|---|
| Year 1 monthly application | 1 cup |
| Year 2 monthly application | 2 cups |
| Year 3 monthly application | 3 cups |
| Year 3 monthly application | 4 cups |

You will also have good results with palm fertilizer, as they have almost identical requirements. Organic products to look for are Citrus-tone from Espoma and Dr. Earth Organic 9 Fruit Tree Fertilizer. Scatter the prescribed amount of organic fertilizer on the ground at least a foot away from the trunk and promptly water it in thoroughly. Notice in the chart below that young citrus will need an increasing amount of fertilizer applied in the first few years you have it in your yard.

Container grown citrus needs to be feed more frequently than trees planted in outdoor ground soil.

## GROWING CITRUS TREES IN COLD CLIMATES

### Selecting the Right Container

You can use just about any type of container that has adequate drainage holes cut in the bottom. Don't worry about potting soil escaping, buy a drainage pan and add a piece of landscape fabric to the bottom before planting to keep the soil from leaching out with your watering. Plastic, resin and wood containers will be far more mobile than heavier clay and ceramic pots. As your tree grows, it is always a good idea to add container wheels for easy moving. Container size selected should never be out of balance with your citrus tree's size. Newly purchased citrus trees with a vigorous root system will normally be in a 6-8″ pot upon arrival. Your initial potting should use a 12″ container for a 1 gallon or 6″ potted tree, and a 16″ container for an 8″ nursery potted tree. You most likely will never pot your indoor citrus tree in anything larger than a 20″ container, but let it grow to need this much root space over time. Too much potting soil around a very young tree's roots will cause you moisture problems.

### Proper Potting Prevents Problems

Start out by installing 1-2 inches of pea gravel over the bottom of your citrus tree's pot to improve drainage. The majority of pre-bagged potting soils have far too much sphagnum peat for good results with citrus. You cannot use garden soil in indoor containers. The easiest way to solve this is to buy a potting soil product that contains perlite or vermiculite and add 1/3 volume of redwood or cedar shavings, blending your two potting mediums well. DO NOT use pine or spruce shavings. Cedar shavings are readily available in the pet section of every big box store. Partially fill your container with the citrus tree potting blend and you are ready to inspect the new tree's root system.

Container grown plants with well developed root systems will most likely be a bit root bound. This is normal and easily corrected at repotting with some careful pruning of some larger roots. Also loosen the rest of the outer roots for better root growth in the tree's new container. Place your citrus tree in the partially filled pot so the depth the plant was previously growing in the original container. Your final soil level should be at least ¼″ lower than the rim of your pot to allow for watering. Finish filling the container and lightly firm up the potting soil surrounding the root ball. Do not fertilize, just as outdoor growing of citrus, you want to wait for new growth to appear before beginning the fertilizer regimen. Water your plant in thoroughly.

**Sunlight Exposure**

Fruit crops tend to be in favor of full sunlight exposure. Obviously, the more hours of sun each day, the heavier your flowering, and ultimately your citrus fruit harvest will be. Most fruit crops grow best in full sunlight, but some will do well in partial shade. So during the cold months of the year your container grown citrus trees should be kept in a spot that they will get the most sunlight possible in your home. When all threat of frost is over, do move your citrus tree outdoors where it will get at least 8 hours of sun a day. However, you will want to allow it to adjust to this vast difference in lighting first. After spending months indoors with rather indirect sunshine, you need to help your citrus plants acclimate to the radical change in lighting. The same is true of preparing your new fruit trees for being moved indoors for the winter. Whichever direction you are headed with light acclimation, you'll want to slowly change from direct sun exposure or part shade exposure in the opposite direction, beginning about 2 weeks prior to the full time move. This will help them to avoid going into shock upon relocation.

**Temperature**

Tropical and subtropical fruit trees cannot tolerate freezing temperatures for very long. Some will be killed back to the soil by mild freezes while only small twigs will be killed on others. Some root damage can occur because the root system is not as well insulated from cold in a container as it would be in the ground. With any type of container grown plant, the roots are much more exposed to cold than those grown in the ground. Since citrus are tropical or subtropical climate plants, exposure of roots in containers to cold can cause root damage. Take care when moving them indoors for cold snaps to keep them away from door drafts and heating vents. You should begin preparing indoor citrus trees for moving indoors once night time temperatures dip into the upper 40's. Don't take a chance with the fickle nature of the weather, it is better to be safe than sorry.

**Water Right for Best Results**

The biggest issue most people have in not succeeding with container grown plants is water – either too much of it or not enough. You should only water as needed, most especially with citrus trees. Plastic, metal and ceramic containers will retain moisture longer than porous clay and wood containers. Unlike growing in ground soil, today's more popular indoor pots of plastic and ceramic don't breathe. This is why you should never use real soil in a container growing situation. To check the soil is best before applying more water. The top surface of your potting medium should be very dry to the touch. Also slowly add the water to your pot to avoid exposure of roots to the air. Do empty the pot's water tray, as with the drainage holes submerged beneath water level will stop the pot's ability to drain well. Cool temperatures slow down the need for water as well as your citrus tree's growth.

**Indoor Citrus Fertilization**

An excellent diet is highly important to your success in container growing of fruit trees. Over fertilization can result in too much foliage growth, poor fruit production and tell tale salt accumulation can also cause foliage to dieback. Either Citrus-tone or Dr. Earth's Organic 9 fertilizers should be used according to label directions. It is easy to see that you are applying adequate fertilizer when the mature foliage on your citrus trees is deep, rich green. Should a white crust form on the surface of your potting mix this is likely due to excess fertilization and/or water containing considerable soluble salts. Citrus trees don't do well in saline soils and you should correct this situation by slowly running water through the container for several minutes. This will carry the excess salts down through the potting medium and out the drainage holes.

**Pruning**

Grown in outdoor soil, these fruit trees will have a naturally full and pleasant shape. Citrus trees grown indoors can become leggy and getting them to be fuller is done by partially cutting back the growth to increase branching and density. Should the canopy of your indoor citrus get way too big for the containerized roots, some falling leaves and fine twig dieback can happen. This is corrected by pruning your citrus plant's canopy back heavily.

## PLANTING AND CARE OF YOUNG CITRUS TREES

Never lift or carry the tree by grasping the trunk or stake, and be sure the tree is lowered into, and correctly set in the planting hole before you slit the poly container and plant the tree. Water the tree right after planting: Your new tree has a large number of active, working leaves which must be kept well supplied with water at all times so as to function and not wilt. Since the ball of the tree contains all of the tree roots it must be kept moist to serve as reservoir for the water. When you remove the plastic sleeve from the ball, you will find that many of the roots are concentrated at the outside vertical surface. It is, therefore, very important that the tree be watered immediately after planting, since these surface roots will otherwise be unable to function properly. If the roots are bound (roots are tangled, matted and growing in circles), score the sides of the ball with a knife.

Leave the upper surface of the ball exposed: The soil in the ball has been specially formulated -- it contains special nutrients and is designed so the ball will readily absorb water that is added directly to its upper surface.

## Planting

Do not dig the planting hole any deeper than the roots of the tree as purchased. The hole may be made somewhat wider than the root mass. The hole can be slightly wider than the root ball, or fork-loosened laterally beyond what is needed for the width of the root system. In most soils, no planting mix is needed. Just backfill with native soil, making sure there are no clods that will cause air pockets around the roots. Consider nutritional amendments that can become part of the backfill at this time. Place NO conventional (commercial) fertilizer in direct contact with roots. Certain organic and naturally occurring materials may be considered with caution. Mulches, compost or organic matter will best serve trees when applied to the surface around trees after planting rather than blending with the backfill.

Make a small cone of soil in the center of the hole and place the tree so that the roots are fanned out around the cone. Add soil until roots are covered by several inches and tamp gently, then add the same amount of phosphate as was placed in the bottom of the hole. Cover with soil to fill the hole using the soil which was removed to make the hole. It is not necessary to add soil amendments. Do not use fertilizer containing any nitrogen when planting.

Plant the tree a little high to allow for tree settling in the hole. The darker color or stain on the trunk below the bud union usually indicates the original soil level. Orient the inside of the trunk bend (the concave bend immediately above the bud union) toward the north or northeast to reduce heat stress and sunburn to the trunk curvature above the union; the sun is less able to impinge directly on the young bark and trunk when so oriented. Using the best, friable, clod-free soil available at the site, backfill leaving no air pockets around the root system. After settling, the soil level next to the trunk should be at or below the level at which the tree grew in the nursery field.

After planting, the soil line on the trunk should be a couple of inches higher than the surrounding ground line, which in turn, should slope downward away from the tree to prevent water from accumulating near the trunk. The graft union should be four to six inches above the soil surface. A basin for water may be made around the base of the tree (not next to the trunk). Fill the basin with water several times to settle the soil around the roots and to remove any air pockets. After planting, irrigate once if necessary, keeping in mind that young trees will not be extracting or depleting soil moisture until growth is well underway (several inches of new growth).

## Watering

Do not allow the ball to dry out. Once your tree has begun to establish a root system, keep the soil damp but not soggy. Water deeply. Apply water according to the needs of the tree or as your soil dries out. Check the soil in the root ball for moisture by probing with a finger to assure yourself that the plant needs water. The tree may be watered by basin for a full year. However, the basin should be broken down during the wet season if water has any tendency to stand in it. After a year you should consider the use of sprinklers or drippers.

**Fertilization**

After trees have made six to eight inches of new growth, three applications over the growing months may be used of one of the following:

| | | |
|---|---|---|
| Ammonium Nitrate | 33-0-0 | 1/2 cup |
| Urea | 46-0-0 | 1/3 cup |
| Ammonium Sulphate | 21-0-0 | 3/4 cup |
| Steer Manure | | 4 pounds |

This amount should be placed six inches to three feet out from trunk and watered in immediately and thoroughly.

**Vertebrate Pests**

Young fruit trees are especially liable to severe damage from gophers, rabbits, voles (mice) and deer. Rodents Pocket gophers eat roots, bulbs and entire young plants. Eradication by trapping is the most efficient method. Two types of traps are commonly used: box traps and Macabee traps. Poison baits are effective, but can cause harm to dogs and cats if eaten. Moles are only present in good soils as they are primarily insect eaters (especially earthworms and larvae). A spear or harpoon trap is usually used for control. Ground Squirrels (Gray Diggers) gnaw roots and bark and will climb trees for fruits and nuts. Controls are baited box-type traps placed near burrows, anticoagulant poison-bait placed inside burrows, and gas bombs placed in the burrow. Metal guards around tree trunks will prevent them from climbing up into trees. Birds The only sure solution to bird depredation is to use screening or netting material. This must be tied about the tree trunk to be effective in keeping the birds out. The use of owl statues, flexible snakes and cat statues will only be effective in gardens if they are repositioned every few days.

Deer They will spend more time looking for food in gardens as wild plants dry out, and will eat foliage and fruit of nearly anything that grows. The only fool-proof protection against their depredation is fencing. On level ground a seven foot woven-wire fence will usually suffice. An "outrigger" extension on top makes it harder to jump. Individual wire cages around plants and cylinders of wire around trees may be effective also. Repellents can work if renewed frequently, especially after each rain, watering or several days of exposure.

### Weeds

If you want good growth, it is imperative that weeds are not allowed to develop near your trees. Keep the space clean for a full six feet from the base of your tree. Allow no weeds or grass to grow in this area and apply no systemic weed killers that may be absorbed by the tree roots. Mulching (keeping it away from the trunk) can be very effective at controlling weeds.

## PLANTING CITRUS TREES?

Floridians absolutely love their Citrus Trees. We are providing some information here to help you plant and grow healthier citrus trees. This information is from our personal experience. We grow citrus trees because we enjoy eating and sharing the fruit with friends and family just like you.

### Soil Requirements:

All citrus trees require deep well drained soil. Your planting site should have good runoff to prevent water standing around the tree. To check and determine the soil is suitable do the following before planting your tree. Using a posthole digger, dig to a depth of 3 to 4 feet and fill the hole with water. All water should drain from the hole within 24 to 36 hours. Soils requiring more than 36 hours to drain completely should be avoided unless raised planting beds are used. Most citrus grows well in a soil pH range from 6 to 8.

### How to Plant Your Citrus Tree

Clear out weeds and grass approximately 3 feet around the diameter of your planting site. After planting your tree you should always keep the site free of weeds and grass. Mulch is not recommended because it can keep soil to moist! Ideally you want good air circulation at the trees base to avoid any fungal diseases caused by excess moisture etc.

### Planting Your Tree

When planting from containers, dig a planting hole in a well drained area of your property. Choose a sunny spot. Your hole should be 4 - 6 inches wider than the root ball and approximately 4 inches deeper. Fill the hole half full of water. Take your tree and roll the container gently on the ground and squeeze the root ball out of the container. Keep the root ball in tact. Don't allow the root ball to fall apart.Position the tree in the center of the planting hole. Gently spread out any circling roots with your hands. Now, add more soil and water the soil and continue backfilling soil to remove any air pockets. The base of your tree should be about 1 inch above the ground. With your hands pack down the soil slightly. Now make a circular basin around the tree trunk to keep the water from running away as you water. This basin can be about 8 - 10 inches away from the trunk.

Planting your tree from a Citus Pot: (see photo on right) Here you will follow all the above instructions with only a few exceptions as descibed here. In this situation you cut away the container with a razor or sharp knife. Do not roll the long citrus pot on the ground as doing so will most likely damage the roots. Place the tree in the hole while gently shaking loose soil into the hole as you move the tree up and down and gently spread out the roots by pulling them loose with your hand. You should have some loose roots and the tap root. The long tap root should go straight to the bottom of the hole. Watering: Young trees need water at least three times per week for the first two weeks. Water twice a week after the third week and keep in mind that during our rainy season you may not need to water at all.

Good Nutriton Keeps Your Tree Healthy! This is the one thing you don't want to ignore. We recommend you fertilize your tree about one month after it has been planted. Be sure to use a citrus fertilizer with minor elements. Spread a small handful around the drip line of the tree at least once a month in spring and summer. Fertilizing around the drip line will encourage the roots to grow out to seek the nutrients. Over feeding is worse than under feeding. Never, ever, ever, ever, use fertilizer spikes! Vigoro 6-4-6 is a good all purpose granular fertilizer for container and landscape grown fruit trees. Read the label prior to application! You can choose to use only organic fertilizers and minerals if you are more experienced in organic methods. Growing citrus is challenging and not for people that are too busy to care for the tree.

**When to fertilize**

All your trees need as a minimum items 1, 2 & 3.

- Vigoro 6-4-6 or similar citrus granuals monthly during spring and summer.
- Citrus nutritional sprays once per month during fall and winter.
- For young trees less than three years of age use a foliar spray applications of 20-20-20 water soluble fertilizer every six weeks, spring and summer. Three year old trees and older can be provided with all the above but switch to a 10-15-10 foliar to decrease nitrogen and increase flowering and fruit production. Container grown trees need to be rinsed at least twice a year to flush out salt residues. As you gain experience you can alter the formula as you see fit. Pay attention to your tree and it will reward you with delicious fruit.

Fertilizer should be applied beginning in late February-early March and ends in late September

**INSECT & DISEASE CONTROL FOR CITRUS GROWING:**

We are going to discuss two major pests and how to control them and even help eradicate them.

Our first villain is enemy #1 The Asian Citrus Psyllid. This trouble maker feeds on your trees new growth flush (emerging new leaves). Evidence of this pests damage is indicated by the curling and distortion of the young leaves. This pest injects toxins during feeding on plant fluids. These psyllids can spread a disease called citrus greening and it's a big headache for all of us in Florida who love enjoying our very own citrus fruit.

The next pain in the neck bug is the Citus Leaf miner: This tiny moth like insect whose populations soar rapidly and develop on the spring and Summer flushes of your citrus trees. As summer approaches and the mercury rises, leaf miner populations increase dramatically causing severe damage to your trees overall growth and health. Leaves are scarred with silver tunnel like trails that spiral inside the leaves. To add insult to injury our summer rains make conditions for citrus canker (bacteria) to spread just perfect! The damaged leaves are extremely susceptible to canker and other diseases. Using copper and other approved fungicides will help keep the canker under control.

Okay now, if you are freaking out at this point may I suggest you relax. With the information you have here you are better informed about growing citrus in Florida than the average person. You can grow healthy citrus trees. You can even do it organically if you choose. Well now that you know, it's up to you! Stop by and say hello so we can share more information and suggest some solutions and products that can help you grow better fruit trees. One product that is often recommended is Sevin. It's not organic and some say it's the only way to stop the blue green root eating beetle. So far I have been fortunate and not had a problem with them. Looks like Neem and other organic products we offer work quite well.

## MACRONUTRIENT FUNCTIONS IN PLANTS

### Nitrogen (N)

Nitrogen is of special importance because plants need it in large amounts. It is also easily lost from soil and fairly expensive to supply. A major factor in successful farming is the grower's ability to manage N efficiently. Nitrogen has numerous functions in plants, and essentially all life processes depend on it. Nitrogen occurs chiefly in amino acids, proteins, and sugars. The most active nitrogenous compounds occur largely in the protoplasm and nuclei of plant cells. Among them are the enzymes that speed up biological processes.

An abundant supply of essential N compounds is required in each plant cell for normal cell division, growth, and respiration. Even the green leaf pigment chlorophyll, which enables plants to use the energy of sunlight to form sugars from carbon dioxide and water, is a nitrogenous compound. A high concentration of N is found in young, tender parts of plant tissues like tips of shoots, buds, and new leaves. The N, present mostly as protein, is constantly moving and undergoing chemical changes. As new cells form, much of the protein moves from older cells to newer ones, especially when the total N content of the plant is low.

The proper functioning of N in plant nutrition requires that the other essential elements, particularly P, K, Ca, and Mg, be present in adequate supply. If the supply of one or more of them is inadequate, the addition of N to most common crops may not produce optimum growth. Such plants often are susceptible to disease, mature late, and produce poor-quality fruit. However, if the nutrient balance and total supply are adequate, significant growth of dark green foliage will occur. Nitrogen is the mineral element used most by citrus trees to produce leaves, flowers, and fruit, although Ca and K are also used in great amounts. Nitrogen is the key component in mineral fertilizers applied to citrus groves; it has more influence on tree growth, appearance, and fruit production/quality than any other element. Nitrogen affects the absorption and distribution of practically all other elements and appears to be particularly important to the tree during flowering and fruit set.

**Phosphorus (P)**

Phosphorus is present in all living tissue. It is particularly concentrated in the younger parts of the plant, in the flowers, and in the seeds. Phosphorus is necessary for many life processes such as photosynthesis, synthesis and breakdown of carbohydrates, and the transfer of energy within the plant. It helps plants store and use energy from photosynthesis to form seeds, develop roots, speed maturity, and resist stresses. Phosphorus is involved in nutrient uptake and translocation. It is a major part of the cytoplasm and the nucleus of cells, where it is involved in the organization of cells and the transfer of heredity characteristics. Phosphorus is also important for cell division and enlargement, thus plant growth is reduced when the supply of P is too low.

**Potassium (K)**

Citrus fruits remove large amounts of K compared with other nutrients. Potassium moves from leaves to fruit and seeds as they develop. Potassium is necessary for several basic physiological functions like the formation of sugars and starch, synthesis of proteins, normal cell division and growth, and neutralization of organic acids. Potassium is important in fruit formation and enhances fruit size, flavor, and color. It helps reduce the influence of adverse weather conditions like drought, cold, and flooding. Potassium is known to influence many enzymatic reactions and is associated with almost every major plant function. Potassium helps regulate the carbon dioxide supply to plants by controlling stomata opening and closing. It improves the efficiency of plant water and sugar use for maintenance and normal growth functions. It moves sugars from the site of photosynthesis to other storage sites. Potassium works with P to stimulate and maintain rapid root growth of plants. It stimulates the synthesis of protein from amino acids. Potassium improves plant health and resistance to disease and tolerance to nematodes and insects. The rate of photosynthesis drops sharply when plants are K deficient. Too much N with too little K can result in a back-up of the protein building blocks, set the stage for disease problems, reduce production of carbohydrates, reduce fruiting, and increase fruit creasing, plugging and drop. A shortage of K can result in lost crop yield and quality. Moderately low plant K concentrations will cause a general reduction in growth without visual deficiency symptoms. The onset of visual deficiency symptoms means that production has already been seriously impaired.

### Calcium (Ca)

Calcium resides mainly in plant leaves. Calcium is an important element for root development and functioning and is an important constituent of cell walls. It is required for chromosome stability and cell division. Calcium activates several enzyme systems and neutralizes organic acids in plants. Plant growth and fruit yield can be reduced by inadequate Ca supply long before deficiency symptoms become evident.

### Magnesium (Mg)

Magnesium is the center of the chlorophyll molecule. It is involved in photosynthesis and plays an important role as an activator of several enzymes. It is also involved in carbohydrate metabolism and synthesis of nucleic acids. Magnesium influences the movement of carbohydrates from the leaves to other parts of the tree and also stimulates P uptake and transport.

### Sulfur (S)

Many plants use about as much S as they do P. Sulfur is an essential constituent of many proteins. Sulfur is important for the production of amino acids, proteins, and chlorophyll, and is a constituent of vitamins and some plant hormones. Protein synthesis is retarded in S-deficient plants. Sulfur enhances the development of nodules and N fixation by legumes. It improves root growth and promotes vigor and hardiness. Sulfur also affects carbohydrate metabolism.

## MICRONUTRIENT FUNCTIONS IN PLANTS

### Functions of iron (Fe)

- Catalyzes the production of chlorophyll.
- Involved in some respiratory and photosynthetic enzyme systems.
- Involved in the reduction of nitrates and sulfates.

**Functions of zinc (Zn)**

- Involved in plant carbon metabolism.
- A necessary component of several enzyme systems that regulate various metabolic activities within plants.
- Part of an enzyme that regulates the equilibrium among carbon dioxide, water, and carbonic acid.
- Part of two enzymes that play a role in protein metabolism.
- Essential for the formation of chlorophyll and function of normal photosynthesis.
- Needed to form auxins, which are growth-promoting substances in plants.
- Associated with water relations in plants and improves water uptake.

**Functions of Manganese (Mn)**

- Involved in the production of amino acids and proteins.
- An activator of several enzymes.
- Plays an essential role in respiration and N metabolism.
- Necessary for the reduction of nitrates and helps make them usable by plants.
- Plays a role in photosynthesis and in the formation of chlorophyll.

**Functions of Boron (B)**

- Important in sugar translocation and carbohydrate metabolism.
- Particularly needed at the location of active cell division.
- Plays an important role in flowering, pollen-tube growth, fruiting processes, N metabolism, and hormone activity.
- Maintains Ca in a soluble form, thus insuring its proper utilization.
- Deficiencies may be aggravated by severe drought conditions, heavy lime applications, or irrigation with alkaline water.

**Functions of Copper (Cu)**

- Part of several enzyme systems.
- Has a role in photosynthesis and chlorophyll formation.
- May have an important function in root metabolism. (Cu appears to be concentrated more in the rootlets of plants than in leaves or other tissues. Cu in citrus fibrous roots may be 5 to 10 times greater than in leaves.)
- Regulates several biochemical processes within the plant.
- Important in the utilization of proteins in the growth processes of plants. (The photosynthesis rate of Cu-deficient plants is abnormally low.)
- May also be involved in oxidation-reduction reactions in plants.

- Heavy fertilization with N tends to increase the severity of Cu deficiency.

**Functions of Molybdenum (Mo)**

- Assists in the formation of plant proteins.
- Helps starch, amino acid, and vitamin formation.
- Considered a catalyst that aids the conversion of gaseous N to usable forms by nitrogen-fixing microorganisms.
- A constituent of the plant enzyme that converts nitrate to ammonia.

**Functions of Chlorine (Cl)**

Although the essentiality of Cl has been established for most higher plants, its need for fruit crops has not yet been demonstrated. The plant requirement for Cl is quite high as compared with other micronutrients, but its exact role in plant metabolism is still obscure. Chlorine is:

- Associated with turgor in the guard cells through the osmotic pressure exerted by imported K ions.
- Involved with oxygen production in photosynthesis.
- Involved in chlorophyll and photosynthesis because its deficiency causes chlorosis, necrosis, unusual bronze discoloration of foliage, and reduction in growth.

**Functions of Nickel (Ni)**

Within the last decade, Ni has been established as an essential element in higher plants. Although well-defined enzymatic functions are known to be associated with Ni in legumes, apparently the need for Ni exists in other plants as well. No one has ever seen a Ni deficiency in soil-grown plants.

**Summary**

Nutrient availability directly affects how well plants grow and produce. A sufficient supply of all nutrients is critical to nutrient management and sustainability. If a single essential element is below the critical level for availability, crop growth and yield will fall even if the other elements are in sufficient supply. Concern for nutrient sufficiency is often confined to N, P, and K because they are needed in relatively large amounts by crops and are most often the limiting factors in crop production. However, nutrient insufficiency can go beyond N, P, and K. A balance of a sufficient supply of nutrients is a key component to profitability. Plant nutrients interact positively when properly balanced. For example, in the case of N fertilization, a shortage of another nutrient could decrease N uptake, reduce N use efficiency and returns on investment, and increase the potential for N loss.

Balanced nutrition of plants should be a high priority management objective for every citrus grower. Plants require a balanced nutrition program formulated to provide specific needs for maintenance and for expected production performance. Properly nourished fruit trees or plants grow stronger, produce more consistently, have better disease resistance, and are more tolerant to stresses.

For most macronutrients, soil application is still recommended because of the large quantities required. However, fertilizer applications to the soil are subject to various fates including leaching, runoff, and fixation to forms not available to plants. Therefore, foliar application should be considered as a possible supplement to soil application for some nutrients. Foliar application of N, K, Mg, Zn, Mn, and B has several positive attributes. It is of significant importance when the root system is unable to keep up with crop demand or when the soil has a history of problems that inhibit normal growth. It is proven to be useful under prolonged spells of wet or dry soil conditions, calcareous soil, or cold weather, which decrease the plant's ability to take up nutrients when there is a demand. Foliar application can reduce overall fertilizer application rates and energy use, and can improve the uptake efficiency of micronutrients because they are directly absorbed into the leaves. Foliar application is the quickest method of getting nutrients into plants over the short term when a nutritional deficiency is diagnosed, but should not be relied upon for long-term tree nutrition. Foliar feeding may also become a best management practice that can help reduce groundwater contamination concerns.

It should be kept in mind that foliar feeding is not intended to completely replace soil-applied macronutrient fertilization. Furthermore, there is a perception that foliar spray of macronutrients is more expensive because several applications are required to satisfy plant needs and maintain high yield. On the other hand, foliar applications of micronutrients are more effective than soil applications with the exception of Fe. Foliar micronutrient (Zn, Mn, Cu, B) sprays provide a more rapid response and are easier to apply, but the effect does not last as long as that of soil application. Soil applications of Fe chelates still offer the most effective means of correcting Fe deficiency.

Fertilization represents a relatively small percentage of the total cost of citrus production, but it has a large effect on potential profitability. Visual evaluation of nutritional status, soil and plant analysis, field history, production experience and economics are all important guidelines to use when making fertilizer rate and source decisions.

## IRRIGATION, NUTRITION, AND CITRUS FRUIT QUALITY

Florida has the highest citrus fruit quality standards in the world. The most important quality factors for Florida citrus growers, production managers, processors, and packers include fruit juice content, soluble solids and acid concentrations, soluble solids-acid ratio, fruit size, and color. Florida citrus growers discern between quality factors for the fresh and processing markets. For example, fruit size, shape, color, and maturity date are most important for fresh fruit, but high juice content and soluble solids are desired for processing fruit. Fruit quality is affected by several factors including cultivar, rootstock, climate, soil, pests, irrigation, and nutrition.

The effects of irrigation and nutrition on fruit quality are important and should be understood and taken into consideration by citrus growers and production managers to increase profitability and enhance sustainability and worldwide competitiveness. In general, excessive irrigation and fertilization reduce fruit quality. Therefore, supplying sufficient nutrition and using sound irrigation scheduling techniques should be high-priority management practices for every grower. Citrus trees require a properly designed, operated, and maintained water management system and a balanced nutrition program formulated to provide specific needs for maintenance and for expected yield and fruit quality.

Irrigation contributes to the efficiency of fertilizer programs. Citrus trees with sufficient water and nutrients grow stronger, better tolerate pests and stresses, yield more consistently, and produce good quality fruit. On the other hand, excessive or deficient irrigation or fertilization may result in poor fruit quality. The most important management practices influencing fruit quality are irrigation and nitrogen, phosphorus, potassium, and magnesium nutrition. Some micronutrients like boron and copper can also affect fruit quality, but only if they are deficient in the tree. In general, when any nutrient element is severely deficient, fruit yield and fruit quality will be negatively affected.

## EFFECTS OF SPECIFIC ELEMENTS

Trends in fruit quality response to increasing nutrient and water availability are described and summarized below:

### Nitrogen (N)

- Increases juice content and color, total soluble solids (TSS), and acid concentration.
- Increases TSS per box and per acre. However, excessive N, particularly with inadequate irrigation, can result in lower yields with lower TSS per acre.
- Decreases fruit size and weight.
- Increases peel thickness and green fruit at harvest.
- Increases incidence of creasing and scab but decreases incidence of peel blemishes like wind scar, mite russeting, and rind plugging.
- Reduces stem-end rot incidence and green mold of fruit in storage.

### Phosphorus (P)

- Reduces acid concentration, which increases TSS-acid ratio. Phosphorus rates have no effect on TSS per box but may increase

TSS per acre due to increase in fruit production in soils that are low in plant-available P.
- Increases number of green fruit but reduces peel thickness.
- Increases expression of wind scar but reduces that of russeted fruit.

**Potassium (K)**

- Potassium produces mostly negative effects on juice quality except for TSS per acre. Potassium increases fruit production, therefore producing more TSS per acre.
- Decreases juice content, TSS, TSS-acid ratio, and juice color.
- Increases acid content.
- Increases fruit size, weight, green fruit and peel thickness.
- Reduces incidence of creasing and fruit plugging. In storage, reduces stem-end rot.

**Magnesium (Mg)**

- Slightly increases TSS per box and per acre, and TSS-acid ratio.
- Slightly increases fruit size and weight, but decreases rind thickness.

**Irrigation**

- Increases juice content and TSS-acid ratio.
- Reduces TSS and acid concentration. TSS per box decreases, but TSS per acre may increase due to yield increase.
- Increases fruit size and weight, increases green fruit at harvest but decreases rind thickness.
- Increases incidence of blemish from wind scar, scab and Alternaria brown spot, but reduces rind plugging.
- Reduces stem-end rot incidence but increases incidence of green mold in storage.

**Summary**

Specific effects on juice and external fruit qualities. This summary is based on numerous field experiments conducted over many years that evaluated the response of oranges to irrigation and fertilization practices. Most of these effects were consistently observed, but some of them appeared to depend on local conditions and growing regions. These observations are useful in developing a strategy to improve fruit quality for a particular variety or location.

## TECHNOLOGIES FOR HIGH QUALITY CITRUS FRUIT PRODUCTION

Citrus cultivation has been one of the most important industries in mountainous areas of western Japan because of well drainage and abundant sunshine in spite of topographic disadvantage conditions. However, novel technologies such as mechanized system have been required to improve fruit quality and productivity, and laborsaving since aging of growers and decrease of successor are remarkable in recent years. In addition, orchards conservation on steep sloping lands is also required. The purpose of this paper is to introduce new systems and technologies which have been researched and developed for high quality fruit production, laborsaving and land conservation in steep lands, (1) the drip irrigation and liquid fertilizing system combined with year-round plastic mulching, (2) new transporting system using contour narrow paths combination of monorail system, and (3) disaster prevention mapping for citrus orchards.

New fruit quality control system using drip irrigation and liquid fertilization technique combined with year-round plastic mulching was developed, expecting high quality and stable fruit production of citrus. Moisture-permeable plastic mulching sheets are covered under citrus trees. Water and/or nutrient solution is automatically supplied through the drip tubes that are laid under the mulching sheets to give adequate water stress to improve sugar and acid content of fruit. Fruit quality was controlled well through the system, e.g. it raised 20-30% of sugar content. This is reflected in grower's income. Furthermore, laborsaving effect of water and fertilization management was observed. Extensive studies regarding to efficient irrigation time and appropriate amount of water or nutritional solution have been conducted.

New transporting system for steep sloping citrus orchard, which is the combination of monorail system and contour narrow paths, was suggested. To excavate narrow paths, the utilization of a small waking tractor was developed and the procedure of narrow path excavation was explained. The efficiency of excavation using the walking-tractor was 20-57 m/h from the investigation in four different steep sloping orchards. To evaluate the improvement effect of growers work by narrow path, the time study survey of main management works of citrus orchard was conducted. After introducing narrow path, the working hours was reduced in the fertilizer and the chemical herbicide application. The working space and the leveled ground provide by narrow path realized the safely work without fal1ing down for all works.

Disaster prevention mapping of citrus orchards on slope lands was developed based on computer-aided seepage estimation and topographic data. The mapping can show the both zones of ascending flow (upward seepage flow) and descending flow (downward seepage flow) of under ground water on heavy rain in citrus orchards. The mapping is considered to be effective for the management of orchards and prevention from erosion on slope lands. Multi-purpose and high-functional farm path foundation on ascending flow area of orchard was efficient for the ability of disaster prevention.

## MANAGING PHOSPHORUS FOR CITRUS YIELD AND FRUIT QUALITY IN DEVELOPING ORCHARDS

Citrus orchards occupy more than 280,000 ha in Florida. Soils planted to citrus require intensive nutrient management to maximize fruit yield and quality. Traditionally, orchards receive ˜40 kg phosphorus (P) per ha at planting followed by applications of up to 100 kg•ha-1 P per year until they enter the fruit-bearing years after age 4. At this point, fertilization diminishes to an average of 20 kg.ha-1 P per year, but some growers apply up to 50 kg•ha-1 P per year (Tucker et al., 1990).

Excessive P fertilization has lowered orange juice soluble solids (Anderson, 1966) and delayed external color development (Koo, 1988). In Florida, the surface layers of some acidic, poorly drained soils planted to citrus are composed of stripped quartz sands that have little capacity to retain soluble P, so the potential for P leaching exists when the orchard is drained. Thus, overapplying P should also be avoided to prevent potential water quality degradation.

Bearing citrus trees rarely respond to P fertilizer (Smith, 1954) except when planted on soils with extreme P fixation capacity or soils nearly devoid of P. For example, Cantarella et al. (1992) and Quaggio et al. (1998) observed positive yield responses of 'Valencia' oranges and lemons [C. limon (L.) Burm.] to annual P fertilizer rates up to 62 kg•ha-1 P on a high P-fixing Brazilian soil. In Florida, Spencer (1963) found that P improved growth of orange trees planted on previously nonfertilized soil but observed a positive response in only 1 of 4 years when 60 kg•ha-1 P per year was applied to bearing trees.Anderson (1966) observed an orange yield response to P fertilizer in 3 of 5 years in central Florida where the Bray P1 soil test was near the accepted minimum adequate value of 40 mg•kg-1 (Koo et al., 1984). This result emphasized the need for a calibrated P soil test for citrus using a common extractant like Mehlich 1.

Soil test data from bearing Florida citrus orchards reflect the ability of many surface soils to retain applied P to a limited extent. For example, the Bray P1 soil test was 44 mg•kg-1 at the beginning of the Anderson (1966) P fertilizer experiment, but it increased to 100 mg•kg-1 after 8 years. Obreza (1993)observed that Mehlich 1 soil test P increased from 15 to 70 mg•kg-1 in a newly planted orchard where a total of 160 kg•ha-1 P was applied during the first 3 years after planting. Anderson (1983), Tucker et al. (1990), and Alva (1993) found that soil test P in a wide range of mature Florida citrus orchards was five to 10 times higher than what would be expected in similar, nonfertilized soils.

No calibrated P soil test exists specifically for Florida citrus orchards. The critical value of Mehlich 1-extractable P currently used (30 mg•kg-1) was determined for agronomic crop production (Obreza and Morgan, 2008).

Environmental concerns about P loss to surface water and grower consideration of nutritional effects on citrus yield and fruit quality emphasize the importance of judiciously using P fertilizer. The objective of this experiment was to establish guidelines for P management in developing Florida grapefruit and orange orchards by determining the effect of P fertilizer rate on soil test P and subsequently calibrating a P soil test for citrus yield and fresh fruit quality.

## MATERIALS AND METHODS

This experiment was conducted in a new citrus orchard planted on a previously noncultivated soil at the University of Florida, Southwest Florida Research and Education Center in Immokalee, FL. The poorly drained Immokalee fine sand (sandy, siliceous, hyperthermic Arenic Haplaquod) at the site is typical of soils used to produce citrus in south Florida (Obreza and Collins, 2002). After clearing native vegetation in 1994, the land was prepared for citrus planting in 1995 by first leveling and then constructing parallel 15.3-m wide beds separated by 0.9-m deep furrows (swales) to enhance surface drainage. After bed formation, organic matter in the top 0.3 m of soil averaged 1.7%. The top 15 cm of soil was limed to pH 6.5 and the site remained fallow for 2 years. In Nov. 1997, 275 'Flame' grapefruit (Citrus paradisi Macf.) and 273 'Hamlin' orange (Citrus sinensis L. Osb.) trees budded to Swingle citrumelo rootstock (Citrus paradisi × Poncirus trifoliata L.) were planted in double rows on the beds. Rows were 6.7 m apart across the top of the beds and in-row tree spacing was 4.6 m (326 trees/ha). Standard nursery-grown grapefruit and orange trees were planted in separate, adjacent orchards. An installed microirrigation system had one microsprinkler (60 L•h-1 flow rate, 3.7-m diameter wetting pattern) per tree placed adjacent to the trunk.

No fertilizer was applied to the orchards during the first 11 months after planting as a result of lack of fertilizer response previously observed with newly planted citrus trees in a similar locale (Obreza, 1990). In Aug. 1998, a general sampling of the top 20 cm of orchard soil revealed that Mehlich 1 soil test P averaged 3 mg•kg, which fell in the very low range according to University of Florida soil test interpretations (Mylavarapu et al., 2007). Nitrogen and potassium (K) fertilizers were applied to assure that nutrient availability was neither deficient nor excessive with respect to tree growth, fruit production, and fruit quality. Application rates and timing were within University of Florida guidelines (Obreza and Morgan, 2008). Nitrogen fertilization began with 49 kg•ha-1 N in Year 2 after planting (1999) and gradually increased to 202 kg•ha-1 N in Year 5 after planting and beyond. Potassium fertilizer rates ranged from 93 to 186 kg•ha-1 K per year.

Four rates of P fertilizer (0, 24, 49, and 98 kg•ha-1 P per year) were applied by plot to establish a range of soil test P throughout the orchard. There were 10 replications of each treatment in a randomized complete block design. A plot consisted of three consecutive trees in a row. Citrus response data were collected from the center tree, whereas the outside two trees buffered the potential influence of adjacent plots. Annual P rates were applied once per year for the first 3 years of the experiment (Oct. 1998, May 1999, and Mar. 2000) using concentrated superphosphate (0N–20P–0K). Fertilizer was spread by hand along the tree row in a continuous band covering the width of the canopy (˜2 m) simulating the pattern of a mechanical fertilizer applicator used in commercial orchards.

Other nutrients applied to avoid potential deficiencies included: 1) soil applications of fritted zinc, iron, manganese, copper, and boron in Dec. 1999 and Mar. 2000; 2) foliar spray application of a nitrogen–iron–manganese–zinc mixture each summer from 2000 onward; 3) foliar spray application of magnesium–manganese–copper–iron–boron–molybdenum in Apr. 2002; and 4) a magnesium sulfate soil application in May 2003. Irrigation was required during dry weather in the spring and fall. The volume of water applied at each irrigation was based on the root zone water-holding capacity of 0.08 cm-3•cm-3 of soil to minimize leaching. Irrigation water was groundwater containing 0.05 mg•L-1 P. Roughly 1 kg•ha-1 P was applied with irrigation water each year, which was considered a negligible P input. A continuous weed-free band ˜4 m wide was maintained beneath the tree row canopy by applying bromacil, diuron, and glyphosate herbicides as needed. Grass growing in bed middles and drainage furrows prevented soil erosion. Grass was mowed as needed and the clippings were left on the surface. Citrus oil 435 was sprayed on the tree canopy once each summer to suppress mites and fungal diseases.

An annual cycle of plot soil sampling began in Summer 1999. A soil sample comprised six cores 2 cm in diameter and 20 cm deep removed in an evenly spaced circular pattern around the canopy edge of the center tree in each plot. The soil was air-dried and passed through a 2-mm screen before determining Mehlich 1 soil test P (Soil and Plant Analysis Council, Inc., 1999). Ten 6-month-old spring growth leaves from nonfruiting twigs were sampled in a circular pattern around each center tree when soil samples were taken. Leaves were dried, ground, and analyzed for total N, P, K, Ca, and Mg concentrations (Southern Extension and Research Activity Group, 1992). In 2005, 75-cm deep soil cores were removed from eight plots within each orchard (two replicates of each of the four P rates). Cores were separated into three 25-cm increments and total P was measured with an ashing procedure (Anderson, 1976).

Tree canopy volume was estimated from measurements of canopy width and tree height [canopy volume = 4/3 (p) (tree height) (canopy radius)2]. Fruit production was measured from 2001 (fourth year after planting) onward. Oranges were harvested in December and grapefruits were harvested 1 month later each season. Fruits from the center tree in each plot were hand-picked and weighed collectively. Juice was extracted from a random 0.018-m3subsample of fruit using a commercial extractor (FMC Corp., Agricultural Machinery Div., Jonesboro, AR). Quality parameters included percentage juice per fruit, total soluble solids (TSS) measured with a Brix hydrometer (Wardowski et al., 1979), acid concentration measured by alkali titration (Wardowski et al., 1979), and TSS-to-acid ratio. For oranges, TSS yield in g•kg-1 fruit and kg/tree were calculated from fruit yield, percentage juice, and TSS in the juice. Because fresh fruit characteristics are important for grapefruit, an additional 10-fruit subsample was taken from each plot to measure size (diameter and height), shape, and peel thickness. Fruit shape (degree of flatness, a preferred characteristic) was evaluated by calculating the height-to-diameter ratio. Fruits were cut in half at the widest point, peel thickness was measured at opposite sides of the grapefruit half, and the two values were averaged.

Response of soil test P to fertilizer rate was evaluated using linear regression (Systat Software, Inc., 2006). Phosphorus soil test calibration was attempted by relating tree size, fruit yield, leaf tissue P concentration, and fruit quality variables measured from each plot in a given year to soil test P measured in late summer of the previous year. The calibration method used was the Cate-Nelson procedure (Dahnke and Olson, 1990), which is an iterative statistical procedure for separating crop response data into two classes based on maximizing the class sum of squares in a one-way analysis of variance. Using this procedure, one finds quantitatively the best division from the point of view of maximizing mean differences among classes (Nelson and Anderson, 1977). The soil test value that separates the two classes is defined as the critical value of soil test P. Previous SectionNext Section

**Results and Discussion**

Annual P fertilizer applications from 1998 to 2000 increased soil test P (Table 1). Linear regression of mean soil test P on annual P fertilizer rate each year resulted in significant positive relationships with R 2 ranging from 0.88 to 0.99 for the grapefruit orchard and from 0.82 to 0.99 for the orange orchard. In the third year after planting, the 24 and 49 kg•ha-1 P annual application rates elevated soil test P to medium (grapefruit) or high (oranges), and the 98 kg•ha-1 P rate raised it to very high in both orchards. At this point, P fertilization was suspended because the objective of establishing a wide range of soil test P among plots was achieved.

Soil test P began to decrease in the fourth year after planting (2001) (Table 1). By the sixth year (2003), soil test P had returned to values almost identical to those measured 4 months after the first P fertilizer application in 1999. Rhue and Everett (1987), Harris et al. (1996), He et al. (1999), Li et al. (1999), andZhang et al. (2002) have shown the limited ability of some south Florida sandy soils to hold P. Thus, residual P fertilizer was either taken up by the trees or was leached below the depth of soil sampling. Indeed, measurements of total P in the top 25 cm of soil in 2005 showed little influence of previous P fertilizer application. The top 25-cm layer contained 31 mg•kg-1 total P in plots not fertilized with P compared with 34 mg•kg-1 where it had been applied. The 50- to 75-cm layer beneath nonfertilized plots contained 21 mg•kg-1 total P compared with 32 mg•kg-1 total P under fertilization, suggesting some accumulation at that depth.

Grapefruit trees grew more rapidly than orange trees during the experiment although all trees were managed the same way. Mean canopy volume of 5-year-old grapefruit trees were approximately three times larger than orange trees (Table 2). Although grapefruit trees are inherently more vigorous, this aspect does not fully explain the size difference. A contributing factor may have been a freeze that occurred on 6 Jan. 1999, when the air temperature fell below 0 °C for 7 h with a minimum of –1.6 °C. Some split bark indicating tree damage was observed across the orchards, but the amount of damage was not quantified because it appeared to be minor and uniformly distributed. However, although the grapefruit trees continued to grow normally during the first season after the freeze, the orange trees grew much less, suggesting that they were more severely affected by cold weather. Mean canopy volumes in Sept. 1999 were 4.30 and 0.75 m3 for the grapefruit and orange trees, respectively. The orange trees began to grow normally during the next season, but they remained 1 to 2 years behind relative to typical tree size. At the end of the orange orchard's seventh year, it contained healthy, bearing trees that were smaller than usual for their age.

Tree canopy volume was unaffected by P availability expressed as soil test P (Table 2). The trees showed a marked lack of sensitivity to P on a previously uncultivated soil considered P-deficient. However, at 5 years, they had grown mostly vegetatively, building a canopy without the P sink of heavy fruit production. The grapefruit trees produced a mean fruit yield of 20 kg/tree in Year 4 and 25 kg/tree in Year 5, which was ˜10% of expected mature tree production (i.e., 10+ years old). Mean orange tree yield in Year 5 was 6 kg/tree, which was ˜5% of the yield mature trees typically produce.

The leaf P–soil test P relationship was the strongest of any observed in the soil test calibration process (Figs. 1 and 2). In the grapefruit orchard, the Cate-Nelson procedure indicated that soil test P 13 mg•kg-1 or greater maintained optimum or higher leaf P in 4- to 7-year-old trees. Of the 42 grapefruit leaf P values that were below optimum, 39 were associated with soil test P below the critical value. The Cate-Nelson model explained 42% of relative leaf P concentration variability, which is reasonably good considering the wide variety of innate and external factors that affect individual citrus tree growth (Jackson, 1991). In the orange orchard, Cate-Nelson critical soil test P was 31 mg•kg-1. Of the nine orange leaf P values that were below optimum, eight were associated with soil test P below the critical value. The

Cate-Nelson model in this case explained only 21% of leaf P variability, most likely because the orange trees were smaller with considerably less P demand compared with the grapefruit trees.

Fruit yield was much less sensitive to soil test P. Correlations of annual fruit yields from 2001 to 2004 (grapefruit) and 2003 to 2004 (oranges) with soil test P from the previous year were not useful to estimate a critical value because Cate-Nelson R 2 was 0.11 or less (Table 2). Part of the reason for this result was the naturally occurring, wide tree-to-tree yield variation that is typical of juvenile citrus trees (Jackson, 1991). A longer-term analysis of yield versus soil test P did not improve the correlation. For example, when cumulative yield from 2001 through 2004 was compared with soil test P from 2000, Cate-Nelson R 2 values were 0.09 for grapefruit and 0.20 for oranges. The highest cumulative grapefruit and orange yields from single trees were 294 and 104 kg, respectively. These yields contained ˜50 and 18 g of P, respectively. Apparently, the amount of P required for the 4- to 7-year-old trees to produce fruit was satisfied by the amount available from the native soil.

Fruit and juice quality measurements were typical of fruit produced by young citrus trees, but little to none of their variation could be attributed to differences in soil test P (Table 2). The largest calculated Cate-Nelson R 2value (0.20) was for the relationship between grapefruit shape and soil test P, suggesting rounder fruits at soil test P below 5 mg•kg-1 and flatter fruits above this value. In a summary of nutritional effects on fruit and juice quality in Florida, Koo (1988) indicated that increasing P nutrition decreases juice acid, increases TSS-to-acid ratio, and decreases peel thickness. No significant relationship between these quality factors and soil test P was found in this study.

The lack of citrus yield or fruit quality response to soil P availability as indexed by the Mehlich 1 soil test was surprising considering that the orchards were planted on a native, low-P south Florida sandy soil. The site was chosen because it provided a reasonable chance of inducing P deficiency in young citrus trees if P fertilization was withheld. The results prompted a subsequent investigation to determine if the native soil contained enough total P to support the observed growth of nonfertilized trees by comparing the total amount of P accumulated by the trees as they grew with the total amount that existed in the root zone.

Annual P distribution in the tree, fruit, and root zone soil were estimated for the largest and highest-yielding grapefruit and orange trees that did not receive P fertilizer, assuming that: 1) P deposited on the soil surface in tree and weed residues (e.g., dropped leaves, twigs, fruit; dead roots) was recycled back to the root zone through soil organic matter; 2) the only P exported from the tree–soil system was in the harvested fruit; and 3) the P in the trees at planting and P from external sources like irrigation water or atmospheric deposition were negligible (Table 3). A linear relationship between citrus tree canopy volume and total P in the tree (including roots) developed by K. T. Morgan (unpublished data) was used to track total P accumulation as the trees grew: total P (g) = (0.94)[canopy volume (m3)] + 15.1. Fruit P was accounted for by multiplying annual yield (kg) by the mean P concentration of fresh citrus fruits (0.017%; Paramasivam et al., 2000). The annual active root zone volume was estimated to be a cylinder of soil with diameter equal to the tree canopy width and depth of 0.25 m. (This estimate was conservative because citrus roots were consistently observed extending laterally beyond the canopy edge.) The total P pool within the root zone was estimated by first determining the soil mass using a bulk density of 1.5 g•cm-3 and then multiplying by 42 mg•kg-1 P, which was the assumed total soil P concentration at planting based on our initial Mehlich 1 soil tests and a relationship we established between Mehlich 1 P and total P. As the trees began to produce fruit, we accounted for the mass of P exported in harvested fruit by subtracting it from the total P pool in the root zone.

Comparison of total tree P plus exported fruit P with the mass of potentially available soil P each year showed that the root zone contained more than enough P to meet the need of nonfertilized trees through 2004 (Table 3), assuming that a substantial portion of it was plant-available. This premise is not difficult to accept considering the perennial root system of citrus trees and the lack of high P-fixing capacity in the root zone soil. It was estimated that nonfertilized grapefruit trees took up roughly 25% of total root zone soil P in the early years. Canopy expansion and fruit production provided additional P sinks that increased soil P uptake to nearly 60% of total root zone P by the end of Year 7. A grapefruit leaf P concentration drawdown observed in nonfertilized trees (orchard-wide means of 0.15, 0.14, 0.12, and 0.11% P from 2001 to 2004) could indicate decreasing availability of remaining soil P and may foreshadow an eventual P deficiency. It was estimated that nonfertilized orange trees took up less than half of total soil P by 2004 as a result of a smaller canopy and less fruit production. Consequently, a leaf P concentration drawdown in these trees was not observed (means of 0.15%, 0.17%, 0.13%, and 0.14% P from 2001 to 2004).

# 3

# Fertilization of Fruit Trees

## INTRODUCTION

Fruit trees are fertilized to ensure continued growth and fruit production. In the backyard orchard, proper pruning in addition to the application of nitrogen in the spring prior to or at bud break helps maintain this productive status. Other than nitrogen and zinc, iron and manganese may limit growth due to our alkaline soil conditions. Apply nutrients based on a soil test analysis conducted by the soil testing lab at Colorado State University or another analytical lab of your choice. The amount of nitrogen to apply can be based on how much shoot growth occurred the previous season or on a soil analysis. Reduced fruiting wood and reduced fruit production results when the growth rate.

**Table. Annual Growth Rate.**

| Nonbearing Trees | Last Year's Annual Growth Rate |
|---|---|
| Apple | 12 to 36 inches |
| Pear | |
| Peach & Nectarine | |
| Tart Cherry | |
| Plum & Sweet Cherry | |
| | **Bearing Trees** |
| Apple Non-Spur | 6 to 18 inches |
| Apple Spur-type | 6 to 10 inches |
| Pear | 12 to 16 inches |
| Peach & Nectarine | 12 to 18 inches |
| Tart Cherry | ~ 8 inches |
| Plum & Sweet Cherry | ~ 8 inches |

## IDENTIFYING ANNUAL GROWTH

Each year of growth can be identified by the ring of bud scale scars remaining when the bud at the tip of the shoot grows. A difference in color and/or bark texture is typically evident above and below the annual growth ring. Check last year's annual growth rate at several points around the tree to determine the average length of last year's growth. Use the average length of annual growth to determine if and how much nitrogen should be applied.

## APPLYING NITROGEN BASED ON ANNUAL GROWTH RATE

Stone fruit trees (i.e. peach, cherry, plum and nectarines) can be fertilized at a maximum rate of 1/8 pound of nitrogen per inch of trunk diameter (measured 1 foot above ground level). Apply this amount if the tree's annual growth. Apply less fertilizer if the previous season's growth rate falls in between the recommended growth increments. If too much nitrogen is applied it can lead to excessive leaf and tree growth over fruit production. The maximum rate of nitrogen to apply to pome fruit trees (apples and pears) is 1/10th pound of nitrogen per inch of trunk diameter (measured 1 foot above ground level). Apply this amount if growth the previous year was at the low end of the recommended rate. As with stone fruits, apply less nitrogen the closer the actual growth rate approximates the recommended growth rate.

Maintain a record on the amount of nitrogen applied each year and the resulting growth. Such records provide a guide for the amount of nitrogen fertilizer to apply to achieve the desired results.

## APPLYING NITROGEN BASED ON A SOIL ANALYSIS

If a soil analysis has been conducted, the recommendations in Table below indicate the amount of nitrogen to apply.

**Table : Nitrogen requirements based on a soil test.**

| Soil Test Value NO-3N ppm | Pounds of Nitrogen to apply per 1,000 square foot area when percent organic matter is: | | | | | |
|---|---|---|---|---|---|---|
| | 0-0.5 | 0.6-1.0 | 1.1-1.5 | 1.6-2.0 | 2.1-2.5 | 2.6+ |
| 0.0-3.0 | 4.1 | 3.8 | 3.4 | 3.1 | 2.8 | 2.1 |
| 3.1-6.0 | 3.9 | 3.6 | 3.2 | 2.9 | 2.5 | 1.8 |
| 6.1-9.0 | 3.7 | 3.3 | 3.0 | 2.6 | 2.3 | 1.6 |
| 9.1-12.0 | 3.4 | 3.1 | 2.8 | 2.4 | 2.1 | 1.4 |
| 12.1-15.0 | 3.2 | 2.9 | 2.5 | 2.2 | 1.8 | 1.1 |

*Note:* The above information is specific to apples and pears. Add 0.2 pounds per 1,000 square foot area for peaches, nectarines, apricots and plums and 0.4 pound per 1,000 square foot area for cherries.

## CALCULATING THE AMOUNT OF FERTILIZER APPLICATIONS

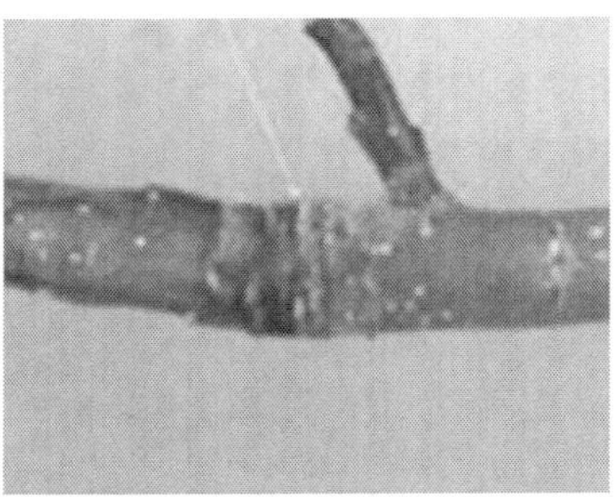

**Fig.** Annual Growth Ring - Apple

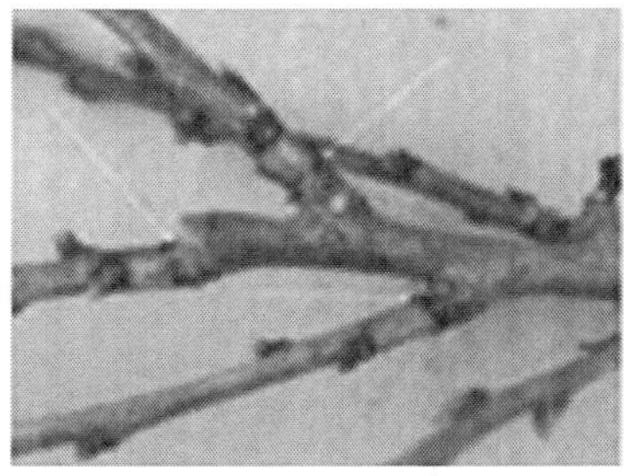

**Fig.** Annual Growth Ring - Pear

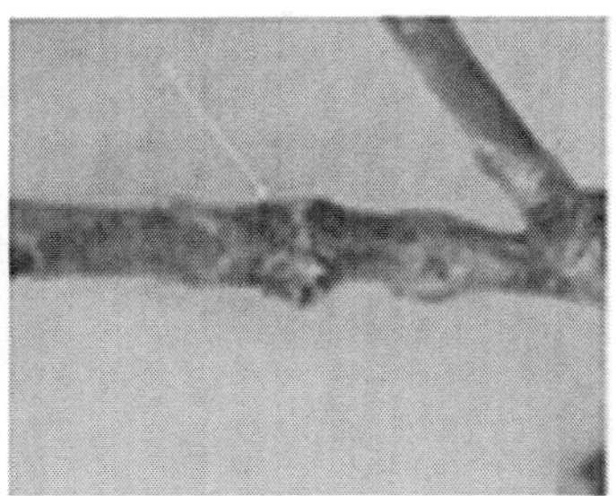

**Fig.** Annual Growth Rings - Apricot

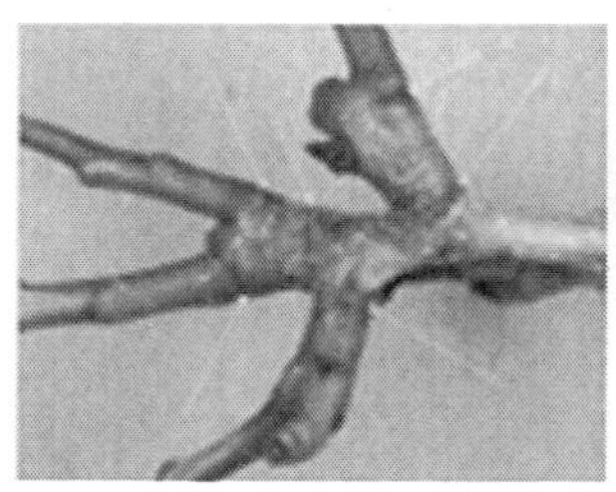

**Fig.** Annual Growth Ring - Sweet cherry

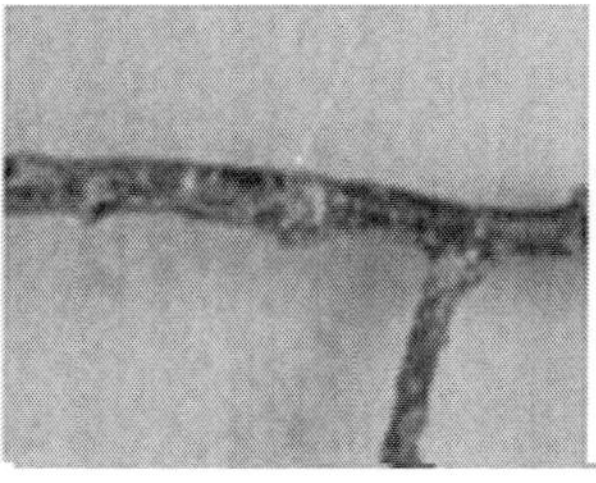

**Fig.** Annual Growth Ring - Peach.

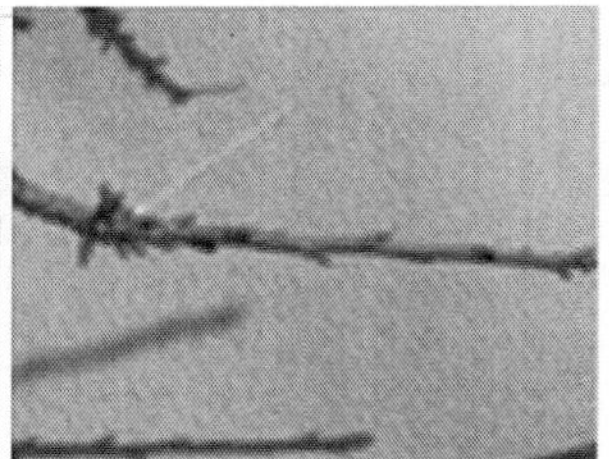

**Fig.** Annual Growth Ring - Italian Plum

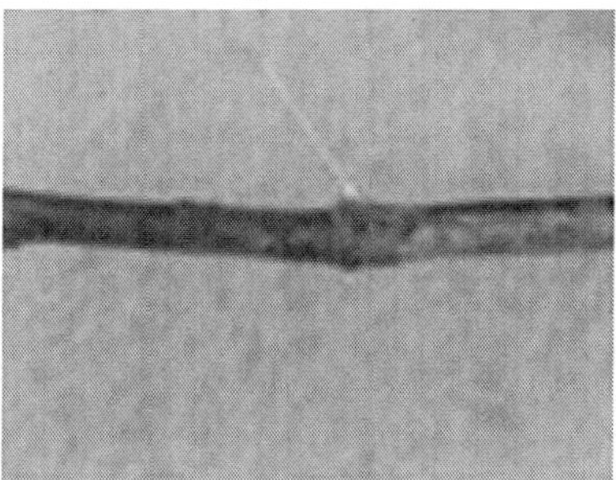

**Fig.** Annual Growth Ring - Santa Rosa Plum.

Fertilizer products contain specific quantities of nutrient based on percentage by weight. This is indicated on the product label such as 15-1-1. The first number (15) is the percentage by weight of nitrogen, the second number is the percentage by weight of phosphorus (P2O5) and the third is the percentage by weight of potassium (K2O). For example, ammonium sulfate (21-0-0) contains 21 percent nitrogen while blood meal contains 12 percent to 13 percent nitrogen. The amount of the fertilizer product needed is calculated by dividing the pounds of nutrient needed by the percent of that nutrient in the product.

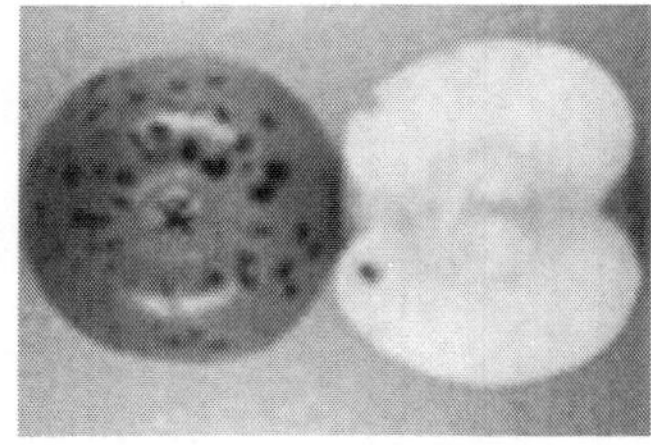

**Fig.** Bitter-pit and cork spot on Honeycrisp apple.

For example, if you need 1/2 pound of nitrogen for a given area and are using a product with 15 percent Nitrogen, divide.5 (one-half pound) by.15 (the percent of N in the product). This tells you 3.33 pounds of this product are needed to apply 1/2 pound of nitrogen.

## PRUNING AND EXCESSIVE IRRIGATION

Pruning the same amount each year will result in the same amount of stimulated growth. If the tree is pruned more severely, apply less nitrogen fertilizer. This will help avoid excessive growth. Irrigating too frequently or too much at one time, as is common with fruit trees planted in turf areas, stimulates growth and subjects fruit trees to possible root rot diseases. Take into account excessive irrigation and correct if possible. Eliminating the grass around the base of a fruit tree and applying a thin layer of mulch is recommended. Kill the grass with a glyphosate product, horticultural vinegar or fatty acid product such as Scythe before applying mulch. It is highly recommended to plant the backyard fruit orchard away from turf where the trees can be watered and fertilized based on their needs and not that of turfgrass

### Calcium

Bitter-pit and cork spot in apples (Figure 8) is a direct result of the inability of the tree to move adequate calcium into the fruit. Small brown lesions up to ½ inch (10 mm) in diameter appear on the fruit. The area below the skin becomes dark and corky. These spots may appear at harvest or during cold storage. Young vigorous trees with few fruit and trees that are over irrigated and over fertilized are more susceptible to this disorder. Calcium sprays are formulated by adding calcium chloride to water at the rate of 3 to 4 pounds per 100 gallons or 0.63 ounces per gallon of water. Apply the first spray about mid-June, a second in mid-July, and a third in mid-August. Trees not affected by this condition will not need this treatment.

### Iron

Iron deficiency symptoms result in the leaves at the end of the branch being yellow or light green with a network of deep green veins. Margin leaf burn may be evident. This deficiency often results in reduced fruit yields and poorly colored fruit with a flat taste. To help correct this problem, avoid over irrigation and improve the organic content of the soil.

### Zinc

Zinc deficiency symptoms are common in Colorado. Since soil applications of zinc have not proven effective, the application of a zinc spray prior to bud break in the spring is recommended. However, applications made within three days (five days for apples) before or after an application of horticultural or dormant oil can cause injury and should be avoided. In this case, except for apricots, apply the zinc in the fall, after October 10. Mix 1 tablespoon of zinc sulfate in a gallon of water. Thoroughly cover the tree with the spray and apply the spray until the bark is no longer able to hold the spray and spray runs off the tree.

## OTHER NUTRIENTS

Other than the application of nitrogen and zinc as described above, base the application of the other nutrients on the results of a soil analysis. Foliar sprays of micronutrients can give remediation for the season if started in April-May and continued until June-July. Miracle-Gro, or a similar water soluble fertilizer can be used for this treatment. Read and follow the label directions on the fertilizer container.

## PLACEMENT OF FERTILIZER

**Table. Phosphorus needs based on the extraction method used.**

| Phosphorus level of sufficiency ppm | AB-DTPA ppm | Bray-1 & Mehlich-II ppm | Mehlich-111 ppm | Olsen ppm | Apply $P_2O$ lbs/$1000_2$ft. |
|---|---|---|---|---|---|
| Very low to Low | 0-7 | 0-12 | <10 | 0-9 | 3.5 |
| Medium | 7.1-13 | 13-25 | 11-31 | 10-16 | 2.9 |
| High | 13-22 | 26-50 | 32-56 | 16-31 | 2.3 |
| Sufficient | <23 | >50.1 | >56.1 | >31.1 | 0 |

*Note:* The Phosphorus extraction method used may be noted on the soil test results. If not, contact the lab to determine what method was used.

Nitrogen (N) and other nutrients, with the exception of zinc, can be broadcast over the ground and watered in, or applied in a band in the irrigation furrows prior to irrigation. Do not apply fertilizer against the trunk as tissue damage may result. Spread the fertilizer evenly and do not dump it in a pile at the base of the tree or injury will result. If the area to be fertilized is more or less than 1,0002 feet, calculate the amount of fertilizer to apply accordingly. Foliar applications can also be used if appropriate materials are chosen.

**Table. Soil test results and recommendations.**

| Potassium ppm | Apply $K_20$ lbs/$1000_2$ ft. | AB-DTPA extractable iron Fe ppm | Chelated Fe * (Sequestrene 138-Fe) lbs/$1000_2$ ft |
|---|---|---|---|
| 0-60 | 1.8 | 1.8 Low -2.0 | 3.5 |
| 61-20 | 0.9 | Marginal 2.0-4.5 | 1.7 |
| 121-200 | 0 | Marginal 2.0-4.5 | 0 |
| 201 + | 0 | | |

*Note:* Sequestrene 138 Fe is 6 percent iron. Other chelated iron products may not be effective in high pH soil

## PROPER FERTILIZATION OF CITRUS TREES FOR OPTIMUM GROWTH AND PRODUCTION

Using the proper fertilizer is also very important. Citrus trees need a balance of micronutrients that are only available in fertilizers specified for citrus and fruit trees. Using a basic 10-10-10 fertilizer won't give them the micronutrients that they need for optimal growth and production ability. It is important to keep all grass removed from the citrus tree trunk out to one foot past the drip line. Mulching is not recommended, but if you must, you need to keep mulch at least a foot away from the citrus tree trunk. Mulching around the trunk can lead to damage from insect and disease that can kill your citrus tree.

### FERTILIZING YOUNG CITRUS TREES

Newly planted citrus trees need to be fed eight times in the first year, at the rate of one pound per tree per month. This is extremely important in helping the tree establish a healthy root system, so that it becomes well established in it's environment. You should start fertilizing in March and end in October. For our winter visitors, this may involve enlisting the aid of a neighbor or friend, or paying a company to fertilize your young citrus trees.

## Fertilizing Bearing Citrus Trees

One very important thing to know is that you never heavily fertilize a citrus tree that is in bloom. A heavy application of fertilizer will cause the buds to drop from the tree. All citrus fertilization should take place before the trees are in bloom (early March) or after the fruit has reached at least the size of a large pea. If you miss your scheduled March citrus fertilization, don't fertilize until the fruit is the proper size, usually around mid-May. Citrus trees that are 4' or larger (bearing trees) should be fertilized three times a year (March, May and October), at a rate of one pound of fertilizer to every foot of limb spread. If you are fertilizing a tree that has been pruned, you must fertilize for the limb spread that existed before the tree was pruned. For example, a tree with a 10 foot limb spread that is pruned back to 8 feet must still received 10 pounds of fertilizer.

## How to fertilize citrus properly

Two cups of granular citrus fertilizer is roughly equal to one pound. Fertilizer should be broadcast starting 10 inches away from the trunk, out to the drip line (where the longest limbs end). For example, a tree with a 10 foot limb spread should be given 10 lbs. (20 cups) of fertilizer. It is also recommended to spray the trees with a good Citrus Nutrional Spray once a month, starting in February and ending in October. This delivers the correct balance of micronutrients in a foliar feed.

# 4

# Fruit Nutrition from Apple Trees

Delicious and crunchy apple fruit is one of the most popular fruits, favorite of health conscious, fitness lovers who believe in the concept "health is wealth." This wonderful fruit is packed with rich phyto-nutrients that in the true senses indispensable for optimal health. The antioxidants in apple have much health promoting and disease prevention properties; thus truly justifying the adage, "an apple a day keeps the doctor away."

**Scientific name: Malus domestica.**

Crimson-red apple. Apples in a store. Apple, york. (Malus domestica)

Apples are obtained from the medium-sized tree belonging to the Rosaceae family. The apple tree is originated in the mineral-rich mountain ranges of Kazakhstan, and now being cultivated in many parts of the world. Apple fruit features oval or pear shape; and the outer skin has different colors depending upon the cultivar type. Internally, the juicy pulp has been off-white to cream in color and has to mix of mild sweet and tart taste. Its seeds are inedible because of their bitter taste. Hundreds of varieties of apples are either eaten as table fruits or as dessert fruit grown in the US and worldwide. Some of the apples are sought-after in cooking and baking too.

**HEALTH BENEFITS OF APPLE**

- Delicious and crunchy apple fruit is notable for its impressive list of phtyto-nutrients, and anti-oxidants. Studies suggest that its components are essential for normal growth, development and overall well-being.

- Apples are low in calories; 100 g of fresh fruit slices provide only 50 calories. They, however, contain no saturated fats or cholesterol. Nonetheless, the fruit is rich in dietary fiber, which helps prevent absorption of dietary-LDL or bad cholesterol in the gut. The fiber also saves the colon mucous membrane from exposure to toxic substances by binding to cancer-causing chemicals inside the colon.
- Apples are rich in antioxidant phyto-nutrients flavonoids and polyphenolics. The total measured anti-oxidant strength (ORAC value) of 100 g apple fruit is 5900 TE. Some of the important flavonoids in apples are quercetin, epicatechin, and procyanidin B2. Additionally, they are also good in tartaric acid that gives tart flavor to them. Altogether, these compounds help the body protect from deleterious effects of free radicals.
- Apple fruit contains good quantities of vitamin-C and beta-carotene. Vitamin C is a powerful natural antioxidant. Consumption of foods rich in vitamin C helps the body develop resistance against infectious agents and scavenge harmful, pro-inflammatory free radicals from the body.
- Further, apple fruit is a good source of B-complex vitamins such as riboflavin, thiamin, and pyridoxine (vitamin B-6). Together these vitamins help as co-factors for enzymes in metabolism as well as in various synthetic functions inside the body.
- Apple also contains a small amount of minerals like potassium, phosphorus, and calcium. Potassium is an important component of cell and body fluids helps controlling heart rate and blood pressure; thus, counters the bad influences of sodium.

**See the table below for in depth analysis of nutrients:**
**Apple fruit (Malus domestica), Fresh, Nutritive value per 100 g, ORAC value-5900**

| **Energy** | **50 Kcal** | **2.5%** |
|---|---|---|
| Carbohydrates | 13.81 g | 11% |
| Protein | 0.26 g | 0.5% |
| Total Fat | 0.17 g | 0.5% |
| Cholesterol | 0 mg | 0% |
| Dietary Fiber | 2.40 g | 6% |
| Vitamins | | |
| Folates | 3 µg | 1% |
| Niacin | 0.091 mg | 1% |
| Pantothenic acid | 0.061 mg | 1% |
| Pyridoxine | 0.041 mg | 3% |
| Riboflavin | 0.026 mg | 2% |

| | | |
|---|---|---|
| Thiamin | 0.017 mg | 1% |
| Vitamin A | 54 IU | 2% |
| Vitamin C | 4.6 mg | 8% |
| Vitamin E | 0.18 mg | 1% |
| Vitamin K | 2.2 μg | 2% |
| Electrolytes | | |
| Sodium | 1 mg | 0% |
| Potassium | 107 mg | 2% |
| Minerals | | |
| Calcium | 6 mg | 0.6% |
| Iron | 0.12 mg | 1% |
| Magnesium | 5 mg | 1% |
| Phosphorus | 11 mg | 2% |
| Zinc | 0.04 mg | 0% |
| Phyto-nutrients | | |
| Carotene-ß | 27 μg | -- |
| Crypto-xanthin-ß | 11 μg | -- |
| Lutein-zeaxanthin | 29 μg | -- |

## Selection and Storage

Fresh apples are readily available in the stores all around the season. Choose fresh, bright, firm textured apples with rich flavor. Avoid fruits with pressure marks over their surface as they indicate underlying mottled of pulp. Fresh apples can be kept at room temperature for few days and stored inside the refrigerator for two to three weeks. Wash them in clean running cold water before use to remove any surface dust and pesticide/fungicide residues.

## Preparation and Serving Tips

Wash apples thoroughly in the running water to remove any surface dust, insecticide/fungicide sprays. Trim off top end using a paring knife and cut the fruit into two equal halves. Take out tiny, centrally placed, bitter seeds. Slice the fruit into desirable cubes or slices.

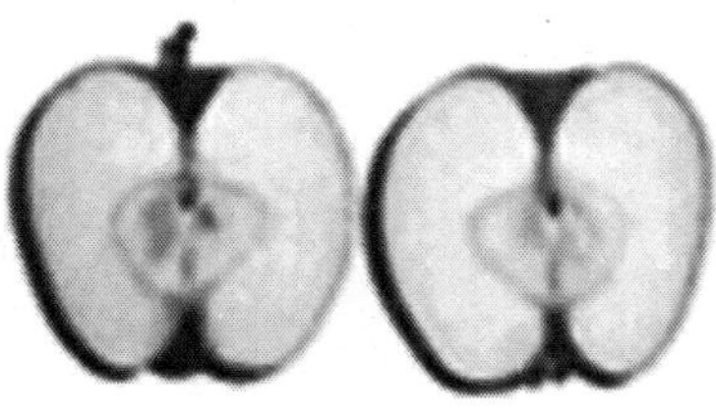

Apple fruit sections.

Apples pie.

Here are some serving tips:

- Eat apple fruit as they are, along with their peel in order to get maximum health-benefits.
- Sliced apple turns brown (enzymatic brownish discoloration) on exposure to air due to conversion in iron form from ferrous oxide to ferric oxide. If you have to serve them sliced, rinse slices in water added with few drops of fresh lemon.
- Cloudy as well as clear apple juice is a healthy alternative drink with dinner.
- Apple fruit is also used in the preparation of fruit jam, pie, and fruit salad.

**Safety Profile**

Good yield demands close attention and supervision of apple crop. According to the environmental-working group reports, apple fruit is one of the heavily pesticide-contaminated produce. The most common pesticides found on apple are organo-phosphorous and organo-chloride pesticides like Permethrin and DDT. Therefore, it is recommended to wash the fruit thoroughly before use.

## GROWING APPLE TREES IN THE HOME GARDEN

Growing apple trees in the home garden can be fun and rewarding. Several factors are important to consider before planting for successful apple production. Apple variety and rootstock, site selection, proper planting, training and pruning, adequate fertility, and pest control all contribute to healthy and productive trees. A brief discussion of these considerations follows. Rootstocks and Tree Spacing - All apple trees sold commercially consist of two parts that are grafted together to form the tree. The "scion" is the top portion that branches and bears fruit and is grafted onto a "rootstock". The type of fruit is determined by the scion variety. The rootstock can be a "seedling", which produces a full size or standard tree, or the rootstock can be "size-controlled" or "dwarfing", which produces a tree that is smaller than full size. The rootstock determines the relative size of the tree but does not affect the type or quality of fruit that the tree bears. Different rootstocks are desirable because they can control the size of the apple tree, reduce the time until the tree reaches fruit-bearing age, and may offer some pest resistance.

Some important characteristics of the rootstocks that are commercially available. Tree size is relative and is shown as a percent of the size that the tree would be on a full size seedling root-stock. Rootstock, soil fertility, and pruning can influence tree size, and therefore influence tree spacing. A range of planting distances with the wider distances for trees planted in good, fertile soils and optimum growing conditions. Trees on the more dwarfing root-stocks must be staked for the life of the tree to obtain optimum growth and to prevent leaning and potential tree breakage. Commonly used stakes consist of a 3-inch diameter wood pole or a 1-inch diameter metal conduit. The stake should be 10 feet high with 2 feet driven into the ground approximately 6 inches from the base of the tree.

**Table. Rootstock Characteristics.**

| Rootstock | Tree Size as Percent of Seedling | Tree Spacing in Row (ft) | Anchorage | Years to Fruit Production |
|---|---|---|---|---|
| Seedling* | 100 | 15-18 | Excellent | 6-10 |
| MM.111 | 85 | 14-18 | Excellent | 4-6 |
| MM.106 | 80 | 12-16 | Excellent | 3-4 |
| M.7a | 70 | 10-14 | Fair | 3-4 |
| M.26** | 50 | 8-12 | Poor | 2-4 |
| Mark** | 35-40 | 6-8 | Good | 2-3 |
| M.9** | 35 | 4-8 | Poor | 2-3 |

* Mature tree is 12-20 feet tall, depending on variety.

** Trees should be staked and tied to the stake at planting.

**Table. Variety characteristics in order of maturity.**

| Variety | Fruit Color | Fruit Use | Relative Bloom Time | Potential Cross-Pollinizers |
|---|---|---|---|---|
| Gala | Yellow-orange to red | Fresh | Early to Midseason | Golden Delicious |
| Empire | Dark red over green background | Fresh, cooking | Early | Golden or Red Delicious, Gala |
| Jonagold* | Yellow with light red stripes | Fresh, cooking | Midseason | Gala, Empire |
| Golden Delicious | Yellow green to light yellow | Fresh, cooking | Midseason to Late | Red Delicious, Gala, Empire |
| Red Delicious | Red | Fresh | Early | Golden Delicious, Gala |
| Stayman* | Blush to red | Fresh, cooking | Midseason | Gala, Golden or Red Delicious |
| Rome | Blush to red | Fresh, cooking | Late | Fuji, Braeburn |
| Braeburn | Green with light red blush | Fresh | Midseason | Rome, Fuji |
| Fuji | Green with red stripes | Fresh | Midseason | Rome, Braeburn |

* Pollen produced by these varieties is sterile.

Varieties–The variety of apple selected should be based on fruit characteristics, bloom time and pollen compatibility. Table 2 (above) shows several popular varieties in North Carolina, listed in order of fruit maturity. Nurseries can also provide varietal information and pollen compatibility suggestions. Crabapple trees can also be used as pollinizers if they bloom at the same time as the desired variety.

Pollination–All apple varieties should be considered self-incompatible, meaning that they cannot pollinate themselves or any flowers of the same apple variety. The highest quality fruit is harvested when cross-pollination occurs with a suitable pollinizer variety. You will need to plant at least two varieties of apple trees together in order to maximize fruit production and quality. Make sure that the varieties you choose have overlapping bloom dates, so that both varieties bloom at the same time. Some varieties, such as Winesap, Mutsu, Jonagold, and Stayman, produce sterile pollen and should never be used as pollinizers. However, pollen from other varieties can be used to pollinate these pollen-sterile varieties. Remember, two trees of the same apple variety cannot be used for cross-pollination. Since the pollen from apple blossoms is transferred primarily by bees, be careful not to spray insecticides during bloom when honey bees are present.

**SITE SELECTION**

Soils - Take a soil test prior to planting your apple trees. Your local County Extension Center can instruct you in collecting the soil sample, help you interpret the results, and provide valuable information about the soil in your county. Results from the soil test will determine the soil amendments necessary to correct nutrient deficiencies and adjust soil pH. The amendments should be worked into the soil to a depth of 12 to 18 inches where the tree will root, not just the planting hole. Apple trees will tolerate a wide range of soils as long as water and nutrients are not limiting and soil pH is adequate. Avoid heavy, poorly-drained soils and low spots, since apple trees cannot survive if water remains standing in the root zone.

Air Drainage - It is important to select a site where the tree will not be in a "frost pocket", where cold air settles in low-lying areas. In a frost pocket, low spring temperatures commonly kill the blossoms or developing fruit because cold air settles around the tree. Good air drainage, especially during early spring frosts, is critical. Choose a higher site with a slope if possible so cold air will flow down away from the trees and will not accumulate around the trees. Do not plant the trees close to a fence row, wooded area, or at the bottom of a hill, as cold air drainage will be impeded.

Other Considerations - Apple trees require full sun and should be planted where the trees will not be shaded from large trees or buildings. Follow tree spacing guidelines that pertain to the rootstock you have chosen from Table 1. Do not plant trees near wooded areas or streams to avoid animal damage.

Prior to planting, remove weeds either manually or with an approved herbicide that will not harm the young tree. If you are planting the tree in a lawn, remove the grass from the planting area in a 4-foot diameter circle. Grass competes with young trees for available water and nutrients and can significantly reduce tree growth and productivity.

## PLANTING THE TREE

Tree Purchase and Preparation - Purchase a healthy 1-year-old nursery tree, 4 to 6 feet tall, with a good root system. A small tree with a good root system will transplant better than a large tree. When you get the tree, protect it from injury, drying out, mouse or vole damage, freezing, and overheating. If the roots have dried somewhat, soak them in water for about 24 hours before planting. If you are unable to plant the tree immediately, there are two options:

- Wrap the roots in plastic along with moist sawdust or newspaper, and place the tree in a refrigerator or cooler at 40 F. Never store the tree with fruit or vegetables, as ethylene gas from ripening foods can kill young trees.
- "Heel-in" the tree. To heel-in a tree, dig a trench and place the tree roots evenly in it, cover the roots with soil, sawdust or peat, and water the tree thoroughly. The tree can be kept for several weeks using this method before permanently planting.

Planting the Tree - In North Carolina, trees can be set from late fall to early spring. To plant the apple tree, first dig a hole approximately twice the diameter of the root system and 2 feet deep. Place some of the loose soil back into the hole and loosen the soil on the walls of the planting hole so the roots can easily penetrate the soil. Spread the tree roots on the loose soil, making sure they are not twisted or crowded in the hole. Continue to replace soil around the roots. As you begin to cover the roots, firm the soil to be sure it surrounds the roots and to remove air pockets. Do not add fertilizer at planting time as the roots can be "burned". Fill the remainder of the hole with the loose soil, and press the soil down well. Important: The graft union must be at least 2 inches above the soil line so that roots do not emerge from the scion. When you have finished planting the tree, water well to eliminate air pockets and provide good contact between the roots and the soil.

## TRAINING AND PRUNING CENTRAL LEADER TREES

Proper training and pruning of fruit trees is essential to the development of a strong tree framework that will support fruit production. Properly shaped trees will yield high-quality fruit much sooner and will live significantly longer. Regular pruning and training will also maximize light penetration to the developing flower buds and fruit. Additionally, a well-shaped tree canopy permits adequate air movement through the tree, which promotes rapid drying to minimize pest problems.

Central Leader Trees - A central leader tree has one main, upright trunk, called the "leader". Branching should begin on the leader 24 to 36 inches above the soil surface to allow work under the tree. The first year, 3 to 4 branches, collectively called a "scaffold whorl", are selected. The selected branches should be spaced uniformly around the trunk, not directly across from or above one another. The major lateral branches are commonly referred to as scaffold branches on which the central leader tree is "built". Above the first scaffold whorl should be an area of 18 to 24 inches, called a "light slot", without any branches to allow light to reach all lower leaves and fruit. This light slot is followed by another whorl of scaffolds. Maintain alternating scaffold whorls and light slots up the leader to the desired maximum tree height. The shape of a properly trained central leader tree is like that of a Christmas tree (except with slots for light).

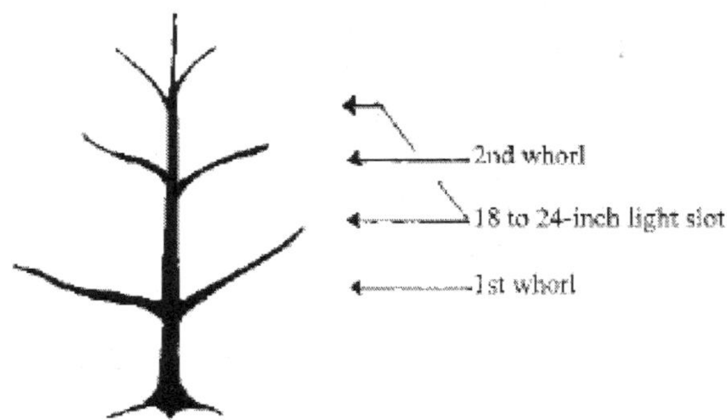


**Fig.** Side view of a central leader tree.

Newly Planted Trees - After early winter planting, wait until just before the buds start to grow in the spring to "head", or cut, the unbranched central leader to 36 inches above the soil surface to encourage new lateral branching. When new growth is 3 to 4 inches long, identify the most upright shoot that will continue to be the central leader. Leave it and remove all new shoots growing 3 to 4 inches immediately below this new terminal to prevent competition. This will also encourage lateral growth in the area 6 to 14 inches below the cut tip of the young tree. Branches that form 6 to 14 inches below the cut tip of the tree are less vigorous, less upright, and easier to train as productive scaffold limbs. When the lateral branches, or scaffold branches, are 3 to 6 inches long, they should be spread to a wider crotch angle that will provide a stronger framework for fruit production. Toothpicks or clothespins can be used to prop the young branches out to a 50 to 60 degree angle. This angle will slow vegetative growth, promote lateral branches, and allow the tree to initiate flowers and produce fruit sooner.

Scaffold Training - Improperly trained fruit trees have very upright branch angles, which result in excessive vigor and serious limb breakage under a heavy fruit load. Larger branches can be spread out using short wooden boards with a notch cut in each end for the branch to fit into. Hanging weights on the branch or tying it down with string wrapped loosely around the limb are other methods for spreading the branches. All upright growth from scaffold branches should be either pulled down to a horizontal position or removed when it is 3 to 4 inches long.

Dormant Pruning vs. Summer Pruning - Pruning the tree during the winter, while it is dormant, will invigorate the tree and cause it to grow and branch more the following season. To promote scaffold branch development, cut the central leader 20 to 28 inches above the highest usable scaffold whorl during the dormant season. It is best to do dormant pruning in late winter or early spring, after the risk of severe freeze is over. Be sure to remove any dead or diseased wood and dried apples at this time as well. After the tree resumes growth in the spring, continue to train the scaffold branches of the tree as you did the previous growing season. Select a new upright shoot to continue the central leader, and remove all new shoots 4 inches below it. Also select the branches to form another whorl of 4 to 5 scaffold branches. Prop all lateral branches out to a 50 to 60 degree angle.

Summer pruning will devigorate the tree and cause it to grow less in that growing season. Remove all undesirable branches directly across from one another on the central leader when they are 3 to 4 inches long. Also, select lateral branches that are spaced uniformly around the leader to prevent crowding as the limbs grow in diameter. Once the tree has filled its allotted space, lateral branches will need to be cut back to their desired length during the summer to devigorate the tree and prevent further growth, not during the dormant season. Ask your County Extension Center for information on the best way to prune your apple tree.

Fruit Thinning - Apple trees often set a heavier crop of fruit than the limbs can withstand. To ensure good fruit size, return bloom for the following year, and to prevent tree breakage, it is necessary to thin the fruit. Every apple blossom results in a bloom cluster of 5 to 6 blossoms. Apples should be thinned when they are about the size of a dime. Cut off enough fruit so that the remaining apples are spaced 4 to 6 inches apart, and leave only one fruit per cluster. It may seem like very few fruit remain, but you will harvest higher-quality fruit, potentially reduce insect and disease problems, and increase the chances for a full crop the next season.

Fertility - Adequate tree nutrition is essential for quality apple production. Determine the nutrient status of your soil by taking a soil sample prior to planting and each year thereafter at the same time of year. Follow the fertilization guidelines provided by the soil test. This will prevent over-fertilization, will be cost-efficient, and will maintain healthy and productive trees. A leaf sample taken in July or August will determine the nutrient status of the tree and can be helpful in conjunction with the soil test. In addition to soil analysis and foliar analysis, regular observation of vegetative growth is a useful indicator of tree fertility. Optimum fertility exists if lateral, outward growth is between 12 and 18 inches per year.

If you are unable to take a soil test, a useful rule is to apply 1 pound of 10-10-10 fertilizer to each tree the first year, 2 pounds the second year, and 3 pounds the third year up to a maximum of 5 to 6 pounds for a mature tree. Always adjust rates of fertilizer application according to annual shoot growth. Apply fertilizer in late winter or early spring before growth begins. Fertilizer should be broadcast on the soil surface around the drip line of the tree. The "drip line" is the circular line at the outer ends of the branches. Avoid getting fertilizer within 6 inches of the trunk as it could burn the tree.

Weed Control - Controlling weeds and gras-ses around young apple trees minimizes competition for soil nutrients and moisture, encourages vigorous tree growth, and increases fruit size. It will also help to minimize damage from pine and meadow voles, and other rodents. Keep all vegetation controlled out to the drip line of the tree where the outer branches end. Do not use weed whips; these will girdle the tree without any visible signs of injury. Avoid mechanical weed culti-vation, such as tilling, as it damages shallow tree roots. Herbicides are effective, but follow the label directions carefully and keep them off the tree. Mulch will also control weeds and conserve soil moisture, however rodents may burrow under the mulch and gnaw tree trunks and roots. When using mulch, place rodent guards around the base of the tree, and pull mulch back in the fall, leaving a 1-foot circle of bare soil surrounding the trunk.

Disease and Insect Control - Diseases and insects can cause serious damage to apple trees and fruit. Good sanitation practices are necessary to control pest problems. Cut out all dead or diseased wood, remove dried apples, and clear leaves and fallen debris away from trees. Disinfect pruning tools with a 10% solution of a household disinfectant (Lysol) or bleach, before and after use and between trees. Household disinfectants, such as Lysol, will not corrode tools or ruin clothing. A regular spray program is essential for high fruit quality and healthy trees. Use a multipurpose fungicide and insecticide labeled for apples. These can be obtained from a garden center and will include application instructions. A spray to control fungus problems should be applied when the first sign of green tissue appears. A horticultural oil should also be sprayed on apple trees at the first sign of green growth in the spring to suffocate scale insects and reduce overwintering mite and aphid eggs. For homeowners with only a few trees, premixed orchard sprays are available from many garden centers. Begin applications after full bloom is over and spray every 10 to 14 days through-out the summer.

If you think you have an insect or disease problem, contact your County Extension Center immediately. It is important to identify the pest accurately so an effective treatment can be selected.

## HARVESTING AND FRUIT STORAGE

Apples reach maturity at different times, depending on variety and climate. There is not a specific date at which you can expect to harvest your apples. Instead, you can observe your apples as they grow and inspect the fruit for certain changes which indicate maturity. The "ground" or base skin color of the apples changes from green to yellow as the fruit matures. Flesh color also loses its greenish tint and turns yellow or white. When you are convinced that the apples look mature, take a bite! A mature fruit will be crisp and juicy. A pleasing taste is the final indicator of fruit maturity.

Proper storage conditions help prolong the shelf-life of your apples. Store apples at 32 oF and maintain high humidity. The crisper drawers of many refrigerators work well, but keep the fruit away from vegetables since ripening fruit gives off gas that may spoil vegetables. Apples can also be stored in plastic bags in the refrigerator to prevent fruit dehydration.

## CUSTARD APPLE NUTRITION FACTS

Have you seen a heart shaped fruit with an unusually scaly rind? If you must know, we are talking about the custard apple! Despite its hideous appearance, this fruit has several benefits. It grows on a small deciduous tree, about 10 meters in height and is known by different names worldwide. At an unripe stage, custard apples showcase a pale green colour and when it is perfectly ripe, brown patches develop on its skin. These fruits are quite rich in flavour and nutritive content. If you are not a fan of the custard apple then you can enjoy the sweetness and aroma of the fruit in the form of juices or ice creams. The custard apple is a wholesome and healthy food. Check out more about the history, heath benefits and storage tips for this fruit.

### History

The native of custard apple is said to be West Indies, but the fruit was carried through Central America to southern Mexico. For a long time, the custard apple was grown and naturalized in countries like Peru and Brazil. Evidence suggests that these fruits have been cultivated in the Bahamas, Bermuda and southern Florida. This fruit quickly spread across tropical Africa around the 17th century and was grown in South Africa as a dooryard fruit tree. In India, the custard apple was prevalent in parts of Kolkata. Eventually, it gained popularity on the east coast of Malaya, south-east Asia and the Philippines. 80 years ago, custard apples were found in abundance in Guam, a region off the western Pacific Ocean.

### Health Benefits of Sharifa (Custard Apple)

- The numerous vitamins and nutrients present in custard apple provide protection from various medical conditions such as heart attacks, strokes and cancer.

- If you have a fluctuating blood pressure then a regular intake of this fruit is effective in keeping it under control.
- Being a rich source of Vitamin A, custard apples give the consumer a glowing and healthy skin and shiny and lustrous hair.
- Looking for a natural way to enhance your eye sight? Then the custard apple acts as a storehouse of various vitamins making your vision sharper and brighter.
- Intake of custard apple helps in curing indigestion problems. Since this is indirectly linked to constipation, these fruits inject copper into the body which ensures the production of stools.
- The high amount of magnesium in custard apples maintains water balance in the body. This promotes the removal of various acids from the joints and lowers the risk of rheumatism and arthritis.
- Custard apples are packed with iron and are the simplest and fastest cure for treating anemia.
- For extremely underweight individuals, custard apples are an ideal way to put on a few kilos. Also, the sugar fibers present in them make the making metabolic system more efficient.
- The skin of custard apple is helpful in fighting against tooth decay and gum pain.
- These fruits are excellent in treating vomiting, gout, lice and vitamin B6 deficiencies.

**Nutritional Value & Calories In Custard Apple**

**Amount:** 1 cup
**Total Weight:** 250 g

| Nutrients | Amount |
|---|---|
| Basic Components | |
| Protein | 5.2 g |
| Water | 183 g |
| Ash | 1.9 g |
| Calories | |
| Total Calories | 235 |
| Calories From Carbohydrate | 213 |
| Calories From Fat | 6.1 |
| Calories From Protein | 17 |
| Carbohydrates | |
| Total Carbohydrates | 59 g |
| Dietary Fiber | 11 g |
| Fats & Fatty Acids | |
| Total Fat | 725 mg |
| Saturated Fat | 120 mg |
| Monounsaturated Fat | 285 mg |
| Polyunsaturated Fat | 100 mg |

| | |
|---|---|
| Omega-6 Fatty Acids | 100 mg |
| Vitamins | |
| Vitamin A | 15 IU |
| Vitamin C | 91 mg |
| Thiamin | 275 mcg |
| Riboflavin | 283 mcg |
| Niacin | 2.2 mg |
| Vitamin B6 | 500 mcg |
| Folate | 35 mcg |
| Pantothenic Acid | 565 mcg |
| Minerals | |
| Calcium | 60 mg |
| Iron | 1.5 mg |
| Magnesium | 53 mg |
| Phosphorus | 80 mg |
| Potassium | 618 mg |
| Sodium | 23 mg |
| Zinc | 250 mcg |
| Copper | 215 mcg |
| Selenium | 1.5 mcg |

How many calories in custard apple (per 100 gm)

Custard appleshave about 94 calories per 100 gm of weight.

How to Buy Custard Apple

- Hold a custard apple in your hand and press it gently. It must feel firm but not hard, as these are still unripe. Soft and squishy custard apples are considered over ripe and are unfit for consumption.
- Observe the colour of the fruit; it should be brownish green or brown but definitely not black.
- Avoid smaller custard apples as they are generally immature. Instead of buying them in bunches, it is advisable to buy single fruits.

Custard Apple Storage Tips

- It is recommended to consume custard apples soon after buying. Meanwhile, wrap them in a plastic bag and place them at room temperature.
- If you have bought a ripe custard apple then you can speed up its ripening process by placing it in a brown paper bag, along with a banana.
- Place the fruits in a refrigerator if you want to store them for longer periods of time.

## FRUIT PRODUCTION AND APPLE NUTRITION

### MANURE FOR ORCHARDS

Manure can pose a food safety risk on many fruit crops. Ensure at least 120 days between manure application and harvest. Manure contains beneficial organic matter and many macro and micronutrients. The organic nitrogen in manure is mineralized over time, providing nitrogen in diminishing quantities for several years. Adjust additional organic and inorganic nitrogen applications accordingly. Excessive nitrogen, particularly in the second half of the growing season, can result in poor fruit colour, reduced storability, excessive growth and delayed cold-hardening of the woody tissue, which makes trees more susceptible to winter injury. Apply no more than 7 tonnes/ha of poultry manure (20 $m^3$ liquid), 40 tonnes/ha of cattle manure (100 $m^3$ liquid), and 35 tonnes/ha hog manure (65 $m^3$ liquid). Since the nutrient content of manure varies greatly, have it tested for nutrients before application (see Manure nitrogen). Broadcast manure at moderate rates and work into the soil in late fall or early spring before planting. Do not put manure around newly planted trees because of potential winter injury. Reduce the rate of nitrogen, phosphorus and potassium fertilizers applied to adjust for the nutrients supplied by manure. Table 3-11. Average Fertilizer Replacement Values for Different Types of Manure, shows the average composition of some manures and suggested reduction of fertilizer when manure is used.

## PH REQUIREMENTS

The pH of a soil is a measure of its acidity and affects nutrient uptake and crop performance. If the soil test recommends a lime application to increase soil pH, add lime at recommended rates prior to planting. For details regarding rates and suggested types of lime to use, refer to Soil pH and Liming. In established orchards, sample soil in the tree row every three years to ensure the pH is satisfactory. If the pH on clay loam soils drops below 5.1 and on sandy soils below 5.6, apply lime to the sod cover in the fall or before spring cultivation. The results will not show immediately because lime moves and reacts slowly in the soil.

## LEAF ANALYSIS

In established plantings, leaf analysis is the best method to determine nutrient needs. The nutrient levels in these plant tissues most accurately reflect the tree's uptake of nutrients. Soil analysis is used in conjunction with leaf analysis to determine the nutrient status of the soil. A combination of both analyses best evaluates fertilizer and lime needs. For more information on these tests, see Plant tissue analysis. Many orchard conditions affect nutrient uptake. Consequently, nutrient levels vary slightly each year depending on the season. To obtain optimum growth and fruit quality, all nutrients must be present in sufficient concentrations.

For leaf analysis to be effective, sample the same trees each year and make adjustments to the fertilizer program on the basis of this leaf analysis. Fertilizer requirements are adjusted to soil management practices, tree age, rootstock, soil type and previous fertilizer applications. Growth, fruit size, colour and storage quality must also be considered to determine the fertilizer required.

**Table. Foliar Nutrient Sufficiency Range of Apple (Mid-Shoot Leaves Taken in Last Two Weeks of July from Mature Trees)**

| Variety | N* (%) | P (%) | K (%) | Ca (%) |
|---|---|---|---|---|
| **Delicious, Crispin** | 2.2-2.7 | 0.15-0.4 | 1.4-2.2 | 0.8-1.5 |
| **Empire, Spy** | 2.1-2.6 | 0.15-0.4 | 1.3-2.1 | 0.7-1.5 |
| **McIntosh, others** | 2-2.5 | 0.15-0.4 | 1.2-2 | 0.8-1.5 |

| Variety (ppm) | Mg (%) | Fe (ppm) | B (ppm) | Zn (ppm) | Mn |
|---|---|---|---|---|---|
| Delicious, Crispin | 0.25-0.4 | 25-200 | 20-60 | 15-100 | 20-200 |
| Empire, Spy | 0.25-0.4 | 25-200 | 20-60 | 15-100 | 20-200 |
| McIntosh, others | 0.25-0.4 | 25-200 | 20-60 | 15-100 | 20-200 |

*Leaf nitrogen in non-bearing trees should be 0.2% higher.

Leaf nitrogen on M.9 or M.26 rootstocks should be 0.2% higher.

## FERTILIZER FOR APPLES

### Fertilizer for Non-Bearing Apples

The best time to effectively incorporate nutrients such as potassium, phosphorous, boron and lime into the soil is prior to planting the orchard. Adequate nutrient levels in the topsoil are 12-20 ppm phosphorus, 120-150 ppm potassium, 100-250 ppm magnesium and 1,000-5,000 ppm calcium. Phosphorus and Potassium Soil Requirements Before Planting Apples, provides information on fertilizer rates prior to planting. On young trees, broadcast the fertilizer under the spread of the branches at least 15 cm from the trunk, since injury can result if placed too close. If the soil was prepared properly through deep cultivation and the addition of organic matter such as manure, there should be an adequate supply of other nutrients to sustain the orchard in its juvenile years. On coarse-textured, infertile soils, it may help to use a starter solution at planting time, such as 10-52-10 or 20-20-20. High nitrogen levels can result in excessive growth and incomplete tree hardening. Cover crops are strongly recommended to check late season growth in cultivated orchards, especially in new plantings. Cover crops such as Italian ryegrass, sown about July 1, take up much of the available nitrogen in the soil and will check the tree growth.

**Table. Phosphorus Soil Requirements Before Planting***

| Soil test(ppm P) | Phosphates ($P_2O_5$)req'dkg/ha | Soil test(ppm K) | Potash ($K_2O$) required kg/ha |
|---|---|---|---|
| 0-3 | 80 HR | 0-15 | 180 HR |
| 4-5 | 60 HR | 16-30 | 170 HR |
| 6-7 | 50 HR | 31-45 | 160 HR |

| | | | |
|---|---|---|---|
| 8-9 | 40 MR | 46-60 | 140 HR |
| 10-12 | 20 MR | 61-80 | 110 HR |
| 13-15 | 0 LR | 81-100 | 70 MR |
| 16-20 | 0 LR | 101-120 | 40 MR |
| 21-25 | 0 RR | 121-150 | 20 MR |
| 26-30 | 0 RR | 151-180 | 0 LR |
| 31-40 | 0 RR | 181-210 | 0 LR |
| 41-50 | 0 RR | 211-250 | 0 RR |
| 51-60 | 0 RR | 250 + | 0 NR |
| 61-80 | 0 NR | | |
| 80 + | 0 NR | | |

*For established apple trees, use leaf analysis to estimate requirmnts of Nitrogen, Phosphorus and Potassium.

HR, MR, LR, RR and NR denote, respectively: High, medium, low, rare and no probabilities of profitable crop response to applied nutrient.

**Table. Nitrogen Required per Tree at Various Tree Densities and Tree Ages, Growing in Sod***

| Tree Density(ha)<br>Tree Age (years) | Tree Density (acre) | 600<br>240<br>(600)* | 800<br>320<br>(800)* | 1,000<br>400<br>(1000)* | 1,200<br>480<br>(1200)* |
|---|---|---|---|---|---|
| | 1 | 30** | 30 | 30 | 30 |
| | 2 | 60 | 60 | 60 | 60 |
| | 3 | 90 | 90 | 90 | *** |
| | 4 | 120 | 120 | 120 | *** |
| | 5 | 150 | 150 | 150 | *** |
| | 6 | 180 | 170 | 168 | *** |
| | 7 | 206 | 190 | 186 | *** |
| | 8 | 232 | 210 | 204 | *** |
| | 9 | 258 | 230 | *** | *** |
| | 10 | 284 | 250 | *** | *** |
| | 11 | 310 | *** | *** | *** |
| | 12 | 336 | *** | *** | *** |
| | 13 or older | *** | *** | *** | *** |

*Trees/ha

*Note:* Reduce nitrogen rate by half if orchard is cultivated without sod between tree rows.

*Note:* Do not exceed 200 kg of actual nitrogen per ha per season regardless of number of trees per ha.

**Approximate grams of actual Nitrogen required per tree - These are approximate values. The exact amount of nitrogen to apply is a function of soil nitrogen level, cultivar, rootstock, soil moisture, etc. The best way to determine nitrogen requirements is with regular leaf analysis.

*** Use leaf analysis to determine nitrogen needs.

**Table. Nitrogen Required per Tree at Various Tree Densities and Tree Ages, Growing in Sod***

| Tree Density(ha) | | 1,400 | 1,600 560 | 1,800 | 2,000 |
|---|---|---|---|---|---|
| Tree Density(acre) | Tree Age (years) | (1400)* | 640(1600)* | 720 (1800)* | *800 (2000)* |
| | 1 | 30 | 30 | | 30 30 |
| | 2 | 60 | 60 | 60 | 60 |
| | 3 | *** | *** | *** | *** |
| | 4 | *** | *** | *** | *** |
| | 5 | *** | *** | *** | *** |
| | 6 | *** | *** | *** | *** |
| | 7 | *** | *** | *** | *** |
| | 8 | *** | *** | *** | *** |
| | 9 | *** | *** | *** | *** |
| | 10 | *** | *** | *** | *** |
| | 11 | *** | *** | *** | *** |
| | 12 | *** | *** | *** | *** |
| | 13 or older | *** | *** | *** | *** |

*Trees/ha

*Note:* Reduce nitrogen rate by half if orchard is cultivated without sod between tree rows.

*Note:* Do not exceed 200 kg of actual nitrogen per ha per season regardless of number of trees per ha.

**Approximate grams of actual Nitrogen required per tree - These are approximate values. The exact amount of nitrogen to apply is a function of soil nitrogen level, cultivar, rootstock, soil moisture, etc. The best way to determine nitrogen requirements is with regular leaf analysis.

***Use leaf analysis to determine nitrogen needs.

**Table. Nitrogen Required per Tree at Various Tree Densities and Tree Ages, Growing in Sod***

| Tree Density(ha) | Tree Age | 2,200 | 2,400 | 2,600 |
|---|---|---|---|---|
| Tree Density(acre) | (years) | 880 (2200)* | 960(2400)* | 1,040(2600)* |
| | 1 | 30 | 30 | 30 |
| | 2 | 60 | 60 | 60 |
| | 3 | *** | *** | *** |
| | 4 | *** | *** | *** |
| | 5 | *** | *** | *** |
| | 6 | *** | *** | *** |
| | 7 | *** | *** | *** |
| | 8 | *** | *** | *** |
| | 9 | *** | *** | *** |
| | 10 | *** | *** | *** |

| | | | |
|---|---|---|---|
| 11 | *** | *** | *** |
| 12 | *** | *** | *** |
| 13 or older | *** | *** | *** |

* Trees/ha

*Note:* Reduce nitrogen rate by half if orchard is cultivated without sod between tree rows.

*Note:* Do not exceed 200 kg of actual nitrogen per ha per season regardless of number of trees per ha.

** Approximate grams of actual Nitrogen required per tree - These are approximate values. The exact amount of nitrogen to apply is a function of soil nitrogen level, cultivar, rootstock, soil moisture, etc. The best way to determine nitrogen requirements is with regular leaf analysis.

*** Use leaf analysis to determine nitrogen needs.

## FERTILIZER FOR BEARING APPLES

Most bearing orchards require annual nitrogen applications. Use a soil test to determine potassium requirments. These two elements significantly affect growth and productivity.

### Nitrogen (N)

Nitrogen is necessary for many tree functions, including growth, fruit bud formation, fruit set and fruit size. Because of the complexity of nitrogen interactions with quality and production, the best guide for nitrogen rates is leaf analysis. Cultivars differ in nitrogen requirements. A cultivar grown for processing could receive more nitrogen than one for the fresh market. In some situations, if fruit tends to be small, more nitrogen may be needed. Rootstocks, spacing and pruning also influence application rates. Tree growth, foliage colour, fruit quality such as colour and storability, nutrient balance in leaves and soil are also important considerations for determining nitrogen rates. Several forms of nitrogen are available, but ammonium nitrate (34-0-0) or calcium ammonium nitrate (27-0-0) are the most economical. If you use blended fertilizers, request ammonium nitrate as the nitrogen source. Do not apply urea (46-0-0) to sod orchards since some nitrogen is lost by volatilization.

### Nitrogen Rates

Given the variety of orchard systems, rootstocks, cultivars and soil types, the exact amount of nitrogen to apply varies. Use leaf analysis to evaluate the nitrogen needs of specific plantings.

- If late winter or early spring pruning is to be severe, reduce or eliminate nitrogen application for that year.
- Do not apply excessive amounts of nitrogen. Late or excessive applications result in poor fruit colour and quality. Available nitrogen late in the season may affect hardening off and increase the possibility of winter injury.

- In cultivated orchards, use cover crops to help lower the nitrogen level in the latter part of the season. Cover crops such as Italian ryegrass, sown about July 1, take up much of the available nitrogen in the soil and limit tree growth
- In orchards with herbicide-treated strips under trees, allow some weed growth late in the season. Weeds will take up extra nitrogen, which helps harden off trees and improve fruit quality.
- For all apple cultivars, do not exceed the maximum rates of 200 kg actual nitrogen per ha per year, even in the case of a severe deficiency.

**Table. Approximate Grams of Muriate of Potash Required per 2.5 cm of Trunk Diameter at various tree densities, grown in sod***

| **Tree Density** | **ha**<br>**acre** | **<500**<br>**<200** | **600**<br>**240** | **800**<br>**320** | **1,000**<br>**400** |
|---|---|---|---|---|---|
| **Tree Age (years)** | 1 | 80 | 80 | 80 | 80 |
| | 2 | 80 | 80 | 80 | 80 |
| | 3 | 80 | 80 | 80 | 80 |
| | 4 | 80 | 80 | 80 | 80 |
| | 5 | 80 | 80 | 80 | 80 |
| | 6 | 80 | 80 | 80 | 80 |
| | 7 | 80 | 70 | 63 | 52 |
| | 8 | 80 | 70 | 63 | 52 |
| | 9 | 80 | 70 | 63 | 52 |
| | 10 | 80 | 70 | 63 | 52 |

Do not exceed 800 kg of 0-0-60 per ha per season regardless of number of trees per ha.
*These are approximate values. The exact amount of muriate of potash to apply is a function of soil potassium level, cultivar, rootstock, soil moisture, etc.

The best way to determine potash requirements is by leaf analysis.

**Table. Approximate Grams of Muriate of Potash Required per 2.5 cm of Trunk Diameter at various tree densities, grown in sod***

| **Tree Density** | **ha**<br>**acre** | **1,200**<br>**480** | **1,400**<br>**560** | **1,600**<br>**640** | **1,800**<br>**720** | **2,000**<br>**800** | **2,200**<br>**880** | **2,400**<br>**960** | **2,600**<br>**1,040** |
|---|---|---|---|---|---|---|---|---|---|
| Tree Age (years) | 1 | 80 | 80 | 80 | 80 | 80 | 80 | 80 | 80 |
| | 2 | 80 | 80 | 80 | 80 | 80 | 80 | 80 | 80 |
| | 3 | | | | | | | | |
| | 4 | | | | | | | | |
| | 5 | | | Use leaf analysis to determine potash needs. | | | | | |
| | 6 | | | | | | | | |
| | 7 | | | | | | | | |
| | 8 | | | | | | | | |
| | 9 | | | | | | | | |
| | 10 | | | | | | | | |

Do not exceed 800 kg of 0-0-60 per ha per season regardless of number of trees per ha.
* These are approximate values. The exact amount of muriate of potash to apply is a function of soil potassium level, cultivar, rootstock, soil moisture, etc. The best way to determine potash requirements is by leaf analysis.

**Nitrogen placement and timing**

Apply nitrogen fertilizer in early April. In cultivated orchards, broadcast nitrogen under the tree canopy. In orchards with sod between the rows, place the nitrogen in a band under the drip line or in the herbicide strip.

**Foliar application of nitrogen**

Foliar applications of urea (46% nitrogen) have been used successfully on apples when weather or crop conditions resulted in the need for additional nitrogen at a critical time. On apples, use no more than 2.7 kg N/1,000 L water (6 kg urea) and apply at least 2,000 L/ha starting 7-10 days after petal fall. Make no more than three applications, about 10 days apart. Do not apply later than the end of July or fruit quality and winter survival of the tree could be adversely affected.

**PHOSPHORUS (P)**

Phosphorus is not required in large amounts by apple trees. With a few exceptions, the level of phosphorus in Ontario soils is generally adequate. Phosphorus may be required for sod or cover crop maintenance. A soil test is the best way to determine if this nutrient needs to be added to the sod cover. If indicated, apply phosphorus before planting so it can be thoroughly incorporated in the soil. Phosphorus soil test values between 12 and 20 ppm are considered adequate for tree establishment and fruit production.

**POTASSIUM (K)**

Potassium is important for fruit colour, winter hardiness, tree growth and disease resistance. Because an excess of potassium can lead to a deficiency of magnesium (Mg), avoid unnecessary potassium applications. Potassium soil test values between 120-150 ppm are considered adequate when planting fruit trees. Muriate of potash (0-0-60) is the most common form of potassium. Apply no more than 3 kg of K2O (5 kg of muriate of potash) per mature standard apple tree in a year, regardless of how severe the potassium deficiency.

**Placement and timing**

The best time to apply potassium, either separately or combined with nitrogen, is the spring. While some growers prefer the fall because of time constraints in the spring, leaching over winter may cause potassium loss. For this reason apply in spring if possible. In orchards with sod between the rows, apply potash in a band around the drip line or in the herbicide strip.

**Calcium (Ca)**

A lack of calcium is associated with fruit problems such as bitter pit of apple. Foliar application of calcium to apples reduces the incidence of bitter pit and cork spot. Where these disorders have been a problem, apply four foliar sprays two weeks apart, beginning in mid-July. SeeTable 4-5. Calcium Foliar Sprays. Where more calcium is required, make additional applications by either beginning earlier around mid-June or by continuing until harvest.

Calcium advances fruit maturity, so the timing of harvest may need to be adjusted as a result of calcium foliar sprays. Some formulations of calcium chloride (CaCl2), if applied too close to harvest, have resulted in poor fruit finish. Calcium sprays must contact the fruit for uptake to be effective. Therefore, water volumes capable of wetting the entire tree are required. The more calcium applied the better the control; however, high concentrations of calcium can cause foliar burn. Do not concentrate CaCl2 (77% flakes) beyond 5 kg/1,000 L of water in mid-July and no more than 7 kg/1,000 L for applications at or beyond mid-August. Do not apply calcium formulations that contain nitrogen after the end of July or fruit quality and storability may suffer. For all formulations, consult the label directions for concentrations and compatibility with pesticides. The product used is not as important as the total amount of actual or elemental calcium applied. For example, calcium chloride (77% flakes) contains 28% actual calcium. For acceptable results, up to 12 kg/ha of actual calcium is often required in a total of four or more sprays.

Calcium sprays may cause foliage and/or fruit injury if applied when low temperature and wet weather delay drying of the spray. Injury can also occur if calcium is applied in hot (over 25°C) or humid weather. Recent studies with calcium sprays on McIntosh failed to show an advantage in fruit firmness and storage quality when fruit was stored in regular controlled atmosphere storage for five and a half months. Limit applications of calcium to fruit that has a known deficiency and/or is prone to bitter pit or cork spot.

**Table. Calcium Foliar Sprays**

| Timing | Product | Rate | Notes |
|---|---|---|---|
| Four sprays spaced two weeks apart beginning in mid-July. Additional sprays can be applied up to harvest. | Calcium chloride* (77% flakes) | 5 kg/1,000 L water | Not recommended on McIntosh or Idared. Wet tree to point of runoff. Pesticides may not be compatible - check label. |
| | Calcium nitrate | 9 kg/1,000 L water | Use only if leaf nitrogen is low. Do not apply later than the end of July. For pesticide compatibility consult labels. |
| | Other formulations including chelates | Consult label | |

*When using calcium chloride, mix required calcium in a pail of water first to be sure all product is dissolved before adding slurry to spray tank.

## Magnesium (Mg)

Magnesium soil test values between 100-250 ppm are considered adequate when you plant fruit trees. Magnesium deficiency has become more evident in orchards, particularly where high rates of potash have been used. Magnesium deficiency can lead to premature drop of fruit at harvest, especially with McIntosh. As magnesium is a part of the chlorophyll molecule, magnesium deficient trees have older leaves that are pale in colour. Leaf analysis is the best way to evaluate magnesium requirements.

**Table. Magnesium Foliar Sprays**

| Timing | Product | Rate | Notes |
|---|---|---|---|
| Three sprays spaced two weeks apart beginning at calyx | Magnesium sulfate (Epsom salts) | 20 kg/1,000 L water | Wet tree to point of runoff. Do not concentrate beyond 40 kg/1,000 L water. |
| | Liquid formulations including chelates* | Consult product label | May be compatible with some pesticides. Consult product label. |

*Use chelates recommended for foliar sprays.

Fruit or foliage injury is possible from a mixture of pesticides with magnesium sulfate, so apply magnesium sulfate separately. Check manufacturer's label about mixing magnesium chelates with pesticides. Use chelates recommended for foliar spays. For long-term corrections, soil applications of magnesium are required; however, crop response is not usually immediate. On some soil types a single, early spring application of soil-applied magnesium is not effective. A second or third application the next spring may be needed before the magnesium level in the tree improves. To avoid early fruit drop in this waiting period, apply foliar magnesium sprays for the first two years, in addition to soil applications. Use dolomitic limestone to supply magnesium and raise the soil pH on acidic soils.

Where lime is not required, apply sulphate of potash magnesia at 5-7 kg per mature standard tree or 3-4 kg per mature dwarf tree. This is a granular fertilizer that contains approximately 21% potash and 11% magnesium. Apply this material in early spring in a band under the tree drip line. It contains potassium (K) and the rate of application depends on potash needs. Other sources of magnesium also work well as a soil application. If magnesium is blended with the fertilizer, apply at least 80 kg of available magnesium per ha when the fertilizer is spread. Warning: Apply nutrient sprays according to recommended rates on the product label. Do not spray at temperatures above 25° C.

## MICRONUTRIENTS FOR APPLES

Deficiencies of micronutrients or trace elements are not widespread in

Ontario apple plantings. Boron deficiency is perhaps the most common. Deficiencies of zinc, manganese and iron appear occasionally, particularly in alkaline or high pH soils. The desirable range for micronutrients is very small. More damage is possible with excess amounts than with deficiencies. Do not apply micronutrients to apples except when deficiency is confirmed by leaf analysis or visible symptoms. Apply only the nutrient that is deficient in sufficient quantities to correct the problem

## FRUIT SET AND NUTRITION OF APPLES

Improvement of fruit set and fruit growth are essential to high yields and consequently to the profitability of apple growers. Fruit yield can be increased by improving pollination, controlling the leaf-to-fruit ratio (achieved by means of pruning and fruit thinning), and by adequate fertilization and irrigation practices. Recently, more attention has been given to the use of plant growth regulators in order to improve fruit set and fruit growth in temperate fruit trees.Cell division and elongation, which affect the final fruit size, are promoted by the growth substances auxin, cytokinin, and gibberellin (Taiz & Zeiger, 1998). Two synthetic substituted phenylureas, CPPU [N-(2-chloro-4-pyridyl)-N'-phenylurea] and thidiazuron (TDZ; N-phenyl-N'-1,2,3-thiadiazol-5-ylureia), show cytokinin-like activity in plant tissues and strong fruit growth promoting effect. CPPU has been very effective in promoting kiwifruit, grape, and apple fruit growth. TDZ promotes apple, kiwifruit, grape, persimmon, and cucumber fruit growth.

In apple trees, TDZ sprayed near the full bloom reduced crop load and, therefore, promoted fruit growth (Elfving & Cline, 1993; Greene, 1995). TDZ sprayed at both, higher doses or later after full bloom, had a more substantial thinning effect on apple trees, with the magnitude of the response being cultivar specific. Greene (1995) reported fruit weight increases greater than 30% for 'McIntosh' and 'Empire' apples sprayed at full bloom with TDZ at doses of 50 mg L-1 and 15 mg L-1, respectively. This effect was reduced when the trees were sprayed 18-22 days after full bloom. For 'Gala' and 'Golden Delicious' apples sprayed at full bloom with 10 mg L-1 of TDZ, the increase on fruit weight was greater than 40%.

However, treatments with TDZ might reduce return bloom in apples, depending on product concentration, time of application, and cultivar. Elfving & Cline (1993) reported substantial effects of TDZ, at doses of 65 and 125 mg L-1, on vegetative growth, flowering behavior, and on mineral content of leaves and fruits in 'Empire' apples. TDZ increased trunk enlargement (possibly as a result of reduced crop load) and reduced shoot growth. Leaves from trees treated with TDZ had higher K and lower N, Ca, and Mg contents. TDZ effects on fruit-flesh nutrient concentration were associated with higher content of K and slightly lower content of calcium. However, the authors did not assess the incidence of physiological disorders and postharvest quality of the fruit. Greene (1995) reported a small increase of bitter pit incidence at harvest when apple trees were sprayed 18 days after full bloom with 15 mg L-1 of TDZ but not when the trees were sprayed at full bloom. The author did not report substantial effects of TDZ on flesh firmness and soluble solids content of the fruits. Besides affecting tree physiology and fruit growth, TDZ also influences other aspects of fruit quality such as fruit shape, maturity and ripening. The published information shows that TDZ has a diversity of physiological effects on treated plants, which depends on product concentration, time of application, plant species and cultivar. This, in association with environmental conditions, might have substantial impacts on the plant physiological response to the product. TDZ has been used in Southern Brazil to improve fruit set and fruit growth in apples (Tagliari, 1997). However, more information is required to characterize the physiological effects of this product on apple trees cultivated in this region. The objective of this work was to evaluate the effects of different doses of TDZ, sprayed at full bloom, on nutrition, vegetative growth, return bloom, fruit set, fruit growth and yield of apple bearing trees.

## MATERIAL AND METHODS

The study was conducted in a commercial orchard in Lages, SC, Southern Brazil. Ten-year-old apple trees, cultivars Gala and Fuji, grafted on Marubakaido rootstock, were sprayed at full bloom, in 1999, with TDZ at doses of 0, 5, 10, or 20 g ha-1 of a.i., with a total spray volume of 1,000 L ha-1. The trees were treated with a tractor-mounted airblast sprayer. Spray barriers, made of polyethylene mounted on wood poles, were used to prevent spray drift between experimental plots. The experiment followed a completely randomized block design with four replicates. Each experimental unity was composed by five rows of eight plants. Only four plants in the center of the middle row were assessed for vegetative and reproductive growth, fruit set, yield, and fruit and leaf nutritional status.

All blossom clusters (at the pink stage of flower development; stage F2) on two representative limbs per tree were counted in 1999 (the year when the trees were treated with TDZ) and in 2000. Return bloom was expressed in terms of percentage of blossom clusters cm-2 limb cross-sectional area in 2000 in relation to 1999. At the end of October drop period, all persisting fruit on each tagged limb were counted. Fruit set was quantified in terms of number of persisting fruit per limb cross-sectional area (PLCA) and per blossom cluster (PBC).

Limb (at a predetermined location, on two representative limbs per tree) and trunk (30 cm above the soil surface) cross-sectional area and shoot growth (on six shoots per tree) were measured at bloom and in the following spring. Eight midshoot leaves from current-season shoots were collected per plant on 15 December 1999 and assessed for length, width, area (with a leaf area integrator LI-COR model LI-3050A), chlorophyll content (with a Minolta chlorophyll meter, model SPAD-502), and dry matter cm-2 (by weighting the material after drying it for 48 hours at 70oC).

Ten midshoot leaves from the current-season shoot and ten fruits per plant were collected thirty days before the start of the commercial fruit harvesting and at commercial harvesting, respectively, and assessed for N, K, Ca, and Mg content. The fruit mineral composition was assessed for the entire fruit (flesh + skin) and for the skin. The dry leaves and the fresh fruit samples were digested with a mixture of sulfuric acid and hydrogen peroxide, as described by Adler & Wilcox (1985). Nitrogen was quantified by steam distillation using a semi-micro'Kjeldahl equipment, as described by Ernani et al. (1997). Calcium and Mg were determined with an inductive coupled plasma spectrophotometer (ICP), and K with a flame emission spectrophotometer.

All fruit on each tagged limb were harvested at commercial maturity and assessed for total weight, total number and average fruit weight. The data were subjected to statistical analysis using SAS (SAS Institute, 1990). Percentage data were transformed to arc sin [(x+5)/100]1/2 before submitting to the ANOVA. The effect of TDZ dose on each attribute assessed was analyzed by orthogonal polynomial contrasts.

**Results and Discussion**

On 'Gala', a cultivar characterized by having low fruit set, TDZ sprayed at full bloom improved fruit set up to the highest dose used. On the other hand, 'Fuji' is characterized by high fruit set and, therefore, requires excessive fruit thinning. In this cultivar, TDZ sprayed at full bloom did not increase fruit set. The untreated 'Fuji' trees had a number of fruit cm-2 of limb cross-sectional area similar to 'Gala' trees treated with TDZ at doses of 10-20 g ha-1. The number of fruit per blossom cluster on untreated 'Fuji' trees was about twice that of 'Gala' trees treated with TDZ at doses of 10-20 g ha-1.

Greene (1995) did not find TDZ effects, sprayed at full bloom, on fruit set of 'McIntosh' at doses of 10 or 50 mg L-1, and of 'Empire', at doses up to 15 mg L-1. However, when applied at both, higher doses or later after full bloom, TDZ might have a thinning effect depending on the apple cultivar (Elfving & Cline, 1993; Greene, 1995). Flowering at the start of the experiment was uniform among treatments, averaging 6.5 and 2.8 blossom clusters cm-2 of limb cross-sectional area on 'Gala' and 'Fuji', respectively. A polynomial orthogonal contrast analysis indicated a highly significant linear effect of TDZ in reducing return bloom in 'Gala'. For this cultivar, the return bloom was reduced to half in trees treated with TDZ at doses of 10-20 g ha-1 in comparison to untreated trees. In 'Fuji', flowering determined in the next spring (spring following TDZ treatment) was reduced in all treatments by 40%, with no effect of TDZ in the return bloom. Some authors have shown that return bloom in apples is affected by TDZ concentration, time of application and cultivar.

Greene (1995) reported a reduced return bloom on 'McIntosh' and 'Double Red Delicious' apples treated with TDZ. In 'McIntosh', more substantial reduction of return bloom occurred with increases on product concentration and with late spraying. At dose of 10 mg L-1, TDZ sprayed at full bloom or 22 days after full bloom reduced return bloom in more than 30% and 70%, respectively. At dose of 50 mg L-1TDZ reduced return bloom more than 70% when sprayed at full bloom, and more than 90% when sprayed 22 days after full bloom. In 'Double Red Delicious', 5 and 10 mg L-1 of TDZ, sprayed at full bloom, reduced return bloom more than 60%. However, return bloom was not affected on 'Empire' trees treated with 1, 5, or 15 mg L-1 of TDZ, either at full bloom or 18 days after full bloom. For this cultivar, concentrations as high as 62 and 125 mg L-1 did not affect return bloom.

The reduced return bloom caused by TDZ in some cultivars might be a direct effect of the product on flower bud formation. Flowering was inhibited on 'Delicious' where there was no treatment effect on fruit set, but on 'McIntosh' there was a general trend to have more inhibition of flowering on trees that thinned the most. In this study, the reduced return bloom in 'Gala' treated with highest doses of TDZ may be the result of higher fruit set, which can cause depletion of reserves for the flower bud differentiation (Faust, 1989), and/or promotion of vegetative shoot growth, that can also affect flower bud differentiation. In 'Fuji', flowering determined in the following spring was substantially reduced in all treatments, and, therefore, TDZ did not affect return bloom. However, this also might reflect low effects of TDZ on flower bud differentiation in this cultivar.

TDZ had no effect on the increment of trunk cross-sectional area, but reduced limb cross-sectional area in 'Gala' (linear effect). On both cultivars, TDZ increased shoot growth. Since the experimental area had high presence of bees, and the weather conditions were favorable for the entomophilic pollination, yield was not affected by TDZ treatments in any cultivar. Average fruit weight did not increase with increases on TDZ dose. This result disagrees with previous published studies, which show substantial increase on fruit size with increases on TDZ doses. However, those studies were performed mainly by dipping the fruit in a solution of TDZ, or by spraying some trees or some limbs per tree with the product to the drip point. This might result in much higher doses of TDZ than those applied in our study, where the plants were treated with a tractor-mounted airblast sprayer, as done commercially. This might have reduced the product effectiveness in promoting fruit growth. Reynolds et al. (1992) also emphasized the importance of field trials to determine the effective concentration of TDZ when the product is sprayed commercially in grapes to promote fruit growth.

TDZ affected the distribution of nutrients between leaves on both cultivars. Treatment with 20 g ha-1of TDZ increased Ca content in the leaf by about 27%, in 'Gala'. The concentration of Ca in the 'Fuji' leaves was about twice that of 'Gala' and it was not affected by TDZ. 'Fuji' and 'Gala' Mg contents on leaf increased by 12% and 17%, respectively, by treating the trees with TDZ at 20 g ha-1. Concomitantly, TDZ decreased the fruit content of Ca and Mg, especially in the skin tissue. Calcium content in the skin reduced 17-20% with 10-20 g ha-1 of TDZ in both cultivars. In 'Gala', the same doses of TDZ reduced Mg and N content in the skin by 14-19% and by 10%, respectively. In 'Fuji', these treatments reduced Mg and N content in the skin by 23-26% and by 20%, respectively. Except for K in the leaves of 'Fuji' and for Ca in the leaves of 'Gala', all the other nutrients are in the normal range for apples, regardless of cultivar and tissue.

The nutritional results obtained in the present study are, however, in contrast to those reported by Elfving & Cline (1993) on 'McIntosh' apples. These authors reported decreases on N, Ca, and Mg and increases on K content in the leaves, and increases on K content in the fruit-flesh. According to them, the increase on K and the reduction on N in the leaves, likely resulting from reduced crop load on trees treated with TDZ, may have induced lower Ca and Mg contents in the leaf. Since the doses of TDZ sprayed at full bloom in the present study increased fruit set and shoot growth, nutritional results different from those reported by Elfving & Cline (1993) are foreseen.

The data show that the intense promoting growth effect of TDZ on vegetative tissues (terminal buds and growing leaves) might increase the accumulation of nutrients into these organs, especially calcium, in detriment of growing fruit.

# 5

# Nutrition and Fertilization from Fruit Trees

As a rule, these soils are fertile and are well-suited to citrus production. Soil pH is alkaline, generally in the range of 7.6 to 8.0, although both higher and lower pH values occur. The ground water table is usually below 5 feet and soil salinity is usually manageable.

## NUTRIENTS

There are at least 15 nutrient elements which are considered essential to citrus growth and development. Three are provided by air and water, the other 12 must be provided by the soil or by supplemental fertilization. Among the macroelements, only nitrogen must be routinely applied to Valley citrus orchards. Long term nutrition studies in Texas citrus have not demonstrated a meaningful response to supplemental applications of either phosphorous or potassium. Among the microelements, only iron and zinc deficiency symptoms have been commonly observed in the field. As a rule, iron deficiency symptoms may appear in early spring, then disappear as soil temperatures increase. The application of iron chelates has not proven to be necessary except for a few plantings of Swingle citrumelo and, to a lesser extent, the citrange rootstocks, especially on soils of high clay content.

Symptoms of zinc deficiency are occasionally noted, but the symptoms are rarely extensive within an orchard or even within a tree. It is uncommon to find zinc deficiency patterns on more than a few leaves or more than a few twigs on a tree. Research into the deficiency has not resulted in increased production as a consequence of foliar zinc applications. Further research is needed, however, especially of a long-term nature.

Iron deficiency of citrus.

Zinc deficiency of citrus.

## SOIL AMENDMENTS

Soil amendments are rarely used in Texas citrus orchards, probably more because of the lack of available materials than any for other reason. Results of research with soil amendments have been inconclusive except in cut areas where the existing topsoil was removed and dispersed over the rest of field during land leveling operations. In such areas, pre-plant incorporation and repeat applications of a suitable organic matter would undoubtedly be beneficial.

### Soil and Leaf Analysis

Citrus can be successfully grown in Texas without using either soil or leaf analyses to direct the fertilization program. Indeed, following the general recommendations for nitrogen fertilization, coupled with close observation of visual symptoms of nutrient deficiency (and the correction thereof), is adequate for average production. For optimal production and quality, however, a routine program of nutrient analysis and soil salinity measurement is essential to correct potential problems before they become limiting to production.

### Soil Testing

Analytical procedures used in soil testing vary considerably among labs, as do the results they obtain. None are capable of reporting available nutrient levels in a sample--they can only report the chemically extractable levels. Moreover, no soil testing extraction procedure has yet been calibrated to correlate the extractable value of any nutrient element with citrus production levels or citrus fruit quality. In other citrus-producing areas, soil testing is widely considered useful to determine soil pH, soil phosphorous, soil calcium and potentially toxic accumulations of soil copper. In our Texas citrus soils, however, alkaline pH is a given, as are more than adequate levels of soil phosphorous and soil calcium. The long-term use of copper fungicides in Texas citrus has not resulted in appreciable increases in soil copper levels. Consequently, routine soil testing in Texas citrus orchards is of limited value.

That being said, however, a standard soil analysis should be conducted every couple of years or so simply to verify pH and soil levels of phosphorus, calcium and copper.

### Soil Salinity

Because salinity is ever-present in Texas citrus soils and irrigation waters and because high salinity dramatically reduces root growth, thereby reducing shoot growth and yield, salinity analysis is critical for optimal citrus production in Texas. The threshold salinity level, i.e., the level at which initial yield declines occur, is considered to be about 1.6 dS/m (deci-Siemens/meter or mmhos/cm), which is about 1120 ppm salts. With irrigation water of greater than 1.0 dS/m (700 ppm), which is about normal for Rio Grande River water, soil salinity can quickly increase to detrimental levels in Texas citrus orchard soils, especially in times of drought and inadequate supplies of irrigation water.

Symptoms of salinity include reduced growth of roots and shoots, decreased flowering and smaller leaves--but these symptoms are not easily recognized, and they usually occur before typical toxicity symptoms appear in the leaves. Probably the most common visible symptoms of excess salinity are chloride toxicity--tip burn and marginal necrosis of the leaves--and sodium toxicity--general bronzing of the leaves. To determine soil salinity, composite samples from several locations in the orchard should be taken in 1-foot increments to a depth of 4 feet. Each separate depth increment should be combined with other samples from the same depth to make a total of 4 composite samples for the orchard. Such sampling will indicate the effectiveness of leaching or the need for additional leaching. In orchards with drip or microsprayer irrigation systems, the samples should be taken from within the wetted zone. Leaching of salts on deep, well-drained soils is accomplished with both rainfall and irrigation. Soils with a high water table or that have internal drainage problems should be abandoned insofar as citrus production is concerned. The alternative to abandonment is the installation of an extensive sub-surface drainage system.

## Leaf Analysis

The use of leaf analysis is based on a given range of concentrations of each nutrient element which correlates well with optimal growth, production and fruit quality--even under different soil types, climatic conditions and rootstock/scion combinations. Indeed, years of research have confirmed the correlation between leaf levels of the various nutrients and citrus productivity. Essentially, there are two schools of thought regarding the type of leaves to be sampled and the optimum concentrations of elements in the leaves. The program developed in Florida is predicated on the sampling of spring-flush leaves from non-fruiting terminals, while that in a few other areas is based on leaves from fruiting terminals. There are pros and cons to both procedures, but the Florida procedure is more widely accepted.

The beauty of leaf analysis is that, properly collected, it indicates the current nutritional status of the orchard, as it measures only the nutrient levels that the trees were capable of extracting from the soil. Recorded over time, along with fertilizer inputs, leaf analysis accurately reflects prior nutritional and production practices in the orchard. Moreover, leaf analysis allows adjustments to the fertilizer program before one or more elements becomes limiting. However, to be of maximum usefulness, leaf sampling procedures must be followed precisely, or the analytical results may be invalid.

The leaf sample consists of at least 100 leaves that are 4 to 6 months old, taken from non-fruiting twigs or terminals of the previous spring's growth flush. Thus, sampling should be conducted from July to September. The orchard to be represented by the leaf sample should consist of only one rootstock/scion combination of uniform-aged trees, within a single soil type.

Select 15 to 20 trees randomly across the orchard from which to collect five or six leaves each. Leaves should be free of damage from insects or disease. Leaves with obvious chlorosis should be excluded, unless the sample is being taken specifically to ascertain a potential nutritional cause of the chlorosis. While still fresh, the leaves should be washed to remove soil or dust, then air dried, packaged and submitted to the laboratory for analysis. When the results are obtained from the laboratory, compare them with the standards. The laboratory will report the levels of major elements as percentage of dry weight, while microelements will be reported as parts per million. If the orchard has been sprayed with any nutrient element at any time since the spring flush leaves emerged, the analytical results for that element or elements should be disregarded, as an indeterminate quantity of foliarly-applied nutrients may still be present on the surface of the leaves, thereby resulting in falsely high readings.

**Table. Citrus leaf analysis standards based on nutrient concentration in 4- to 6-month-old spring-flush leaves taken from non-fruiting terminals from July to September (based on University of Florida recommendations).**

| Element | Symbol | Unit | Deficient | Low | Optimum | High | Excess |
|---|---|---|---|---|---|---|---|
| Nitrogen | N | % | <2.2 | 2.2-2.4 | 2.5-2.8 | 2.9-3.2 | >3.3 |
| Phosphorus | P | % | <0.09 | 0.09-0.11 | 0.12-0.17 | 0.18-0.29 | >0.30 |
| Potassium | K | % | <0.7 | 0.7-1.1 | 1.2-1.7 | 1.8-2.3 | >2.4 |
| Calcium | Ca | % | <1.5 | 1.5-2.9 | 3.0-5.0 | 5.1-6.9 | >7.0 |
| Magnesium | Mg | % | <0.20 | 0.20-0.29 | 0.30-0.50 | 0.51-0.70 | >0.80 |
| Sulfur | S | % | <0.14 | 0.14-0.19 | 0.20-0.40 | 0.41-0.60 | >0.60 |
| Chlorine | Cl | % | --- | --- | <0.5 | 0.5-0.7 | >0.7 |
| Sodium | Na | % | --- | --- | <0.2 | 0.2-0.5 | >0.5 |
| Iron | Fe | ppm | <35 | 35-59 | 60-120 | 121-200 | >250 |
| Boron | B | ppm | <20 | 20-35 | 36-100 | 101-200 | >250 |
| Manganese | Mn | ppm | <18 | 18-24 | 25-100 | 101-300 | >500 |
| Zinc | Zn | ppm | <18 | 18-24 | 25-100 | 101-300 | >300 |
| Copper | Cu | ppm | <4 | 4-5 | 6-16 | 17-20 | >20 |
| Molybdenum | Mo | ppm | <0.06 | 0.06-0.09 | 0.1-1.0 | 2-50 | >50 |

## NITROGEN FERTILIZATION-GENERAL CONSIDERATIONS

Ammonium sulfate is the most commonly used dry fertilizer in Texas citrus, as it is among the lowest of nitrogen materials in terms of relative salt index, and its sulfur content is useful in attenuating alkalinity. N32 or urea ammonium nitrate is perhaps the most common liquid nitrogen source. Obviously, other materials and formulations are available and are used. Because yield differences have not been correlated with different formulations, the grower's choice is usually based on the method of application, i.e. broadcast dry or injected through the irrigation system, and the price per unit of nitrogen (not the price per ton of fertilizer!).

Nitrogen fertilization is commonly based on tree age, tree size or prior yields. General recommendations for citrus trees of different ages. During the first three years of orchard establishment, the recommendation is in pounds of actual nitrogen per tree, since young tree fertilization is usually accomplished by hand application. Fertilization should be split equally into several applications during the spring-to-summer months, as young trees are better able to utilize small applications frequently applied.

For bearing trees, the rates are presented as pounds of nitrogen per acre per year, as applications are normally accomplished by use of mechanical spreaders which broadcast the fertilizer over the entire orchard floor. Higher applications than those shown are rarely justified in Texas citrus, and then only on the basis of leaf analysis.

**Table. General nitrogen fertilization rates for citrus in Texas.**

| Tree Age | Pounds of N per tree per year | Pounds of N per acre per year |
|---|---|---|
| 1 | 1/8 | -- |
| 2 | ¼ | -- |
| 3 | ½ | -- |
| 4 | -- | 50 |
| 5 | -- | 75 |
| 6 | -- | 100 |
| 7 | -- | 100 |
| 8 | -- | 125 |
| 9 | -- | 125 |
| 10+ | -- | 150 |

In other production areas, the amount of nitrogen to be applied is calculated on the basis of prior production. Under that scenario, the rate to be applied is 8.9 pounds of nitrogen per ton of fruit produced. Since harvest of the current crop is rarely completed before the need to apply fertilizer for the coming crop, this method to determine nitrogen rates would be better calculated on the basis of the last completed season's production, i.e., the 2003-04 crop should be fertilized on the basis of 2001-02 production.

Neither the timing of application nor the number of applications of nitrogen appear to matter as much as the consistency of applications from year to year. Most Texas growers make a single application annually, usually two or three months prior to bloom. A few growers utilize split applications, with two-thirds of the rate applied pre-bloom and the balance in May. Growers who fertilize through the irrigation system commonly make injections at least monthly through the year.

## SPECIAL SITUATIONS

### Alternate Bearing

Over the last several seasons, Texas citrus production has been characterized by higher levels during seasons which bloomed in even-numbered years (e.g., the 2000-01 season) followed by lower production in odd-numbered seasons (e.g., the 2001-02 season). This alternating tendency exists across the board for all varieties of citrus--grapefruit, navels, early oranges and Valencias. While some orchards do not exhibit alternation, having fairly consistent production from year to year, others that do exhibit alternation are out of phase with the industry, being up when the industry is down and being down when the industry is up.

The Texas citrus grower has only two options with which to try to attenuate alternation--fertilization and pruning, with the latter being discussed in anotherpublication. Alternate bearing persists because the tree uses almost all of the nutrients and reserves available to it to mature a large crop of fruit, with little energy being left to provide sufficient growth for the next crop. Consequently, the next crop is lighter, so the tree produces more growth than average, setting up yet another year of higher production.

In order to begin to break the cycle, nitrogen fertilization rates should be increased during high-production years and deceased during years of low production. To accomplish this with a single fertilization, the grower should apply 8.9 pounds of nitrogen per ton of fruit produced in the season preceding the current one. In other words, fertilizer the 2003-04 crop based upon production in 2001-02. An alternative solution is to utilize split applications of nitrogen. Apply 2/3 of the expected annual rate pre-bloom, then assess fruit set in early May. If set is deemed to be heavy, apply the remaining balance of nitrogen plus an additional 50 percent of that. If set is light, skip the May application entirely.

Annual leaf analyses may indicate the existence of alternate bearing when the same quantity of nitrogen is being applied every year. In years of higher production, leaf nitrogen levels should be a little higher because the absorbed nitrogen is concentrated in fewer leaves. Conversely, in years of lower production, leaf nitrogen levels should also be lower because the nitrogen is diluted over considerably more leaves.

### Topworking and Rehabilitative Pruning

When bearing orchards must be severely pruned in order to topwork the trees to a different variety or to rehabilitate the trees following significant freeze damage, nitrogen fertilization rates should be reduced to levels consistent with the reduced tree size. Indeed, no fertilizer should be applied during the first season after such severe pruning, as regrowth will be more than sufficiently vigorous without it. In the second season after severe pruning, use the general rate recommended for either 4-year-old or 5-year-old trees. In the third and subsequent seasons, switch back to the normal fertilization program.

**Sheepnosing**

Sheepnosing of grapefruit is apparently caused by climatic factors occurring at some specific but undetermined time during flower induction-differentiation-initiation and early fruit development. It has never been considered to be a common occurrence, as most grapefruit varieties may not exhibit sheepnosing more often than once every 8 to 12 years or more. Rio Red grapefruit, however, apparently has a genetic proclivity to sheepnosing, inasmuch as sheepnosed fruit can be found in every Rio Red orchard in Texas every year--only the percentage of affected fruit and the severity of the disorder differ from orchard to orchard and year to year.

Neither nitrogen fertilization, overall nutrition, irrigation system, amount of irrigation nor other cultural practices cause sheepnosing, so its occurrence cannot be induced by anything the grower does or does not do in orchard management. Production inputs can, however, influence the severity of the disorder--any production practice that induces greater vigor than is necessary will increase the severity of sheepnosing, i.e., excess vigor will cause an otherwise moderately sheepnosed fruit to become severely sheepnosed.

Consequently, growers must try to limit tree vigor, particularly during seasons in which the overall percentage of sheepnosed fruit is expected to be higher than normal. Current work to try to predict, in relative terms, how bad sheepnosing will be is still underway. Many growers have routinely reduced nitrogen fertilization levels for Rio Red orchards by about 25 percent. Others have opted for split applications, with the idea of eliminating the second application if it is determined that the incidence of sheepnosing is high.

Such practice may or may not coincide with the need to adjust nitrogen fertilization levels to attenuate alternate bearing. If it does not, it may be preferable to concentrate first on alternate bearing and then work on the attenuation of sheepnosing.

**Urea and Nutriphite Sprays**

Winter-applied foliar sprays of 53 to 60 pounds per acre of low-biuret urea have increased navel orange production in California. Similar results have been reported for Valencia oranges in Florida, both from winter-applied urea and from 2.5 quarts per acre of Nutriphite applied either in winter or at the initiation of open bloom. Both products can be phytotoxic to citrus, which may be a factor in causing enhanced fruit set. Neither product has been tested under Texas conditions, nor am I aware of plans to do so. The timing of application of urea is critical to success--and such timing is not easily determined. For example, urea sprays must be applied after some natural flower induction has occurred but before most flower bud differentiation begins. In the Florida work, the critical period for application occurred in late December in some years, early January in others. Because the critical window for the application of urea is likely to be different each year, we have to develop a means to determine that window before we can test urea sprays and expect meaningful results. In the Florida work, Nutriphite was applied either at the same time as urea or, alternatively, at 5 to 10 percent open bloom. The latter application timing is one which we can use in coming seasons to test Nutriphite under Texas conditions. Yield increases averaged 13 to 17 percent over the four years of the test.

### Micronutrients

Although micronutrient sprays are sometimes routinely included in post-bloom or later pest management sprays, there is no supporting research for this practice. The inclusion of any micronutrient in the spray program should be based on low or deficient levels as determined by annual leaf analyses. While some growers might argue that the presence of visual sypmptoms of deficiency should be sufficient, the fact is that growth and productivity may have already been affected by the time visual symptoms of deficiency appear.

### Fertigation

Most growers who have drip or microsprayer irrigation systems are using them for nitrogen fertilization, typically injecting liquid nitrogen at least monthly. Injection should occur near the end of the irrigation cycle, as earlier injection could push the fertilizer lower in the root zone and soil profile than is desirable. Following injection, the system should be run long enough to clear the fertilizer solution out of the lines.

Because of the greater efficiency of fertilizer application and utilization via fertigation, nitrogen rates should be reduced to about 80 percent of the level which would be broadcast as dry material. Too, it doesn't happen often, but rains can occasionally occur with such frequency that the irrigation system may not be used for several weeks--without irrigation, there is no fertilization. Consequently, some growers broadcast about one-third to one-half of the annual nitrogen requirement during the months pre-bloom--and then only in alternate middles. As a final note, growers should understand that while nutrition inputs in the current season may impact the quality of the current crop (e.g., sheepnosing in Rio Red) as well as overall set and fruit size, the applied nutrients must also be sufficient to provide the optimal vegetative growth which will be needed to produce next season's crop.

## FRUITS WITH HIGH NUTRITION VALUE

'Taste every fruit of every tree in the garden at least once. It is an insult to creation not to experience fully' said famous British actor Stephen Fry. Tropical countries like India are blessed with a variety of fruits and vegetables that they naturally grow and produce.

Fruits like mango, apples, banana, guava, citrus fruits, are available in plenty due to their distinct flavor and taste. These fruits also constantly enjoy the attention of organic scientists and horticulturists and are further improved in their quality. However fruits lovers and the common man should also remember that there are other fruits too that are sometimes termed as "underutilized fruits" that are hundred times more nutritious than the popular fruits in terms of their vitamin C, cartenoids, and antioxidants. And added to this they are inexpensive and are easily available.

Some of these too have a distinct flavor and taste and these are easy to grow even under harsh conditions as compared to exotic fruits like apples, mangoes, kiwis, cherries, grapes and their hybrids.

The underutilized fruit plants are also naturally disease-tolerant and are adapted to hot, hardy climate conditions very much in sync with equatorial climate conditions. In India they are found locally in the dry regions of Rajasthan, Madhya Pradesh, and Gujarat. Some of them are confined to natural wild and semi-wild arid zones.

Although the nutritional value of majority of the forest foods is not precisely known, but most of them which have been researched have been found to be a storehouse of nourishment.

There are various obvious reasons for the under consumption of underutilized fruits. -

- Most indigenous fruit trees are not commonly cultivated on the farm and there is scant and dispersed knowledge about their fruits and their nutritive value.
- People prefer to have fruits with good taste since wild fruits have high levels of the unpleasant tasting tannins and glycosides.
- Indigenous fruits are neither large nor fleshy and contain lots of seeds.
- The fruits are highly perishable and difficult to store in the fresh form.
- Most wild fruits are not really easy and handy to eat.
- Some fruit species are not acceptable as fresh fruit because of high acidity and/or strong astringent taste.
- Most of them are often available only in the local markets and are rarely known in other parts of the country.

On the contrary, benefits of such underutilized fruits far exceed their limitations.

- They are cheap and highly nutritious.
- They have known medicinal and therapeutic properties and are used by the local people to cure various diseases.
- Many of them, the fruit, seed, leaf of the plant are used as curative foods in the traditional Indian Medicine and Ayurveda. For example, amla is used for treating diabetes, bael fruit for beating the heat, bael leaf for diabetes, and ber and phalsa being highly rich in vitamin C are used in cases of vitamin C deficiency.
- In addition some have an excellent flavor and taste and are used for preparing delicacies at home. eg. Ber, matira, pickles made of lasora, or dishes made with ker, sangri, kachri, and kumat in Rajasthan.

## Nutrient Value of Some Underutilized Fruits (per 100 grams)

| Fruit | Protein (g) | Fibre (g) | Calcium (mg) | Phospho - rous (mg) | Iron (mg) | Vitamin A (IU) | Vitamin C |
|---|---|---|---|---|---|---|---|
| *Underutilized Fruits* | | | | | | | |
| Hingota (Balanites aegyptiaca) Ripe fruits eaten raw or sun dried and stored as dates | 4.9 | 3.5 | 147 | 58 | 4 | - | 46 |
| Ker (Capparis deciduas) Shrub; fresh dried tiny unripe fruits used as vegetable, fresh red fruits eaten raw | 5.9 | 12.3 | 153.8 | 50.8 | 2 | - | 133 |
| Lasora (Cordia spp.) Fruit used as a veg or in pickles, ripe fruits freshly eaten | 2 | 2 | 55 | 275 | 6 | - | - |
| Sangri (Prosopis cineraria) Pods are locally called sangria; chocolate colour, dried pods are eaten as a vegetable | 23.2 | 20 | 414 | 400 | 19 | | 523 |
| Meetha Pilu (Salvadora oleoides) Fruits are red brown when ripe; eaten raw when ripe; can be cooked; dried and stored | 6 | 2 | 6 | 76 | 8 | - | - |
| Bael (Aegle marmelos) Fruit are small to large with a hard cover; pulp eaten or made in a fresh sherbet, bael leaves good for diabetics | 1.8 | 2.9 | 85 | 31.8 | 0.6 | 91.6 | 1.1 |
| Chironji (Buchanania lanzan) Fruit eaten raw; sweetish sub acidic flavor | 7.3 | 5.2 | - | - | - | - | 2 |

| | | | | | | |
|---|---|---|---|---|---|---|
| Karonda (Carissa spp.) Immature fruits used as vegetable, mature fruits eaten raw. Fruits are processed as pickle, jam, marmalade | 1.1 | 1.8 | 21 | 28 | - | 1619 | 9-11 |
| Aonla, Amla (Emblica officinalis) Fruit eaten fresh, processed and preserved | 0.5 | 1.9 - 3.4 | 20 | 26 | 0.48 | 17 | 500-625 |
| Phalsa (Grewia subinaequalis) Small tiny brown fruits; used to prepare a sherbet | 1.3 | 1.2 | 129 | 39 | 3.1 | 800 | 22 |
| Mahua (Mahua indica) Ripe fruits eaten raw or cooked | 1.3 | - | 45 | 22 | 1.1 | 512 | 40.5 |
| Khirni (Manilkara hexandra) Fresh fruits are very sweet and eaten raw as well as after drying | 0.48 | - | 83 | 17 | 0.92 | 675 | 15.67 |
| Manila tamarind (Pithecellobium dulce) Pulp consumed raw | 2-3.3 | 1.1 | 13 | 42 | 0.5 | 25 | 138 |
| Jamun (Syzygium cumini) Eaten as fresh fruit Jamun seeds contain various alkaloids such as jambosin and glycoside which inhibits the conversion of starch in to sugars | 0.7 | 0.9 | 15 | 15 | 1.2-1.6 | 80 | 18 |
| Tamarind (Tamarindus indica) | 2 | 2.9 | 34-94 | 34-78 | 0.2-0.9 | - | 44 |
| Ber (Zizyphus spp.) Eaten as a fruit | 2 | 2 | 60 | 120 | 7 | - | 88 |
| Common Fruits | | | | | | | |
| Apple | 0.2 | 1 | 10 | 14 | 1 | - | 1 |
| Banana | 1.2 | 0.4 | 17 | 36 | 0.9 | 78 | 7 |
| Orange | 0.7 | 0.3 | 26 | 20 | 0.3 | 1104 | 30 |
| Papaya | 0.6 | 0.8 | 17 | 13 | 0.5 | 666 | 57 |
| Grapes | 0.5 | 2.9 | 20 | 30 | 0.5 | - | 1 |
| Mango | 0.6 | 0.7 | 14 | 16 | 1.3 | 2743 | 16 |
| Dates (dried) | 2.5 | 3.9 | 120 | 50 | 7.3 | 26 | 3 |

## DATES HEALTH BENEFITS AND FRUIT NUTRITION FACTS

Dates are classified as the fruit of the date palm. These are darkish reddish brown, oblong, and approximately 1 1/2 inches lengthy. Date skin is old and wrinkly as well as covered together with a sticky, wax-like film. Dates mature within groups underneath the fronds on the date palm tree. An individual group holds 600 to 1,700 dates. Date palms can easily develop as tall as 100 ft and remain in production for more than Sixty years. The fronds of the date palm are usually featherlike as well as Ten to twenty feet lengthy. A complete developed date palm or even Phoenix dactylifera could achieve 82 feet (25 meters) tall, having a distributing crown of greenish leaves.

The date fruit develops in large groupings hanging underneath the leaves, and they are generally yellow during the early maturing stage, or even kimri, the Arabic term for unripe. Several customers take pleasure in date fruit within the next phase, khalal, which means full-sized yet crispy, although some look forward to dates to achieve rutab, ripe as well as soft. As soon as the date fruit is permitted to sun dry about the tree, it really is regarded as being within the last tamr phase of maturing. Dates are generally gathered within February as well as March in Arizona along with California the location where the U.S. crop is cultivated. In case you have never had a newly collected date, they're worth trying to find for their huge fairly sweet, caramel-like flavor as well as gentle structure. Producers markets in Southern California as well as Arizona might have fresh new dates in season; also, they are accessible mail order from several farmers and may even be discovered in Middle Eastern marketplaces.

It's got various titles in accordance with the various nations on the earth just like:

- In French, it really is dattier
- In German, it can be dattel
- In Italian, datteri or even dattero
- In Spanish, datil
- In Dutch, dadel
- The Portuguese term is tamara.

## HEALTH BENEFITS OF DATES

A dry fruit date each day does miracles to body of a human: beside various other health advantages, dates provide power as well as control food cravings. Dates are generally fairly sweet dry fruit, as well as grow on palm tree referred to as phoenix dactylifera and provide darkish brown drupes, oblong clusters in dimensions. A single group has about 600-1700 clusters on a single. The fruit includes a wax-like as well as tacky cover and it has dry as well as old and wrinkly overall look. Dates are essential component of Middle Eastern as well as Mediterranean cooking: dates are generally sliced up and therefore are adornment on cakes or even pudding or perhaps combined with milk, butter or even yogurt. The fruit has got numerous dietary valuations, are usually packed with all-natural fibers, vitamins and minerals: also really lower with calories with no cholesterol in any way. Frequent consumption of dates curbs stomach cancer. The main nutrition just like sculpture, magnesium, potassium, iron, fiber, calcium, manganese, copper, amino acids, phosphorous as well as fats, vitamins A1, B1, B2, B3, B5, along with dietary fiber encourages various health improvements.

### Cures Constipation

Bowel problems is often times understood to be having a bowel motion a lot less than Three times each week. It always is associated to hard stools or even trouble passing stools. You might have discomfort whilst passing stools or even might be not able to have a bowel motion right after pushing or even pressing in excess of Ten minutes. Date is referred to as to become a healthy laxative food items. This is why that dates are very therapeutic for individuals struggling with constipation. To get the laxative impact through dates, you have to dip dates for just one complete night in water. You need to eat these soaked dates each and every morning just like syrup to obtain their genuine benefits.

### Prevents from Intestinal Disorders

"In anatomy, the intestine is definitely the section of the alimentary tube stretching through the stomach towards the anus as well as, in humans along with other mammals, includes two sections, the small intestine as well as the large intestine. In human beings, the small intestine is additional divided in the duodenum, jejunum as well as ileum whilst the large intestine is divided into the cecum as well as colon. The problem within this intestine may cause various illness which usually damages extremely terribly to the overall health.

If you suffer from any kind of intestinal problem then making dates a significant part of the meals are one of several actions that you could decide to try deal with this kind of difficulty. The nicotinic content which is contained in the dates is considered to be very therapeutic for treating just about any forms of intestinal problems. Constant consumption of dates helps you to sustain a check on the development of the pathological microorganisms and therefore, helps with the increase of helpful bacteria within the intestinal tract.

**For Healthy Heart**

The heart is really a hollowed out muscle which pumps blood through the entire blood vessels simply by repetitive, stroking contractions. It really is present in just about all animals with a circulatory system (which includes all vertebrates. The word cardiac (as in cardiology) indicates "related to the heart" as well as originates from the Greek ?a?d??, kardia, for "heart".

The vertebrate heart is primarily made up of cardiac muscle mass as well as connective tissue. Cardiac muscle is surely an unconscious striated muscle tissue discovered just within this organ as well as accountable for the capability of the heart to pump blood. The typical human heart, beating at 72 beats per minute, will beat roughly 2.5 billion occasions throughout an typical 66 year life expectancy. They weigh around 250 to 300 grams (9 to 11 oz) in women as well as 300 to 350 grams (11 to 12 oz) in men.

Dates are very useful when you are preserving your heart within a healthy and balanced situation. Whenever drenched for the night time as well as smashed the next day as well as ingested, they turn out to be really beneficial for vulnerable hearts. They assist in building up the heart, if consumed two times in one week. Manage your usage of dates to align with one of these heart-healthy diet suggestions: saturated fats, 16 g to 22 g daily; trans-fat, as much as 2 g daily; calories from sugar, Five percent to fifteen percent of the everyday total, as well as fructose, 50 g to 100 g each day. Dates consist of virtually no fat, but some varieties contain high amounts of fruit sugar. Among dates, deglet noor varieties make a better choice than medjool dates.

**Prevents from Diarrhea**

Diarrhea is really a situation which involves the regular passing of loose or even watery stools – it will be the complete opposite of bowel problems and may have numerous reasons, which might be contagious or even non-infectious. Ripe dates consist of potassium. Potassium is helpful for managing looseness of the bowels. They may be very easily digestible, which will help to deal up with the difficulties of diarrhea. Potassium can also be required to sustain a healthy and balanced central nervous system.

**Cures Intoxication**

Intoxication is definitely the severe state of bodily as well as intellectual disability brought on by alcohol consumption or just being confronted with a psychoactive medication. Intoxication usually possesses both mental and physical side effects, which may consist of problems with managing motion, mood changes, disability of cognitive skills, impaired common sense, impaired impulse control, as well as alterations in interpretation your surroundings or even the actions of other individuals. Even though the phrase "intoxication" is most often utilized to make reference to the serious (instant or even short-term) negative effects of alcoholic beverages, intoxication could also be used to consult the effects of various other drugs, which includes amphetamines, caffeine, cannabis, cocaine, hallucinogens, inhalants, opioids, phencyclidine, sedatives, hypnotics as well as anxiolytics. Dates are referred to as an outstanding treatment for alcoholic intoxication. Dates offer fast comfort when it comes to alcoholic intoxication. They must be rubbed as well as drenched over night to get a lot more nourishing valuations from them.

**Boost Energy**

The high content of sugars just like glucose, sucrose as well as fructose causes it to be a power house providing instantaneous energy. Dates usually are power filled with 29 grams of natural sugars, 3 grams of soluble fiber as well as 31 grams of carbohydrates. As a result, the small date fruit could be effectively referred to as an energy improving dry fruit.

**Maintain a healthy Nervous System**

The central nervous system is really a complicated network of nerves as well as cells which carry messages back and forth from the brain as well as spinal-cord to numerous areas of the body. The nervous system consists of both the Central nervous system as well as Peripheral nervous system. The Central nervous system comprises of the brain as well as spinal cord as well as The Peripheral nervous system consists of the Somatic as well as the Autonomic nervous systems. Dates usually are abundant with potassium content, that happen to be important for a proper nervous system. Potassium in dates is a vital element of cells and the body fluids which help control heart rate as well as blood pressure level, therefore offering safety against coronary heart disease as well as heart stroke. Study discloses that greater potassium consumption of about 400 mg can help to eliminate the potential risk of heart strokes simply by 40%.

**Treat Anemia**

Anemia is described as an unusually lower quantity of red blood cells within the circulating blood. It often affects sufferers along with cancer. In reality, in lots of cancer healthcare determinations just like multiple myeloma as well as serious leukemia, the existence of anemia might be what initially encourages a doctor to suspect an actual tumor (neoplasm). Whether or not anemia grows depends upon the kind of cancer discovered, the treatment used, as well as the existence or even deficiency of various other underlying medical problems. Dates have got higher iron content within them, which will make them excellent natural home remedies for decreasing a deficiency of iron. Anemia patients are encouraged to consume numerous dates on a regular basis to boost their amount of iron within the blood.

**Strengthen Tooth Enamel**

Tooth enamel is definitely the tough external surface coating of the teeth which acts to safeguard from dental cairies. In reality, tooth enamel is regarded as the toughest mineral substance within your body, even more powerful as compared to bone. Inspite of its power, everyday acids which develop from particular foods as well as beverages, especially those which are sweet or even include starch, can easily place your enamel at an increased risk. Acids can easily attack as well as soften the tooth surface area. Plaque bacteria produce acids that may deteriorate as well as ruin the enamel. And when your enamel is finished, it may be gone for good. Dates may also help avoid tooth decay due to its fluorine content, an important chemical element which slackens the growth of tooth decay simply by shutting out plaque as well as making the tooth enamel powerful. Fluorine is known to keep plaque away simply by conditioning the tooth enamel. However, so many people are of the misunderstanding that tooth enamel can easily amazingly change tooth decay. This is simply not true. Tooth enamel is made up of hydroxyapatites, that when they are introduced contact with fluorine forms hydroxy fluoroapatites, which can be resistant against decay. This obviously implies that fluorine helps in avoiding even more tooth decay, yet are not able to turn back decay procedure.

**For Pregnant Women**

Pregnancy is really a time period whenever a lady suffers from excessive bloatedness, bowel problems along with other intestinal tract difficulties. Dates are generally an outstanding treatment for these kinds of intestinal disruptions. In addition, being abundant with iron, dates avoid anemia in mother as well as child. Dates assist improve the uterus muscles, allowing it to enlarge in the smooth manner in the course of delivery. Dates in addition improve the mother's milk by giving it along with nutrition very therapeutic for the newborn's overall health. Dates in addition avoid hemorrhage soon after childbirth. Additionally it is considered to provide excellent memory as well as intelligence to the unborn child. Dates is additionally seen to enhance strengthen the uterine muscles as well as allows for childbirth.

### Stops Night Blindness

The American Cancer Society declares that usage of one date on a daily basis assists in maintaining healthy eyes for one's life time. One date each day can prevent eyes blindness. 20-35 grams of soluble fiber every day is sufficient.

### Abdominal Cancer

Abdominal cancer will come in numerous types, all relating to the irregular, uncontrolled development of cells within a specific portion of the entire body. Kinds of this cancer consist of stomach cancer, liver cancer, renal cancer, pancreatic cancer, bowel or even colorectal cancer, uterine cancer, and ovarian cancer. There are numerous of signs and symptoms of abdominal cancer. These include, but aren't restricted to, exhaustion, fat reduction, appetite loss, nausea and vomiting, night sweating, itchy skin, a fever, jaundice, difficult or even painful ingesting, lack of sex drive or even trouble carrying out sexually, anemia, shortness of breath, bowel problems, looseness of the bowels, extended or even heavy periods, pale skin, stomach pain, as well as lumps, masses or even swelling within the abdomen. Dates are very therapeutic for treating stomach cancer. They act as beneficial tonic for all those age groups. They act as better as the medications, and therefore are all-natural and don't bear any kind of negative effects within the body. They may be effortlessly ingested as well as utilized for offering additional as well as required energy.

### Sexual Weakness

A sexual weakness may be covered up along with dates, simply because the fruit increases sexual endurance. A secret potion for sexual endurance will be a hand of drenched dates combined with goat's milk throughout the night, with a few cardamom powder as well as honey. Dates are also very therapeutic for raising sexual stamina within your body. A number of dates, whenever soaked within fresh goat's milk for the night and after that grinded within the very same milk together with the combination of cardamom powder as well as honey, turns into a very beneficial tonic to increase sex endurance as well as sterility brought on because of functional problems.

### How to store Dates

Store dates at room temperature within an air-tight pot for a number of months or even refrigerate for as much as 1 year.

## DISADVANTAGE OF DATES

Dates are definitely the fruit of date palms. You are able to consume them fresh new or even dried as snacks or perhaps in recipes just like cakes or even stews. Dates are usually nutrient-dense plus they count toward your everyday suggested portions of fruit, yet a lot of dates may cause negative effects. Consume them moderately as an element of a well-balanced all round diet.

### Weight gain

Dates have got 415 calories in the cup, which can be just 147 g, or even about 5 1/4 oz. You will gain weight if you consume a lot more calories as compared to you expend, or even burn, as well as weight problems raises your chance for long-term illnesses just like heart disease as well as diabetes. It is possible to consume a lot of calories from calorie-dense foods for instance dates. Fresh fruits usually are lower-calorie options. A tiny apple, pear or even cup of blackberries or perhaps raspberries offers no greater than 80 calories, in accordance with the 2010 Dietary Guidelines from the U.S. Department of Health and Human Services.

### Diabetes

A lot of dates can easily improve your chance for type 2 diabetes simply because they have a great glycemic index, in accordance with the Linus Pauling Institute Micronutrient Information Center. The glycemic index signifies a food's potential impact on your blood sugar levels. High-glycemic foods just like dates may cause sharp raises within your blood sugar levels. By consuming a high-glycemic diet and therefore are constantly causing your blood sugar levels to increase, you might ultimately develop type 2 diabetes.

### Intolerance

In case you have fructose intolerance, date fruit can easily result in a response simply because every cup offers 29 g fructose. Fructose intolerance, or even fructose mal absorption, implies that the body is not able to effectively absorb fructose, and you might experience gas, bloating as well as diarrhea, in accordance with the Mayo Clinic. If you possess the condition, prevent fruits, fruit drinks, sugars, honey as well as sweetened products just like soft drinks as well as condensed milk.

### Tooth decay

The teeth may rot when you eat dates simply because every cup offers a lot more than 100 g carbohydrate. Whenever bacteria within your mouth ferment, your teeth can easily rot due to the acid. Any kind of food along with carbohydrates can result in tooth decay in case the food stays in your teeth for too much time. To decrease your danger for cavities, wash the mouth along with water as well as brush the teeth following a meal, and don't drink sugary drinks in between meals.

## HISTORY OF DATES

The date palm is considered to possess originated from the lands round the Persian Gulf as well as in medieval times was particularly plentiful between the Nile as well as Euphrates rivers. Alphonse de Candolle stated that this ranged in the prehistoric era from Senegal to the basin of the Indus River within northern India, particularly in between latitudes 15 as well as 30. There is certainly ancient proof of farming in eastern Arabia within 4,000 B.C. It absolutely was much adored as well as thought to be symbolic of fertility, as well as shown in bas relief as well as on coins. Literature dedicated to its historical background and romance is spacious. Nomads planted the date at oases within the deserts as well as Arabs introduced it into Spain. It's got long been cultivated for the French Riviera, in southern Italy, Sicily as well as Greece, although the fruit doesn't reach excellence within these locations. Probably it costs better within the Cape Verde Islands, for a program of date improvement was introduced there within the later 1950's. Iraq has usually led the entire world in date manufacturing. At present, there are actually 22 million date palms in that nation providing just about 600,000 tons of dates on a yearly basis. The Basra area is acknowledged for its cultivars of remarkable quality. The dates have been typically a staple food items in Algeria, Morocco, Tunisia, Egypt, the Sudan, Arabia as well as Iran. Blatter quotes the writer, Vogel, as declaring: "When Abdel-Gelil besieged Suckna in 1824, he reduce no less than 43,000 trees, to force the city to give up; even so, you will still find a minimum of 70,000 still left."

In 1980, creation in Saudi Arabia was delivered to almost a half-million tons through 11 million palms due to government subsidies, enhanced technology, along with a regal decree which dates be contained in foods in government as well as civic institutions and that hygienically-packed dates be frequently obtainable in the marketplaces. Growers obtain financial rewards for every offshoot of the high-quality date planted in a recommended spacing. The Ministry of Agriculture has generated training programs through the entire nation to show modern day agricultural techniques, which includes mechanization of all probable procedures in date culture, as well as acknowledgement and also particular roles of the numerous nearby cultivars. In West Africa, close to the Sahara, just dry, sugary types could be developed.

Bonavia launched seeds of 26 types of dates through the Near East into northern India and Pakistan in 1869; and, in 1909, D. Milne, the Economic Botanist for the Punjab, launched offshoots as well as established the date like a grown crop in Pakistan. The fruits mature effectively within northwestern India and also at the Fruit Research Center in Saharanpur. In southern India, the weather is undesirable for date creation. Several trees around Bohol within the Philippines have been demonstrated to deal with a good amount of fresh fruits of excellent quality. The date palm has become presented into Australia, as well as into northeastern Argentina as well as Brazil exactly where it might thrive in dry zones. Several dates are usually offering fruits for the market on the little island of Margarita off the coast through northern Venezuela. Seed-propagated dates are located in several tropical as well as sub-tropical regions exactly where they're appreciated as ornamentals yet the location where the weather conditions are unacceptable for fruit manufacturing.

In November 1899, 75 plants were delivered from Algiers to Jamaica. These were saved in a nursery till February 1901 and after that 69 were grown at Hope Gardens. The feminine palms eventually carried huge bunches of fresh fruits however they were ready to fully developed in October throughout the rainy period as well as, appropriately, the fruits rotted as well as fell. Very rarely have date palms carried normal fruits within the Bahamas as well as South Florida.

Spanish visitors introduced the date into Mexico, about Sonora as well as Sinaloa, and Baja California. The palms were just seedlings. Still, the fruits had excellent attractiveness as well as appeared to be released from Baja California in 1837. The very first date palms within California were seeds grown by Franciscan as well as Jesuit missionaries within 1769. Potted offshoots from Egypt arrived at California within 1890 and various other introductions have already been converted to that state as well as in the drier regions of southern Arizona about Tempe as well as Phoenix. In 1912, Paul as well as Wilson Popenoe bought a total of 16,000 offshoots of chosen cultivars within Algeria, eastern Arabia and Iraq and also transferred these to California for distribution by their father, F.O. Popenoe who was simply an innovator in motivating date culture in California. It had been a lucrative crop, especially in the Coachella Valley. Nowadays there are about a quarter of the million bearing trees in California as well as Arizona.

## TYPES OF DATES

It might be not practical to manage in depth together with date cultivars here. Paul Popenoe outlined 1,500 as well as supplied explanations of the fruit as well as palm, along with the background and importance, of the most basic, country simply by nation, in 90 pages of his book, The Date Palm, developed in 1924 yet released in 1973 as well as easily available. In Iraq, there are actually at present 450 female cultivars, the most crucial of which are usually: 'Zahdi' (43% of the crop; lower in price); 'Sayer' 23% of the crop as well as high-priced); 'Halawi' (13% of the crop as well as high-priced); 'Khadrawi (6% of the crop as well as high-priced); additionally 'Khastawi, 'Brem', as well as 'Chipchap'. Sawaya and colleagues (1983) have documented on the sugars, tannins as well as vitamins in 55 main date cultivars of Saudi Arabia.

The following, together with short remarks, are definitely the dates most often cultivated: 'Barhi'- released into California within 1913 from Basra, Iraq; almost round, mild amber to darkish brown whenever ripe; gentle, together with thicker flesh as well as abundant taste; of outstanding quality. For shipment requires refrigeration once selected, after that curing as well as special packing. 'Dayri' (the "Monastery Date")-introduced into California from convent grounds in Dayri, Iraq, in 1913; lengthy, slim, almost black, gentle. Palm needs additional care. Not developed thoroughly within California. 'Deglet Noor'-a top date within Algeria as well as Tunisia; as well as in the second country it really is developed in inland oases and is also the main export cultivar. It absolutely was launched into California within 1900 and today comprises 75% of the California crop. It really is semi-dry, not so sweet; retains effectively; is moisturized just before delivery. Significantly utilized for cooking. The palm is higher producing although not very resistant of rain as well as environmental moisture. 'Halawy' ('Halawi')-introduced into California through Iraq; gentle, incredibly sweet, small to moderate; might wilt throughout maturing until the palm is well-watered. It really is particularly resistant of moisture. 'Hayany' ('Hayani')-the cultivar most substantially grown in Egypt; however, not exported. Introduced into California in 1901, and is also distributed fresh new; isn't easy to cure. The fruit is dark-red to almost black; gentle. The palm is probably the most cold-tolerant.

'Khadrawy' ('Khadrawi')-important within Iraq as well as Saudi Arabia, and is also developed to some degree within California as well as Arizona. It's the cultivar most in-demand simply by Arabs yet too dark in color to become well-known around the American marketplace, even though it is really a gentle date of the highest quality. It really is early-ripening; doesn't keep too well. This particular cultivar is definitely the tiniest delicious date palm developed in the usa and it's also pretty resistant of rain as well as moisture. 'Khastawi' ('Khustawi'; 'Kustawy')-the top soft date in Iraq; sirupy, tiny in dimensions; valued for treat; maintains well. The palm is big as well as energetic and also generates its offshoots high in trunk area in California. The fruit is resistant against moisture.

'Maktoom'-introduced into California from Iraq in 1902; huge, red-brown; thick-skinned, gentle, mealy, moderate sweet; resistant against moisture. 'Medjool'-formerly released from Morocco; 11 off-shoots imported into California from Bou Denib oases in French Morocco in 1927; is currently promoted like a luxurious date within California; is big, gentle, as well as delicious yet ships properly. 'Saidy' ('Saidi')-highly prized in Libya; gentle, extremely sweet; palm is actually a hefty bearer; requires a hot weather conditions. 'Sayer' ('Sayir')-the most generally developed cultivar within the Old World and a lot released to Europe as well as the Orient; dark orange-brown, of moderate size, gentle, sirupy, and infrequently a few of the sirup is exhausted out as well as sold individually; not of top quality however the palm is probably the most resistant of salt along with other undesirable elements.

'Thoory'('Thuri')-popular within Algeria; really does very well within California. Fruit is usually dry; whenever treated is brown-red together with bluish bloom with extremely old and wrinkly skin as well as the flesh is oftentimes tough as well as breakable however the taste is nice, fairly sweet as well as nutty. Keeps well; usually carried on trips. The palm is strong together with short, stiff leaves; contains seriously, as well as groups are extremely big; fairly resistant of dampness.

'Zahdi'('Zahidi')-the oldest-known cultivar, eaten in great quantity in the centre East; launched into California about 1900. Of moderate dimension, round, mild golden-brown; semi-dry but gathered and traded in 3 phases: gentle, medium-hard, and hard: extremely sugary; keeps well for a few months; much utilized for cooking functions. The palm is strong, rapid developing, heavy bearing; drought tolerant; has very little tolerance of mugginess. Among the less well-known cultivars within California usually are:

'Amir Hajj'-introduced from Mandali Oasis within Iraq in 1929. The fruit is gentle, together with thin skin as well as thicker flesh; of good quality yet very little cultivated within theUnited States of America. 'Iteema'-offshoots from Algeria were launched into California in 1900. The fruit is big, oblong, mild amber, gentle, extremely sweet. Much developed in Algeria although not rain resistant as well as little developed in California. 'Migraf' ('Mejraf)-a extremely popular cultivar within Southern Yemen. Fruit is mild golden-amber, big; of excellent quality. In inland oases of Tunisia, along with the 'Deglet Noor', there is certainly 'Ftimi' ('Alligue') which can be likewise susceptible to moisture, significantly less fruitful and much less disease-resistant. 'Manakbir' features a big fruit as well as ripens sooner yet has got the drawback that the palm delivers very few offshoots and its particular multiplication is restricted.

## NUTRITIONAL VALUE OF DATES

Dates usually are free of fat, cholesterol free, and therefore are an excellent source of dietary fiber. One particular date consists of close to 23 calories, which makes it a nutritious low-calorie snack food. They will consist of vitamins A, Vitamin B-complex, C and over 20 distinct amino acids that will help us absorb as well as deal with carbohydrates along with manage blood sugar as well as fatty acids levels. Dates may also be excellent options for carbohydrate, fiber, as well as potassium, calcium and iron together with minerals within small quantities. Selenium present in dates reduces the chance of cancer as well as heart diseases and enables us keep our immunity processes healthy and balanced.

| Nutrient | Unit | Value per 100.0g |
|---|---|---|
| Proximates | | |
| Carbohydrate, by difference | g | 75.03 |
| Energy | kcal | 282 |
| Fiber, total dietary | g | 8.0 |
| Protein | g | 2.45 |
| Sugars, total | g | 63.35 |
| Total lipid (fat) | g | 0.39 |
| Water | g | 20.53 |
| Minerals | | |
| Calcium, Ca | mg | 39 |
| Iron, Fe | mg | 1.02 |
| Magnesium, Mg | mg | 43 |
| Phosphorus, P | mg | 62 |
| Potassium, K | mg | 656 |
| Sodium, Na | mg | 2 |
| Zinc, Zn | mg | 0.29 |
| Vitamins | | |
| Folate, DFE | µg | 19 |
| Niacin | mg | 1.274 |
| Riboflavin | mg | 0.066 |
| Thiamin | mg | 0.052 |
| Vitamin A, IU | IU | 10 |
| Vitamin B-6 | mg | 0.165 |
| Vitamin C, total ascorbic acid | mg | 0.4 |
| Vitamin E (alpha-tocopherol) | mg | 0.05 |
| Vitamin K (phylloquinone) | µg | 2.7 |
| Lipids | | |
| Fatty acids, total monounsaturated | g | 0.036 |
| Fatty acids, total polyunsaturated | g | 0.019 |
| Fatty acids, total saturated | g | 0.032 |

## WAYS TO EAT DATES

Eat as it is. Whilst it's healthy to be a part of several red dates every single day, like several dried fruits, they must be consumed moderately to prevent gloating. 3 to 5 are generally sufficient for just one day. Diabetics, however, and individuals along with digestive problems ought to stay away from red dates. Chinese herbal medication states that red dates usually are not ideal for individuals with an excessive amount of internal humidity. Listed here are several typical red date dishes, that you simply must be able to pick up at most of the non-regional or even zhou

- Red dates, white fungus as well as lotus seed soup (hóng zao yín'er iánzi geng)
- Longan, red dates and glutinous rice porridge (guìyuán hóng zao nuòmi zhou)
- Osmanthus as well as red dates cake (guìhua hóng zao gao)
- Red date, Chinese yam as well as pork rib soup ( hóng zao shanyào páigu tang)
- Black rice, red date as well as soy bean paste (Hei mi hóng zao dòusha)

## FOOD AND NUTRITION

### FOOD FROM THE FOREST

Fruit of the Elaeis guineensis (oil palm), a palm that provides an oil used in the manufacture of soap, candles and lubricants, as well as edible palm hearts. Millions of households in the developing world depend on food and fodder from forests to supplement their own and their livestock's diets. Although forest foods do not usually provide a complete diet, they do make a critical contribution to the food supply. Forest foods increase the nutritional quality of rural diets; supplement other sources of food -particularly agricultural crops that are only seasonally available; and are used as emergency food supplies during drought, famine and war.

Forest foods are nutritionally important and are traditionally used as supplements to the staple diet. Leafy vegetables and wild animals add diversity, flavour, vitamins and minerals to characteristically grain-dominated diets. Forest foods are often collected and stored for later use. Forest foods can thus raise rural peoples' nutritional intake by providing a year-round supply of food.

The most important and well documented use of forest foods is in meeting seasonal shortfalls such as the `hunger periods' at the beginning of the rainy season before crops are ready for harvest. During hunger periods, the practices of digging for roots and tubers and gathering fruit and nuts are almost universal. Other gaps in food supply are caused by sudden needs for cash-for school fees, for example -which force families to sell crops. Such demands leave families short of both food and the money they need to buy food.

Trees are also important in emergencies such as drought and famine. They provide food when crops fail and products such as gum arabic to sell for cash income with which to buy food. Energy-rich foods such as the roots of the baobab tree are most sought after in times of famine, while other foods, such as the baobab's fruits, are often eaten during periodic shortages of staple foods. Several wild foods are used only in times of scarcity and famine, among them fibres, seeds, tubers, leaves and stems. Many of these famine foods have a high protein and energy content. In India, Malaysia and Thailand, about 150 plant species, representing nearly one-fifth of the wild species consumed as food in these countries, have been identified as sources of emergency food crops, among them bark, kernels and tubers. The role forests play in providing and maintaining suitable conditions for wildlife should not be under-estimated. Forested areas, mangroves, streams and fallow agricultural areas within the forest provide a habitat for many wild animals and fish. For, example, forests maintain suitable conditions for fresh water fish by helping to reduce the sediment in streams, while mangroves provide an important habitat for fish. Birds, their eggs, insects, rodents and other larger mammals are sometimes the only source of animal protein for rural people. The naturally abundant small animals-rats, squirrels, mice, porcupine and grasscutters - are the most important species for subsistence consumption.

**Fig.** The cashew nut -another nutritive forest product; one that is easy to collect and roast.

In some regions, wild game consumption is extremely important. For example, in West Africa the consumption of bushmeat depends largely on availability. In Nigeria, people living near forest reserves consume as much as 84 percent of their animal protein in the form of game whereas, in areas of Nigeria with no access to forests, bushmeat makes up only 7 percent of total meat consumption. Similarly, in some areas of the Peruvian Amazon, people obtain more than 85 percent of their animal protein from wild animals, including fish.

## Food from cultivated trees

Homegardens are producing an increasingy important supply of food in many countries, as population pressures reduce the amount of land available to each household for food crops. Homegardens support the cultivation of multi- purpose trees and shrubs, often in association with annual and perennial agricultural crops and livestock, within the household compound. Such gardens are found in most regions of the tropics and sub-tropics, particularly in lowland areas with high population densities. Many homegardens resemble those of Java or southeastern Nigeria, with an intensive combination of trees, crops and livestock. In other cases, however, a single mango tree provides a source of food at a time of the year when few other foodstuffs are available or when the need to plant the next season's crop means that there is little spare time or labour available for gathering and preparing food.

The average size of a homegarden is usually much less than one hectare, yet in many parts of the world the fruit, nuts, edible leaves and other foodstuffs grown in homegardens provide a substantial part of the household food requirement. In some areas of Java, homegardens provide more than 40 percent of the total calorific intake of farming communities.

**Fig.** The mango-typical fruit found in homegardens in Asia, Africa and South America.

Gardens within a household compound can produce food all year round with a relatively low labour input. When intensively managed, the yield from a compound, in monetary terms, can be five to ten times as much per hectare as that from traditional field cropping systems, and returns on labour are typically four to eight times higher. Many homegardens support very large numbers of different species. In southeastern Nigeria permanently cultivated compounds around the household contain trees including the oil palm, coconut, banana and plantain, intercropped with cassava, gums and other arable crops. Studies have shown that households with homegardens have higher than average nutrition levels. In Puerto Rico, for instance, food from gardens tended by women significantly contributes to the total food supply and is an important source of both betacarotene (converted to Vitamin A in the body) and Vitamin C, especially for children. Trees are also part of traditional shifting cultivation systems practised within forest areas by more than 300 million people world-wide. Shifting cultivation can involve clearing forest areas to develop agroforestry systems similar to those found in homegardens. Trees are maintained or grown to provide a range of fruits, seeds, nuts and leaves for food as well as to maintain suitable soil conditions for food production.

## FOOD FOR LIVESTOCK

An adequate supply of livestock fodder is a crucial part of food production for millions of people. Good fodder increases livestock productivity, making more milk and meat available for rural families; it helps maintain animals used in working the land; and it provides manure for fertilizer. A reliable supply of fodder is critical to the 30 to 40 million pastoralists world-wide. There are 20 to 25 million pastoralists in Africa alone, many of them roaming the dry sub-Saharan belt that stretches from Mauritania to Ethiopia.

These pastoralists manage to keep their herds alive on arid and semi-arid land by supplying them with twigs, leaves, small branches, seed pods and fruit from trees and shrubs. Pastoralists in parts of northern Senegal, for example, rely on high-quality supplements of browse-leaves, fruits and seed-pods from trees and shrubs-for at least six months of every year, to keep their herds alive and healthy. Probably as much as three-quarters of Africa's nearly 10 000 tree species are used for browsing.

During the dry season, browse can comprise as much as 60 percent of goats' fodder and 30 percent of cattle fodder. Fodder from trees and shrubs is particularly important during this season when the nutritional quality of grasses is markedly reduced. Dry grasses contain only minimal levels of energy and protein, and lack essential minerals, especially phosphorus. For this reason, herders in Sahelian West Africa try to avoid pastures with no shrubs or trees. Unfortunately, fodder supplies are coming increasingly under threat as pressures on tree fodder cause overbrowsing, reduce the regeneration of trees and shrubs, and in many places actually eradicate them. In the Sahel, overbrowsing has led to a decline of the fodder species Acacia seyal (Whistling Thorn) and Acacia senegal (Gum Arabic tree). In the central rangelands of Somalia, Cordeauxia edulis (yicib), the main dry season food of camel and goat herds, is being overbrowsed and progressively eliminated.

The management of fodder resources on farms, in rangelands and in forest areas can help reduce overbrowsing and provide a sustainable source of animal fodder. A ready supply of fodder means that women, often the principal fodder providers, do not have to travel so far to find this necessity. The extra time and energy thus acquired are highly valuable, especially as women also prepare the family meals and often have a working day that lasts from four in the morning to eleven at night.

Many Nepalese farmers have now started to grow fodder trees on their farms, and they are popular on farm and fallow land in southeast Nigeria. Cattle, sheep and goats are not the only animals to benefit. In western Java, trees provide up to 15 percent of the fodder used to feed small stall-fed ruminants.

**Fig.** Prosopis chilensis, (Chilean Mesquite) originally imported from South America, but now commonly grown in Africa. Not only does it provide nutritious fodder but the wood can be used to make poles, stakes and other agricultural items.

Livestock productivity can also be increased on rangelands through greater use of tree and shrub species. Species with particular potential in arid areas of Africa areOpuntia and Atriplex nummalaria; in Latin America, Prosopis species (Mesquites) have shown their worth in maintaining livestock health and productivity. Managing the growth of naturally occurring fodder species or planting more palatable plants to replace woody shrubs and trees is another way to improve rangeland production. In the Sahel, many degraded pastures could be improved by controlling the unpalatable Calotropis procera, now especially common around boreholes. In large areas of Turkana in northern Kenya, Acacia reficiens has rendered the rangeland impenetrable to animals. In both regions, control of the unpalatable species, linked with re-planting of more useful species, would enhance conditions on the rangeland and enable these lands to support a greater number of animals.

Agroforestry can also play an important role in providing fodder for livestock, whether from hedges and windbreaks established to stabilize sand dunes and allow more cropping, or from the variety of trees cultivated in homegardens. In either case, incorporating th provision of fodder into the agricultural syster has significant spin-offs for the welfare of the household.

## NUTRITION AND HEALTH

Food from trees in forests homegardens, and the plants and wildlife supported by forests are often nutritionally essential to the diet of rural people. The medicinal and other properties of tree products also play an important role in keeping people healthy. Many forest foods are higher in vitamins and other important nutrients than domesticated varieties. While the vitamin C content of an orange is famously high at 57 mg/100 g, the fruit of the baobab tree has 360 mg/100 g and Ziziphus jujube var. spinosa 1000 mg/100 g. Similarly, on a weight-for-weight basis, wild leaf vegetables contain more riboflavin-another vitamin necessary for good health -than eggs, milk, nuts and fish. It is estimated that some 250 000 children go blind in south-east Asia every year because of lack of vitamin A. Many forest fruits and their leaves are good sources of beta-carotene, which the body converts into vitamin A. Riboflavin deficiency, responsible for several eye and skin disorders, can be corrected by many forest foods, especially leaves. Iron, needed to produce blood haemoglobin, is abundantly available in many forest foods. One of the most common causes of dietary deficiencies is the decreasing diversity of diets. Medical surveys have shown that Pacific Islanders, for example, began to consume fewer minerals and vitamins as they became more dependent on imported cereals and developed a preference for introduced vegetables; there was a significant decline in health as a result.

In rural Bangladesh it was found that, although people living in ''modern' villages had calorific, protein-rich rice and wheat available all year round, a greater number of people suffered from malnutrition than in `traditional' villages where food was not as readily available. Because of the availability of rice and wheat, the modern villagers were eating less of other foods, and the diet of traditional villagers, which contained more roots, tubers, pulses and vegetables, had a higher mineral and vitamin content.

As well as making a direct contribution to health by improving nutrition, tree products provide the only medicines available to many people in developing countries. Some plants contain high concentrations of chemicals used as the base for modern drugs. Many plants are used traditionally for these medicinal qualities, and others undoubtedly depend on effects not yet exploited in Western medicine. These medicines are important nutritionally. By keeping the body healthy, they not only help it to absorb food efficiently but also increase its ability to fight off infections that might otherwise impair digestion and the ability to eat. Some trees have specific properties that can improve the quality of water supplies. Moringa species, for example, are used by women in Egypt to clarify turbid water. The tree's seeds contain natural coagulants which can clear water to tap water quality in one to two hours, and eliminate up to 99 percent of bacteria. The fruits of other species are lethal to the snails that act as intermediary hosts of bilharzia, and the water fleas that harbour the guinea worm.

## LEAVES AND STEMS

Wild leaves, either fresh or dried, are one of the most widely consumed forest foods. As the base for soups, stews and relishes they add flavour to otherwise bland staples such as rice or maize, making them more palatable and thus encouraging consumption. One study in Lushoto, Tanzania, found that people who consumed wild leaf relishes favoured the taste of wild leaves over introduced cultivated vegetables. Leaves from wild and cultivated trees are often boiled fresh in stews. They can also be dried and powdered, or fermented, to preserve them; they can later be made into a paste which is used in stews and soups as a meat substitute.

The carotene, vitamin C, calcium and iron content of leaves varies greatly. One study in Swaziland found that the nutrient content of wild leaves compared favourably with that of the leaves of cultivated plants. In Swaziland, the leaves of 48 different species are used; and at least half the adults eat meals which include wild leaves more than twice a week. The stem of the sago palm contains starch, a valuable carbohydrate, commonly used in cooking in Indonesia. It contains 352 calories/ 100 g and provides 85 percent of the energy intake of people in the rural area of Upper Sepik, in Papua New Guinea.

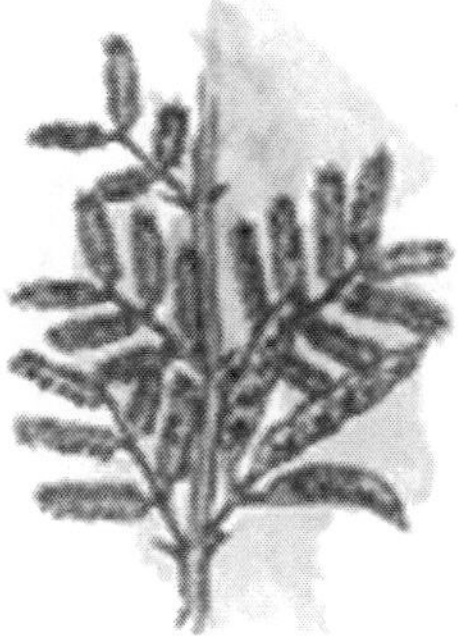

**Fig.** Leaves are widely used in rural society as spicy supplements to stews and soups and often have a high nutritional value.

## SEEDS AND NUTS

The nuts of the coconut, oil palm and babassu palm are at the forefront of nutritionally important nuts and seeds, adding substantial calories, oil and protein to the diet. (Nutritionally, fats and oils are important for several reasons, not least because they facilitate the absorption of vitamins A, D and E.) Coconuts are of central dietary importa in many cultures, and account for 7 percent the world's fat consumption. In the areas of northeastern Brazil where the babassu palm grows, its kernels provide oil for an average percent of households. In Sierra Leone, oil from the kernel and fruits of the oil palm is consumed by 96 percent of rural households. Among other important oil nut trees are, the shea butternut, cashew nut, African breadfruit, the mongongo nut and the Park species. The seeds of Parkia form an import part of the diet in most parts of the Sahel. Fermenting Parkia improves the digestibility the protein and increases the vitamin content of the seeds, providing a nutritious protein fat-rich food known as dawadawa. It is an important ingredient in side dishes, soups a stews made to accompany porridges in northern and western Africa.

**Fig.** The baobab tree is heavily used in Africa as a source of fuelwood, food, oil and even dug-out canoes. The fruit pulp can be dried to produce cream of tartar, and the leaves are added to stews and soups as a nutritious spice.

## ROOTS AND TUBERS

Roots and tubers provide carbohydrates an( some minerals, and are often important ingredients in traditional medicines. They used as drought and famine foods, not only because they survive low rainfall periods, b because they can be an important source of water. However, they require time to find a dig up, and often involve extensive process such as soaking and prolonged cooking.

### Mushrooms

Mushrooms, eaten as meat substitutes and i flavouring, are good sources of protein and minerals. In an analysis of the nutritional values of 30 edible mushrooms from Upper Shaba, Zaire, the mean protein content was found to be 22.7 g/100 g dry weight, with a high calcium and iron content.

**Fig.** Mushrooms are nutritious as well as tasty-one study found that 30 species had a high protein value of 22.7 g/100 g as well as high calcium and iron contents.

## Gums and Sap

Sap, high in sugars and minerals, is tapped beverages. Gum, used as a food supplement good source of energy; and both saps and g, have many medicinal uses. Palm wines froc fermented sap are an important cultural beverage in many areas. For example, in southern Cameroon the wine is consumed in most households several times a week. The gum of Sterculia species, a good source of beta-carotene and vitamin C, is added to soups and stews in northern Senegal. Gum arabic (from Acacia senegal) is a traditionally important food for pastoralists, agriculturalists and hunter-gatherers; more recently it became an important source of food for gum collectors. Less than 200 g have been estimated as sufficient to feed a person for a day.

## Honey

Honey is highly valued almost everywhere for its high energy content: 100 g of honey contain more than 280 calories. Tree blossoms provide a year round food supply for bees and, in turn, the fertilizing action of bees during their hunt for nectar can increase the yields of oilseed, pulses and fruit trees.

## Fish

Forests help to maintain the conditions in rivers necessary for fish to live and breed. In coastal areas, mangrove forests are home to important breeding areas for fish and molluscs, which together provide significant animal protein for many rural communities. Average figures for Nigeria, for instance, show that more than three times as much fish as beef is eaten. In Sarawak, Malaysia and the Peruvian Amazon, 50-60 percent of animal protein comes from fish.

## Animals

The protein content of wild meat-often 20-25 percent by weight-is comparable and sometimes higher than that of meat from domestic animals. Wild meat is an important source of animal protein in many parts of the world, though availability depends on conditions in the forest. Although wild meat is also a good source of iron and vitamins A and B, which are commonly lacking in diets, often only small amounts of meat are consumed because of shortages, high prices and legal restrictions on the hunting of animals. The protein and vitamin content of insects such as caterpillars has been compared to that of vitamin pills. Studies found that 100 g of termites provide 561 calories, and that bee larvae contain 10 times more vitamin D than fish liver oil and twice as much vitamin A as egg yolk.

# 6

# Landscaping with Fruit Trees

## LANDSCAPING TIPS FOR FRUIT TREES

Fruit trees come in all shapes and sizes, so that there is one for every garden. Once established, they should provide a plentiful crop and need remarkably little maintenance.

First Steps:

- Inspect smaller fruit trees and remove lower sideshoots and suckers (vigorous, unproductive shoots).
- Cut away old and congested branches and shoots and clear away the cut branches.

### SMALL VARIETIES FOR SMALL GARDENS

Low-growing fruit tree cultivars are best for most domestic gardens. With a crown not exceeding 2.5m in diameter, and a stem height of about 60cm, apples and pears sold as spindlebushes are easy to look after and to harvest. Half or full standards, on the other hand, have stem heights of 1–2m and mighty 10–12m wide crowns that would burst the boundaries of many gardens.

There are small cultivars of sour (Prunus cerasus) and sweet cherry trees (Prunus avium), plums (Prunus domestica) and damsons ( Prunus damascena) with a stem height of 60cm and a 3m wide crown. The crowns of apricot ( Prunus armeniaca) and peach trees (Prunus persica) usually measure about 5m across. They prefer warmer spots, but both can be trained as an espalier against a sheltered house wall.

#### Watch Out For...

Never leave fruit trees unpruned. Columnar trees will remain in shape, but without corrective pruning all other varieties will deteriorate quickly and will soon cease to bear fruit and become susceptible to disease. Fruit trees are available as bare-root, root-ball or container-grown plants. Bare-root and root-ball trees are best planted in the cooler months of spring or autumn, while container-grown plants can be planted out all year round as long as the ground is not frozen.

Whatever the root type, soak the roots in a bucket of water while you prepare the hole. It should be about 1m wide and 60cm deep so that there is plenty of room for the roots. Trim the roots of bare-root trees a little before planting. Always plant the tree so that the grafting site, which is visible as a thickening in the trunk, remains above ground. Tread down the soil around the trunk to firm it and tie the tree to a stake so that it will withstand strong winds.

**Pruning Fruit Trees**

Immediately after planting, remove competing branches, leaving only the central branch and a maximum of four strong leader branches. Over the next few years, prune formatively so that the numerous shoots coming from the leader branches do not get in each other's way and are not too congested. This will also shape the crown nicely.

In subsequent years, pruning keeps the crown in shape for a good yield and thins out branches that are too congested. Remove any stray shoots sprouting from the base of the trunk together with any long 'water' shoots on the branches that take nutrients from the tree without producing any fruit.

If necessary, prune large branches right back to the trunk. To ensure that the crown gets sufficient light, regularly cut away congested growth that prevents stronger branches from developing. Cut old fruit branches back to the point where young shoots are growing, which will in turn set fruit.

**Help with Aphids**

Creating a circular bed around a young fruit tree promotes healthy growth.

- Remove weeds, grass or other plant matter from an area measuring about 1.5m across around the tree.
- Mulch the bed with compost in autumn, then sow nasturtiums ( Tropaeolum majus) in spring to provide shade for the soil and prevent it from drying out. Nasturtiums are a magnet for blackfly so will help keep the tree free of this pest.
- Put up nesting boxes to attract blue tits. The birds feed on insects living in the branches. Alternatively, hang a clay plant pot upside down in the branches and fill it with straw to attract earwigs, which prey on aphids.

## APPLE TREES

No landscape should be devoid of fruit trees. Aside from the obvious benefit of having said fruit available for consumption, fruit trees are beautiful.

Few things are more attractive than an apple tree blooming in spring. The delicate blossoms fill the air with a sweet smell and look like lace decorating the branches. As the season changes and the year moves toward summer, dark green shiny leaves create the canopy providing a sanctuary for birds and other small animals.

While it is true that apple trees need some tending throughout the year, the effort is worth the returns. If you wish to harvest your fruit you will need to spray for mites and other types of pests and fungi. You will also need to invest in some nets to keep your fruit from being eaten by birds or deer. Imagine baking pies in the fall from apples you grew yourself. Or putting up jars of applesauce, knowing exactly what is in the jars, unlike the ones you pick up at the supermarket. Fruit trees, while lovely in their own right, have a fairly specific payoff.

Apples can be grown for food or beauty. If you are desire only beauty, consider planting crabapple trees instead of traditional fruit trees. With crabapple you still get the benefits of the spring bloom, but don't need to provide the careful tending that a fruit producing tree needs. Crab apples, when large enough, can still be used for food, by the way, but generally speaking, crab apples are left for the animals.

Birds, deer, bears and squirrels are but a few of the animals that enjoy the fall production of crab apples. Crab apple trees don't tend to grow as large as full sized apple trees, which makes them ideal for a smaller yard as well. While some folks complain about the fruit lying upon their grass after the tree goes dormant in the fall, most of that fruit will be gone by spring, eaten by animals searching for nutrition during the cold, snowy New England winter. What doesn't get eaten decomposes, feeding the soil.

Apple trees cross pollinate, so if you do invest in apple trees and want fruit, make sure you get at least two trees and keep them close enough to each other to ensure pollination. If you don't want fruit, but still enjoy the sight of an apple tree in spring, get only one, or space them far enough apart that pollination is unlikely; for example, one in the front yard, and one in the back. You can have the best of either world.

When planting fruit trees in New England remember that they need to be planted in the spring. Don't plant them in waterlogged soil or they will likely succumb to some sort of fungus or rot. Pick a site with direct sunlight and leave enough room for the trees when they are fully grown. If you have deer in your area, remember that hungry deer can cause significant damage to young trees, so protect them with a fence; nothing else will work as well. Then, sit back and enjoy your annual apple tree display.

## CHERRY TREES

Cherry trees are magnificent fruit trees that provide four seasons of visual beauty in your landscaping. In the spring the cherry tree puts on its spectacular bloom. Come the summer you will be treated to dark green leaves and plenty of fruit. In the fall the cherry leaves turn beautiful colors and in winter the stark, frosty branches stand out against the sky. No cherry trees like wet soil, so plant in the drier areas of your yard. If you plant a row of cherry trees along a path or property edge you will be treated to a marvelous sight when the trees bloom in spring.

Growing cherry trees in New England may seem a bit odd, but the fact is that these hardy trees do very well in northern climates. Cherries come in sweet and sour varieties, and for ease of growing, sour cherries are better. Sour cherry trees will self pollinate, removing the need to have more than one tree in your yard. On the other hand, if you don't want the fruit, a sweet cherry tree, all by itself isn't likely to pollinate; the exception to this is the Stella cherry. If you are not interested in fruit at all you can always select a flowering cherry tree. These trees produce only blooms and fall into the ornamental category.

Tart cherry trees are hardy into the northern reaches of New England. The "Montmorency" cherry is the industry standard producing copious amounts of fruit for minimal effort. Tart cherries are particularly wonderful in pies, cobblers and preserves. If you purchase stock that is certified virus free your tree should flourish through the humid summer months and survive the cold winter very nicely.

Sweet cherry trees are susceptible to more problems than their tart cousins. They are only suitable for the southern regions of New England, but can still do well with regular care. Depending on the stock you chose some varieties are less attractive to birds and others are smaller, coming in dwarf versions. Dwarf varieties are easier to care for since you can access all parts of the tree with less effort, making spraying, covering and harvesting simpler.

Given the opportunity birds will try to share in your cherry crop. About the time you are ready to start harvesting the birds will be doing the same. The most effective way to keep the birds out of your trees is to drape them with nets. Nets for fruit trees are usually available at garden supply stores. If you can't find them at a garden supply store you might find them at a pond supply store, they are very similar to nets used to cover koi ponds.

Including fruit trees like the cherry in your landscaping is a wonderful way to add beauty and a little something extra to your yard. While it is true that fruit trees require a bit more effort, they pay off your work tangibly. There are few things as satisfying as enjoying the fruits of your labors; literally.

## PEAR TREES

Pear trees, like so many other fruit trees, make wonderful additions to your landscaping. Although New England isn't associated with the growing of fruit trees, beyond apples, the fact is that many types of fruit trees thrive in the cooler climate. Longer, cooler springs give fruit trees time to blossom and set fruit before the hot months of summer arrive. Set in the center of a planting bed as the main feature pear trees add height and visual appeal to any yard.

Pear trees cannot be planted in the autumn in New England. The cold winters are too difficult for a newly planted tree, and you are likely to lose trees to frost and damp. Wait until the early spring when the ground has thawed and drained a bit. Dig an appropriate hole, add some plant food and center the tree inside taking care not to cover the graft site. Cover with the dirt you removed to create the hole. Bare root stock is cheapest, but small trees will produce fruit between a year or two sooner.

The Bartlett pear is particularly well adapted to the New England Climate. The cool, humid climate found up in the northern reaches of New England help this pear variety thrive where-as it does very poorly in warmer areas and is prone to blight. If you don't succeed with the Bartlett pear don't give up, other pears are also exceptionally good matches for New England. Check with your local gardening store. Pear trees can be trained to trellises or allowed to grow naturally, but they do need regular pruning and maintenance or they will quickly produce lots of additional small branches. If you want quality fruit you must invest some time in regular care and feeding.

Pears are best eaten out of hand, but they can be canned and made into a spread. Some people enjoy drying their pears and using them for pies or other baked goods. Birds like to next in pear trees, but aren't likely to eat the fruit; you will have more problems with insects in this regard.

Adding fruit trees to your landscaping may seem a bit odd, but in fact, fruit trees provide at least three seasons of beauty on top of the benefit of the fruit itself. Fruit trees need to be cut back and maintained just like other trees, but the blossoms they provide each spring can't be topped by traditional hardwood trees. As pears set in the summer the growing fruit looks so beautiful among the leaves. If you are growing a red or brown pear the startling color peeking out from below the branches adds another visual aspect to your garden. The fall brings colorful leaves and even in the winter, the sight of the branches arcing up against the sky is striking.

Incorporating fruit trees into your landscape design just makes good sense. Not only do you get to enjoy the fruit, your children have the opportunity to make the connection between trees and food. We have become so disassociated from our food production cycle; anything we do to put ourselves back in touch is a good thing.

## PEACH TREES

Peach trees are usually associated with Georgia, a southern state. You can however, use peach trees when landscaping your yard, even in New England. Many varieties of peach are hardy even into the northern areas of New England. Peaches require more care than most fruit trees, but few things compare to the taste of fresh peaches.

Peach trees set blooms early in the spring. Their flowers run the gamut of colors from pink to red to stripes. If you are fortunately and the spring isn't windy you will have many days of beautiful blooms. If, however, your spring is a windy or rainy one, the flowers will fall off prematurely. Either way the summer brings dark green, long leaves that are lovely. Of course, you also get peaches.

One of the hardest things to do if you are growing a peach tree for landscaping purposes is to cull the fruit. Estimates suggest that you want to remove up to 90% of your fruit to allow the remaining fruit to mature properly. In practical terms that translates to leaving one fruit every 8 inches, more or less. The fruit will grow larger and be more likely to ripen in time for harvesting. If fruit falls off the tree, which it is likely to do during the "June Fall" that most peaches experience, make sure to remove it unless you enjoy yellow jacket incursions.

A vase shaped pruning method is recommended for peach trees. To do this you prune out branches from the center of the tree, leaving a vee. This allows for good sun exposure in the middle of the tree and more fruit. You will also need to spray the tree for a variety of insect infestations if you want it to thrive. There are also a number of diseases that affect peach trees and need to be treated regularly. Birds love peaches just as much as you do, so net your trees if you want to keep the fruit for yourself.

The best part of having your own peach tree is what you can do with the fruit once it ripens. Peach cobbler is a summertime staple. Peaches can be canned, made into jam or, with a little work and creativity, make fabulous liquors. Of course, you can always eat them out of hand, still warm from the sun.

Adding a peach tree to a home based orchard is a wonderful choice. Most peaches will self pollinate, removing the need to invest in multiple trees just for fruit. Some varieties are reported to be ornamental, which means they don't set fruit, but why get one then? Ornamental cherry trees are easier to manage if all you want is flowers. A small orchard with some self pollinating cherry trees, a couple of apple trees a pear tree and a peach tree is perfect for the homeowner looking to beautify their landscape and enjoy home grown fruit. If you choose dwarf rootstock you can get the whole orchard set up in a fairly small yard. Fruit trees are beautiful year round and they contribute to the landscape in so many ways.

## FRUITS GALORE

Summertime, and the living is easy. This is the moment of succulence, maximum build-up of sucrose, other sugars and minerals in edible fruits. What can ever taste better than a fresh off the vine homegrown June strawberry, July blueberry, August tomato and September peach? Oh, the glories of plant based nutrition every summer, and it is summer every new day somewhere on earth. Indeed, it is summertime in the heart anytime fruit is eaten, regardless of the calendar. Dried berries bring sunshine energies into an igloo when Eskimos winter feast on preserved summer berries. In the tropics, where night is the only winter and summer always prevails, fresh fruits fall off trees continually and people never lack vitamin C. A fruit is the fleshy seed pod formed behind the flower of an edible plant - and most plants are edible, a biologist reports. By this definition, asparagus may be considered a fruit and so may wheat. Look in dictionaries, encyclopedias, biology and botany and horticulture textbooks to find traditional definitions and classifications. None of these matter, however, when one is in a cornfield in August sampling succulent sweet milky kernels, chewing them off the cob and enjoying everything about the experience. Nobody is happier than a produce grower standing among plants sampling the fruits of whatever angiosperms - flowering plants - are being grown. Literature on forests advise harvesters to look for fruiting bodies at the bases of trees. Edible fungi, friendly mushrooms, are technically fruits. To fruit is to store seeds, which contain genes and chromosomes, to have another chance to live. The plant kingdom fruits lavishly, angiosperms flamboyantly.

If you want to maximize nutrition, nothing seems to beat the good old fashioned prune plum. A blue-purple plum tree is king of the home orchard, for no other fruiting tree has its lust for survival in hot cold wet dry seasons nor its amazing resistance to attacking bacteria, fungi, viruses and insects. It is even fairly well pollution tolerant, marvelously resistant compared to its weaker relatives too numerous to name. Prune plum trees can be grown in every state, with tender loving care, appropriate placement and proper soil nutrition. Scientific researchers report over and again that plums top the scales in terms of antioxidants. For more than several centuries, a Belgian fruit processor has been slowly simmering annual batches of plum purees at low temperatures in copper kettles to produce a unique 100% plum jell paste which is delicious spread on bread, crackers, carrot slices or anything. It is a perfect mate for American peanut butter. According to company experts, plums contain self preserving substances which allow plum products to be handled easily and without refrigeration.

This plum jell is distributed globally and has a long shelf life before and after opening. Eastern Europeans brought a similar version of plum puree to America where this levkar is a standard item in delicatessens, traditionally packaged in wax paper lined small wooden barrels and scooped out by customers using a communal wooden paddle. A pectin rich extract of California plum production is used as both dough conditioning binder and sweetener in bakery products. Because of prolific productivity of cultivars, ease of cultivation and processing - plums have been developed into diverse products over recent centuries and remain very popular today. Prunes are good for people and everyone knows it. Plums grow on bushes and trees, some fruits are blue, others purple, black or yellow and both sizes and shapes vary. The Italian purple prune plum tree was brought by 19th century Italian immigrants to California spawning a largely veganic organic agriculture based industry which continues to contribute to American health improvements. Native or imported, wild or domestic, little or big, sweet or sour, fresh or dry, any color, plums are superb human food. It is common sense to eat plums and plum products regularly.

The same goes for figs. Superior human food. You can grow a fig tree in a large pot near your house and when you move, take the fruit bearing family member along. Pick and dry them in sun or oven, store in jars for years. Nutritional powerhouse figs are basic. Figs are fantastic.

Grapes are extremely nutritious. Lately grapeseed oil has been rediscovered, outside of France where it has reigned for a thousand years, and crushed ground grapeseeds are sold for healthful antioxidant properties pretty much everywhere. Some say red grape peelings are the most potent carriers of health supportive phytonutrients, others claim grape seeds lead. The safest path through this maze is to eat the whole fruit, fresh or dried and often. Chew a few seeds? Why not? Or grind and mix with other foods. Grapeseed oil makes superb salad dressings. Grape leaves are useful wrappers and when young might be served as steamed greens. Raisins delight.

Don't forget red and black currants, gooseberries, elderberries, and those hardiest of scraggly wonder trees the mulberries which some gardeners have reported cannot be killed and never fail to fruit. Sour cherries are delicious and so healthy, the trees are fairly shade tolerant, Michigan is the leading producer of both sours and Maraschinos. Sweet cherries are more temperamental, Utah's amazingly fine when a crop makes every other year and California's excellent every year.

Apples have miraculous health benefits and there are more kinds than anyone can grow or know. Apples, applesauce, apple butter, apple juice and cider and vinegar, dried apples, pectin extracted from processed apples and used in many jelled products, all these remind that nothing is better to revive appetite and stimulate memories of sweet aromas and Grandma's and Mom's apple cobblers and pies. No wonder Pilgrims and Puritans did not leave home without them, they brought apple seeds and probably apple tree cuttings to root and re-plant in the New World. A Pennsylvania Christian Communalist missionary grew up on the Johnny Appleseed stories, which are true, and a century ago himself spawned India's vast apple industry which is prolific.

Whomever smells and tastes a native American paw paw fruit need never again depend totally on bananas; to swoon is the appropriate response to this overwhelming aromatic sweet custard in a thin skin. Wild in Appalachia they are well known and oft sought out at their seasonal peak. Every County Agent in the wooded east knows where they are if any exist in the area. Paw paws do not ship well, so grow some in a niche in the yard plantings and share with neighbors. Also try persimmons, Asian favorites underestimated by Americans. Too soft when ripe to ship.

Cucumbers are fruits which love growing on trellises and fences. Bitter melons and winter melons appreciate trellises. So do roses which produce hips that make nice rosehip soup or dry and grind into vitamin C rich powder. Decorate summer soups and salads with surplus flowers from bountiful cucumbers and other cucurbits such as squashes. Also use nasturtiums – blossoms, leaves, stems, roots, and the tiny caper-like seedpods - in lieu of capers fresh or pickled, in salads, rose and dahlia petals, along with marigold petals strewn over every dish.

Cattails can be eaten, from roots to flowers. The seed pod atop makes a nice flour. Oh, daylilies, every part edible, use the flowers and then eat the seed pods which remind both of asparagus and green beans. Golden St. John's Wort flowers are colorful in salads, maybe along with yellow dandelion flowers and leaves. Early spring, eat violas and violets, using flowers and leaves. Okra flowers are decorative and delicious, and the seed pods, fruits of this cousin of cotton, are loaded with protein and vitamins and minerals galore. To not know okra is to have never lived, or so okra devotees maintain. In West Africa, okra seeds are dried and ground into highly nutritious flours rich in proteins. No okra, no gumbo. Okra deserves wider utilization.

The many types of pears provide months of happy eating and are easily dried and canned for cold season use. Delicate and a delicacy, whether eaten fresh in the home orchard, supplied by a friendly farmer, wrapped in paper appearing from an opposite hemisphere on the planet, or preserved in plastic bag, jar or can, pears are indescribably delicious and demonstrably nutritious. Fiber and pectin rich, pears are revered around the world as stool stabilizers. So are apricots, which also may be eaten fresh off the tree - or Mongolian bush - and shipped about from continent to continent, pureed and dried and vacuum preserved through steam or hot water bath canning processed widely used by survivalists here, there and everywhere. Both happy and unhappy Hunzas swear by them – and walnuts, just in case you haven't read all the Hunza books or had a chance yet to visit these sturdy folk in a so-called protected state north of Pakistan.

Say "fruit" and images of bananas and mangos and papayas are likely to come into the mental images formed. Bananas of all sizes are sweet, similar looking plantains are not. In the tropics, breadfruits are common and once were island survivalist staples. Papayas. Coconuts are fruits as well, discard the husks, shed the shell and eat only the fibrous and pulpy layers and liquid. Pomegranates are underestimated and too scarce. Messy and red staining, they are also deliciously sweet and tart and memorable as well as being Biblical. Tamarind is another seedpod as is carob, both from legumous trees which thrive in hot dry climes and alkaline soils. Mexican native vanilla bean orchid attracts by aroma, just a little goes a long way in flavoring dishes exquisitely. Avocados from Mexico are called the Haas variety of alligator pears. They are the black rough skinned ones which generally bring the highest price. Grown around the world in sub-tropics and tropical areas, avocados are uniquely rich in healthful oils. There is apparently no limit to the creative ways these fruits may be used to make meals outstanding – guacamole is only one. Star fruits appear regularly in urban supermarkets. Little known outside of the tropics, where summer never departs and night is the only winter, popular fruits include durians, lychees, mangosteins, rambutans and so many more too fragile to become commercialized.

Just where do the definitions of fruits and seeds and vegetables lead? Are not zucchinis in all these categories? Why not add all beans and lentils? Tomato cousins, tomatillos and eggplants surely are fruits. What else? If it is an angiosperm, it has flowers and makes fruits. Fungi?

Mid latitude and mid-altitude fruits include all the berries, those small bush rhibes including blackberries and raspberries and all their cross-bred kin like the gigantic boysenberries. In these so-called temperate regions elderberry bushes or trees can proliferate outside every ground level kitchen door and in apartment balcony pots. Great antiviral fruits. But, perhaps all the berries are antiviral, only years of further research can tell and the environment for securing plant based nutrition and nutraceutical research funding is parsimonious. No matter, herbivores, herbalists, native medicine men and women since time immemorial have been recommending fruits for whatever ails. Sour cherries in Germany for gout. Elderberries in Poland for colds and flue. But for scientific curiosity wondering why Hungarians stayed well so often, vitamin C would not have been extracted from paprika, eureka, and the whole pepper family green, red, yellow, large and small might not yet have become so highly regarded.

Durian and saw palmetto for libido and related plumbing system maintenance were discovered, respectively, by Native Southeast Asians and Native Americans. Cranberries for urinary tract infections, Amerindians taught Europeans. European bilberries and American blueberries both were recognized for strengthening eyes and ophthalmologists now recommend them for daily consumption. Apples have long been respected as good for practically everything. Apples with the peelings on, current research insists...straight from Cornell University nutrition science laboratories and field tests...are anticarcinogenic and have preventive effects with regard to heart diseases. Apples are documented as diversely beneficial to humans - organic and non-organic apples well washed.

Which brings up a related point of interest in addition to the reminder that all fruits should be well washed, whether grown organically, certified organic or conventionally. What does it mean to say "organic" with regard to foods? Prior to the 1940s and World War Two, essentially every food grown was produced organically and met or exceeded contemporary standards for that agricultural style. Today, only approximately one percent (1%) of commercial produce can legally be labeled "organic" or "organically grown." No one keeps track of the statistics for homegrown and local small produce plots and community gardens, though they are numerous. Relatively few of the 280 million Americans now living consume any organic produce and it is statistically improbable that anyone today eats only organic produce no matter how big their garden is. Given that around. 9% of Americans are vegans, folks who eat only plants and products made exclusively from plants and the occasional mineral, it must be the case that not even vegans can manage a 100% organic diet, nevermind the even more desirable 100% veganic organic food protocol. A fortunate few, mostly educated and relatively affluent or wise and insightful if poor, consume some organic produce during the year – most in summer, perhaps more the closer they live to a major college community, and much in the form of easily grown fruits such as greens and tomatoes and cucumbers produced in large quantities by astute specialty crop growers who for their personal home use buy most of the foods they eat in the same supermarkets as almost everyone else. Is an organic apple still organic after being stored alongside non-organic apples and sprayed by produce vendors with chlorinated and sometimes fluoridated water? Wanting to eat only organically produced food is not the same as actually doing it 100% of the time.

That is a standard essentially no one can meet. It would be really hard to do, even if one had the greatest garden in the world. To limit food intake to 100% veganic organic produce is still more difficult and practically impossible. Anyone living in a zero pollution area where water and air are perfectly clean with no traces of radiation, spent or otherwise, and no remnant dioxin or DDT can be found? That's where the next IPBN Demonstration Veganic Organic Garden should be planted. Where is there such a place? There is no such place anymore. But neither is mass starvation so prevalent globally as it was before World War Two, despite pockets of politically induced hunger and malnutrition such as in Chad and North Korea. Modern people live with non-organic chemicals and not all of these are bad. People just have to do the best they can and sanctify their food consumption however they like, and, if growing produce for sale, meeting popular standards of the day and squeezing profit margins from whatever they think others will like. Even in an impure post-Eden world, however, fruits are quite healthful, amazingly prolific and cleansing.

Fruits and other edible plants rarely contain pollutants inside. Consumer Union recently studied fresh produce at point-of-purchase marketplaces and found pesticide residues on only approximately 23% of organic and 73% of conventional products. These are good statistics, probably better than in any previous years since 1945. Things are getting better. Produce growers and distributors care and are striving to reduce toxicities of every sort. Clean water spray, soaking, scrubbing if appropriate will remove essentially all such surface pollutants. Grapes warrant special care, double sprays and soaks. A little baking soda in wash water neutralizes many contaminants. So do salt and vinegar. Dr. Bronner's Liquid Peppermint soap has surfactant qualities which will reduce the footprint of any hitchikers. It is these, not synthetic chemical fertilizers and pesticides which are likeliest to cause immediate health problems, as for long term effects...research is underway on all fronts. Health Food Stores offer numerous produce wash concentrates, each designed to dissolve and detach adhering pesticides, but when even organic produce has been waxed and possibly shellacked, maybe the best thing to do is peel it. The chlorine in most tap water can reduce 100% organic pathogens such as e coli, salmonella and their troublemaking friends. Washed, drained, dried and displayed or served in luscious food combinations, fruits are super healthful. Who does not smile after eating an apple? Who does not prefer it first be well washed?

The Dean of Cook College, the Agricultural School of Rutgers University, the New Jersey State University at Brunswick, has recently returned from Central Asia with seeds from hundreds of melons grown traditionally in the countries north of Afghanistan. These will be field tested to see which varieties may have commercial potential in the United States. Undoubtedly, he will be sharing seeds with colleagues in other agricultural research universities, and, surely, also with friends and sponsors at the USDA Agricultural Research Station northeast of Washington, D.C. surrounding Greenbelt, Maryland. Sooner or later, Americans will be eating some Tajikistan melons and making comparisons with their earlier arriving cantaloupe, casaba, honeydew, muskmelon, Persian and other plant kingdom melon cousins. On hot days what tastes better?

"Can you help me find a supplier of organic peaches?" asked the produce manager of a time honored healthfood store in Columbia, Missouri, near the University of Missouri campus. "Sure," answered the IPBN traveling volunteer, "grow them yourself." A health conscious Pennsylvanian advises, "Dad has grown peaches all his life in South Carolina and just can't make a crop organically." A Georgia peach grower in production nearly half the year offers no organically grown specimens. To cheer the Missouri produce man, it was reported that "In California and a few other places, however, there are valleys where nature and farmers get together and have considerable success with organic peaches. These are worth more dried, though, because of pricing and profit margins and shipping of fragile produce requires so much handling that peaches suffer and nobody will buy bruised, spotted, bird or worm damaged soft fruits." He agreed that nobody wants less than perfect fruits or anything else. "Go down to Louisiana, Missouri," he was encouraged, "and talk with the people at Stark Brothers Nurseries. They are huge peach tree sellers to orchardists and homegrowers in most every state. If anyone knows peaches, they do, and perhaps someone nearby, here in Missouri, will see the opportunity and become the world expert on veganic organic peach production. Check the internet. And, oh, for sure, contact your County Agent for every county in America has one or more – even New York City and San Francisco. The United States Department of Agriculture Research Stations may have good stories regarding peaches to meet your standards. Ask about nectarines and apricots too. The County Agent may know a farmer in your area who is already committed to meeting this need. If there is, find and help him build a market and survive." Any professional horticultural supplier can advise regarding how to grow fruit trees and bushes in any climate and soil situation on the continent and in Hawaii and Territories as well. So can the fruit specialist of every State University Agricultural College. Garden supply catalogs are marvelous textbooks.

Oranges, grapefruits, lemons, tangerines and all the other citrus fruits thrive in semi-tropical climates. Picked early winter through early spring, they store well in controlled atmosphere low oxygen high carbon dioxide humid chambers chilled to 34 degrees Fahrenheit. That's why the once-and-done storing-shipping-display boxes have holes around the sides and provide excellent plant based cellulose corrugated paperboard insulation. They breathe. They stack. They secure so that people can enjoy the benefits of fresh picked or freshly displayed citrus fruits, primarily oranges, all year around. Unbeautiful citrus get peeled and their juices squeezed and dehydrated to make frozen pulp for commercial and home re-hydration in future seasons. What is a summer salad without citrus, whether from cold storage, jar or can. And in winter when citrus are fresh harvested, why shouldn't every salad contain some? If only for its essential folic acid, citrus juice should be treasured, but there are also the vitamin C and precious sugars uniquely present in these beautiful and versatile fruits. Kumquats anyone? Eat the next ones you find. Ecstasy. Colorful. Tangerines? Superb. Mandarins? Tangelos? So very special. Nearby grow olives and cacti such as prickly pear fruits (napalitos) and perhaps saguaro (tuna) and cholla (buds).

Dates. What tastes better? Often organically grown. High priced and in short supply. "We are planting new trees fast as land and water and financing can be obtained, but can't keep up with the market. The Chinese want them. Koreans. "We could sell everything we grow for export to Asia, but we choose to keep most for the American market. It is tough, because we can get a better price elsewhere. Don't write about them or encourage people to contact us, we don't need more business and never will the way this thing is going. We don't need anything but more land and there isn't any, we are making a patchwork quilt set of small farms wherever we can find soil and water that are suitable. Dates are very particular, but if they like a location will grow there for many years. They are tough and hardy, but diseases and insects are creeping in and giving us trouble. Find me more land dates like." The Southern

California desert valley date grower was expounding in response to an IPBN volunteer question, "How might we help you?" The solution? Love the dates you get. Love all those who get dates to you. Appreciate every date and learn to grow some alternative fruit on your homestead or in a friendly location nearby.

Palm oil is squeezed from fruits of Southeast Asian tropical oil palms. Colonialists planted these in Brazil as well. For all the railing against saturated fats, which are present in palm oil and coconut oil and many other plant based foods including peanuts, it appears that these relatively solid fats can have some healthful benefits when consumed in small quantities. At their worst, plant based saturated fats appear to be friendly when compared with others.... Worst of all healthwise are the hydrogenated and semi-hydrogenated fats. Southeast Asians who eat palm and coconut oils are rarely fat from them, but rather modern lifestyles which have people sit continually before computers and televisions, ride automobiles, rarely exercise, with addictions to contemporary fast foods containing copious quantities of demonstrably bad fats - not from plants - which are truly unhealthful, commonly overeaten to the carcinogen and coronary and diabetes and stroke proliferating level. Obviously, life offers alternative choices. If tropical plant oils are not ever fit to eat, as some claim, they make excellent diesel engine fuels and can fire steam turbines which generate electricity. Fruits are useful, whatever the best uses may be.

Spanish and Moroccan Clementines will be arriving in tens of thousands of containers on ships docking in Philadelphia in September or early fall and thousands of trucks, on highways and railroad cars, will distribute this Mediterranean citrus harvest throughout North America until the inventory runs out sometime before winter. Kiwis - those successfully marketed Chinese gooseberries - come in from New Zealand regularly, supply never exceeding demands so that prices fall precipitously. California growers supplement this inflow, but the kiwi bin of every produce vendor is sometimes empty, no matter the season.

Before summer has been forgotten, Argentine pears and apples will be arriving, some organic and all beautifully packed, at Port of Philadelphia - the nation's largest fresh produce handler as it has been since the 1600s when Caribbean pineapples and jackfruits, mangos and papayas, bananas and maybe even guava were regular fare year around. American tomatoes will keep coming until frostbitten in the usual north to south succession to be replaced by Mexican harvests and then, after the bottom of winter passes by, Florida crops and then south to north every latitude will fruit in turn all the way up to Nova Scotia. All year long, Canadian greenhouse tomatoes will be filtering in, typically hydroponic but also grown in ground under cover and mostly in the Ontario Niagara fruit belt, but also in other provinces from Labrador to British Columbia. Dutch greenhouse tomatoes will be flown in, reds and yellows, then the exquisite small tasty tomatoes from Israel. For those with money to spend, there will be fresh tomatoes every day, from somewhere, and for everyone peak moment picked canned peaches will be pulled off grocer shelves until fresh tomatoes once again ripen on vines throughout the American homeland. As the various peaks of tomato seasons pass (cherry tomatoes early and beefsteak tomatoes late) and finally summer tomato consciousness abates or fades, summer squash have gone to seed and hardened to protect their genetic futures in a sturdy encasement. The last peaches and pears hand deliciously awaiting eaters who become seed spreaders. Frost. Cranberries appear, and fall squashes – more fruits sustaining human survival - will dry up their leaves signaling winter is coming and be ready for harvest and storage, providing fruity pulpy sweet golden bisques through fall and winter when again flowers and the seed pod fruits they produce appear and renew the earthship food cycle.

Want to get rich? Grow black currants in your backyard. They are quite trouble free and nutritionally and medicinally super fruits. Maybe alternate rows of plum trees and black currant bushes as the berries are relatively shade tolerant. Also try black elderberries. All these high profit potential fruits are thorn freer reliable croppers relatively trouble free in terms of diseases and insects. Birds, though, may target and savage your crop. Think netting, just do it rationally and don't fall off any ladders. Berries will make money while the plum trees are limbing getting ready to flower and fruit in a few years. Mulch with straw and rotten hay, composted leaf and wood chip and sawdust mulch. Black currants were illegalized by well intentioned misinformed governmental bureaucrats who paid a dollar a day to U.S.D.A. Civilian Conservation Corps volunteers who rid forty-eight states of these rhibes which Europeans and Asians relish and pay at high prices. If black currant crops seem too daring, consider earning moderate wealth through mulberry or sumac cultivation. Anyone can grow these two plants. Dry the mulberries or boil down their juice or eat and sell them fresh. Sumac berries, are reddish with little hairs. Boiled and strained produce a beverage like pink lemonade. (Avoid blue berried poison sumac.) Sumac ground becomes "fatoush" and retails for five to $15.00 a pound. Sour. Vitamin C rich. Try it.

Nature must love fruits, it made so many of them. Sweet, sour, hot, starchy, oily, juicy, dry, hard, soft. What but fruits are peapods and green beans? Soybeans? All legumes? Citron and quince? Amaranth, barley, corn, kaffir corn maize, flax, oats, millet, quinoa, rice, rye, sesame, sorghum, teff, wheat and its relatives the large grained kamut and half rye triticale? Aren't these all seed pod fruits of grasses? Aren't nuts fruits? Acorns, almonds, black walnuts, chestnuts, hazelnuts, hickories including pecans, pistachios, macadamias and walnuts are called nuts, but botanically they are fruit seedpods from which flowers shriveled and outer layers separated.

Who thinks of an almond as a sister of the peach and mango cousin, only with a non-edible pulpy pod around the edible seed, instead of the other way around? Even people silly enough to say they never eat fruits actually do. The boundaries between fruits and vegetables and melons and nuts and seeds bear exploration.

They overlap curiously and seemingly illogically. Some wish to argue to their deaths over distinctions they know are true but lack constituencies to impose their definitions on others. Names, boundaries and definitions vary among the many countries and cultures, and these have varied throughout written history as well. In fact, a bean is an edible seed encased in a fleshy pod, as is a banana and a walnut, but the bean seed shell is thin, soft and edible, whereas banana seeds are very small and soft inside an edible fleshy pod of which the peeling is indigestible, and both peeling - or outer husk - and hard shell of a walnut are inedible at the mature stage. Picked earlier and pickled whole, green walnuts are delicious. Trying to classify all nature imposing human logic is futile. Suffice it to say that edible fruits have proliferated around the planet and whatever are available locally probably provide adequate nutrition. Unaware of plant diversity and complexity, Native Americans were happily eating popcorn over 5,000 years ago, caring not a whit about scientific classification of plants. They just knew what to eat, when and why. Migrating to a new area, they figured out what was best to eat and what to avoid. Anyone who inquires into food plant origins will become fascinated and enjoy a lifetime of surprises, never learning all there is to know. So it goes. How can cashews and mangos be cousins? Coffee, chocolate and papaya? The plant kingdom fruits gloriously.

What is not fruit? Flowers. Leaves. Stems and stalks. Roots. Tubers. Edible bacteria such as chlorella and spirulina. What about "vegetables"? Vegetables are an indefinite category which includes edible plants or plant parts which may or may not be fruits also - such as tomatoes and zucchini for example. Perhaps all this is what fruitarians have been trying to communicate for centuries. Given this broad inclusive comprehension and definition of fruits as edible seedpods of angiosperms, an individual could happily and healthily live a long lifetime on fruits alone. Possibly fruits and leafy greens alone? Add edible flowers? Also some stems, stalks, roots, tubers? Bacteria? Fungi? Why not eat a little from each of the plant based nutrition categories and utilize the benefits each may provide while reaping advantages from interactions between and among them their many forms - fresh and dried, raw and cooked? Could this be the core and essence of plant based nutrition and cuisine logic and philosophy? Time will tell. Honest science and practical experience reveal the truth that people need to eat diverse fruits daily..

# TREE PLANTING ON PRIVATE LANDS

## NEED FOR AFFORESTATION

With increasing population and growing consumerism, there has been severe pressure on food production and employment generation, particularly in agro-based development countries like India. This has had a direct impact on deforestation, increase in soil erosion and run off of rain water, resulting in depletion of natural resources and environmental pollution. Simultaneously, increasing use of fossil fuel for industrial production, power generation and automobiles has accelerated the emission of green house gases (GHGs) which are responsible for global warming and climate change. The negative impacts of global warming are far more serious in India due to prolonged droughts, rising sea level and melting of snow caps of the Himalayas thereby affecting steady supply of water to major rivers of North India and our food security. To reverse this trend, it is necessary to reduce the emission of GHGs, while taking up massive afforestation to serve as carbon dioxide sink and for supporting rural livelihood.

Trees have a significant role in keeping the environment clean, while supporting the livelihood. Over 43% of the cooking energy in the world is met from biomass. In rural areas, where 65% of the population lives, biomass is the only accessible and affordable source of energy. In the developing countries, the average per capita consumption of biomass in rural areas is equivalent to a ton of wood per annum and 50% of the wood cut in the world is used for fuel (Hall and de Groot, 1985). More than 2 billion people in the Third World are dependent on biomass to meet their energy needs which is equivalent to 22 million barrels of oil every day. In 1979, about 68.5% of the total rural energy was met from wood in India. In 2000, the annual demand for wood in the Indian rural sector was 192.6 million tons while it was difficult to meet even 50% of it from the available sources. This indicates the extent of damage caused to the natural forests. Most of the rural people have been dependent on Government-owned forests and community woodlots, which are under severe pressure and vanishing rapidly. Thus, deforestation has been severely affecting the ecosystem and economy in India as well.

Presently, only about 12% of the geographical area in the country is under close forests as against the stipulated forest cover of 33% required for maintaining an ecological balance (Anonymous 1989). To solve the energy crisis, the strategy is to promote production, and ensure judicious use of wood energy which will indirectly conserve our forests. This calls for improving the existing forests through people's participation and increasing the area under forest cover even on non-forest lands, by bringing available barren and wastelands under afforestation. It is estimated that India has about 80-100 million ha of denuded forests and wastelands, which are neither put to any productive use nor considered for conservation. As such, these denuded lands have been accelerating soil erosion, wastage of rainwater and loss of biodiversity, contributing to global warming.

## SOCIAL FORESTRY – A DRIVE FOR PEOPLE'S PARTICIPATION

With the background of developing private and non-forest public wastelands under afforestation while protecting the natural forests, the Government of India introduced several people-oriented afforestation schemes in the early 1950s. However, the activity gained significance only during the Sixth Five Year Plan (1980-85) under social forestry, as a powerful tool to generate sustainable livelihood for rural people. To support this programme, afforestation was introduced under various development schemes such as Rural Landless Employment Guarantee Scheme (RLEGS), National Rural Employment Programme (NREP), Employment Guarantee Scheme (EGS), Drought Prone Programme (DPAP), Western Ghats Development Scheme, etc.

In the 1980s, all the State Governments had set up a separate wing either in the Forest Department or in the Department of Social Forestry to operate social forestry schemes. Various schemes such as development of community woodlots on public lands, establishment of decentralised kisan nurseries for free distribution of seedlings and promotion of tree plantations on marginal agricultural lands were launched with huge financial support. The objective was to increase the supply of fuelwood, bamboo, small timbers and fodder, while generating rural employment and maintaining environmental stability. The strategy was to grow fuelwood closer to the consumption points. Among various afforestation activities under social forestry, raising block plantations in the form of village woodlots on community lands, degraded forests and wastelands was the major programme. Such plantations aimed at reducing the hardship of women and children, who travel long distances in search of fuelwood. The remaining 30% of the programme included raising of seedlings for free distribution to farmers and schools. The focus was on fodder and fuelwood production.

Based on the recommendations of the National Commission on Agriculture (1976), the Sixth Five Year Plan focussed on the production of fodder, fuel, small timber, minor forests produce and industrial raw materials under the social forestry programme. The theme was "Development without destruction". The overall target of the Sixth Plan was to bring 1.65 million ha under tree cover. In addition to afforestation on community wastelands, 37.25 million seedlings were distributed free for establishing farm forestry. Simultaneously, production forestry was also given a boost to bring 0.62 million ha under industrial wood plantations. Except for farm forestry, the other programmes could not make a significant impact because the objective of meeting the basic needs of fodder, fuel and timber of the local participants could not be fulfilled.

Looking at the drawbacks of the Sixth Plan, the Seventh Plan (1985-89) introduced a new programme with the theme "Forest for Survival" for expanding rural fuelwood plantations with the involvement of NGOs. In 1985, the National Wastelands Development Board (NWDB) was established to increase tree and other green cover on wastelands with a target of 5 million ha coverage every year, while promoting fuel and fodder plantation to meet local needs. NWDB intended to coordinate soil and water conservation, dryland farming, fodder development and conservation of land resources to prevent desertification as well. To popularise afforestation, various innovative schemes were also launched and significant among them were decentralised plant nurseries for distributing among small farmers, cultivation of fodder, fuelwood and round timber species through Forest Development Corporations, fuelwood plantation on urban wastelands, production of industrial raw materials on Government-owned wastelands and leasing revenue wastelands to poor for growing trees, etc. In 1988-89, after observing the performance for 4 years, the programme was restructured to cover activities such as reclamation of wastelands through agro-forestry, silvipasture and farm forestry, involving small farmers.

Under NWDB, 7.18 million ha wastelands were brought under afforestation during the first 4 years of the Seventh Plan with a survival rate ranging from 43.6 to 70.4% (GoI, 1989). Many NGOs, public sector undertakings and cooperatives such as NDDB and IFFCO initiated various innovative schemes to promote afforestation on degraded lands. All these programmes heavily depended on People's Participation for their success. Apart from planting and maintenance, cooperation from local people was also expected for protection of plantations from stray animals and illicit felling, until harvesting for ensuring equal distribution among the participant families. However, the outcome was not very encouraging. Only 9% villages were covered under the programme. Community wastelands were not easily available in many villages because either the Panchayats were not prepared to spare the lands or there were many encroachments.

The outputs from fodder and fuel were very low as compared to the local needs. Hence, the benefits were not significant (NCAER, 1988). Poor rate of survival of saplings, poor growth, poor protection from livestock and trespassers were other factors which contributed to the failure. In many States, the focus was shifted from fodder and fuelwood species to species such as eucalyptus because of local demand, fast growth and protection from stray animals. With regard to employment generation, fodder and fuelwood plantations generated only 60-80 mandays/ha while farm forestry generated 600-800 days/ha. The wage rate under these schemes being low compared to the local wage rate, there was no enthusiasm among the poor to take active part in the fuelwood production programme (Saxena, 1988; 1989; Sathe, 1990). To enhance people's participation in tree planting programme, the schemes were modified to integrate livestock with forage production and shift from fuelwood production to income generation by introducing short rotation species with long rotation trees and timber species with fuelwood. Emphasis was laid on extension programmes to motivate local families to take active part in afforestation (Shingi, 1988, Deshpande et al, 1990, Singh, 1990). During 1980-88, the State Forest Departments across the country claimed that 20 billion seedlings were distributed for planting. This meant an average of 35,000 plants per village but there were hardly any villages with such significant number of trees in the country. This reflected the poor performance of social forestry schemes in the country. This failure could have been avoided if suitable tree species had been selected. Most of the farmers would have taken good care if species of their choice had been provided (Hegde, 1987). Lack of marketing arrangements was another weakness of the social forestry programme.

Interestingly, farm forestry scheme which was promoted by Forest Development Corporations and private paper and pulp mills by distributing seedlings of eucalyptus and other commercial species, had exceeded the target as the participants were motivated by the prospects of economic gains. Fast growth, high value for the produce, sustained demand from industries and existence of an easily accessible market were the reasons for the popularity of eucalyptus plantation (NCAER, 1988). Higher profitability due to higher rate of survival, short gestation period, higher yield, ready market, high value products like round wood, remunerative price, negligible impact on seasonal crops, easy management of labour, ease in protection and favourable Tax and Land Ceiling Acts were the other reasons for acceptance of eucalyptus by farmers (Gupta, 1990). Farm forestry introduced on agricultural lands as a substitute for low yielding food crops with species such as eucalyptus was also successful (Muranjan, 1988). Farmers also preferred farm forestry as it demanded only 120-140 mandays of labour ha/year while bajra, gram and other rainfed crops demanded about 120-130 mandays/ha crop in 3-4 months (Singh, 1985).

## Choice of Tree Species for Private Lands

Based on various social forestry projects implemented over the last 2-3 decades, it is clear that choice of species is the key to the success of any afforestation programme. When it comes to tree plantation on private lands, profitability is the main factor followed by other minor factors such as gestation period, demand for produce, level of investment, access to market, availability of planting material and specific local uses, which influence the farmers to select tree species for planting on their lands.

Selection of suitable species is the most important motivating factor for people's participation as it influences the profitability. Tree planting on private lands is being carried out by the land owners either because they are convinced or motivated by some of the agencies engaged in promoting afforestation.

There are very few farmers who take initiative in establishing plantations of new species, based on the information they have gathered about the utility and profitability of these species. However, most of the small farmers are driven by the publicity and attractive benefits as highlighted by the promoters, while selecting tree species for planting on their lands. The popularity of the species also varied from region to region, based on the demand for produce, marketing infrastructure, agro-climatic conditions, available inputs and the extent of awareness and publicity created by the programme implementing agencies.

Tree Species for Income Generation: In a study conducted in Pune and Nashik districts of Maharashtra, where multiple agencies were independently promoting tree planting, a majority of the farmers preferred growing fruit trees on their marginal and wastelands. This was followed by timber and round wood species. Among 35 most popular tree species promoted in the state, 18 species were grown for food, 8 for timber, 3 for fuelwood, 2 each for oil and ornamental purpose and 1 each for fodder and fibre. The most preferred among them were eucalyptus (Eucalyptus hybrid), mango (Mangifera indica), teak (Tectona grandis), custard apple (Annona squamosa) and jujubee (Zizyphus mauritiana). The list of these species with their popularity rank is presented in Table 1 (Hegde, 1991). This preference is based on the profitability as well as market demand for the produce and field publicity. However, eucalyptus was the most popular species because of reasons other than high returns. First of all, eucalyptus had good demand as round wood in the local market. Any wood that was not sold as pole was purchased by paper and pulp mills at the site. In addition to assured demand, easy marketability and an attractive price, eucalyptus is a fast growing, non-browsing, coppicing species with a short harvesting cycle and well adapted to adverse agro-climatic conditions. Being one of the few species promoted by the wood-based industries, it has received wide publicity. Other tree species cultivated in India on a commercial scale under farm forestry by farmers are casuarina in coastal areas and poplar (Populus deltoides) which is confined to Northern India, beyond latitude 28o N.

In interior areas, where marketing facilities for wood were inadequate, farmers preferred to grow fruit crops and used existing market outlets for selling their produce. Thus, about 50% of tree selected were fruit species. Among fruit trees, seedlings of custard apple, jujubee (ber), tamarind, jambolina, drumstick, jackfruit, cashew, Indian gooseberry, wood apple and bullock's heart (Annona raticulata) were raised by farmers in decentralised rural nurseries while other species such as mango, guava, pomegranate, coconut, mandarin, orange and sapota were raised in commercial nurseries promoted by the Horticulture Department. Most of the farmers did not mention any other species although there were many with high income potential because there was neither any publicity nor availability of the planting materials locally. The farmers were also confident of selling timber of eight species, namely eucalyptus, teak, chinaberry (melia), leucaena (subabul), portia (bhindi), casuarina, bamboo and shishum in local markets. Three fuelwood species preferred by the farmers were Ramkathi acacia (Acacia nilotica var. cupressiformis), Gum acacia (Acacia nilotica var. telia) and Australian acacia (Acacia auriculiformis). The former two species are native to Maharashtra and used as fuel and timber, while the latter has been introduced only recently. Five other species selected by the respondents were gulmohar (Delonix regia), Ashoka (Polyalthia pendula), neem (Azadirachta indica), sandalwood (Santalum album) and agave (Agave sisalana). Of these, the former two were planted as ornamental, the other two for oil and agave shrub for fibre. Sesbania (Sesbania sesban) is the only fodder species, although other multipurpose species like subabul, and Shishum could yield fodder to some extent. Neem is an excellent species as a source of oilseed and biopesticide and drought tolerant. However, farmers did not prefer neem as there was no attractive market for selling the seeds and they also lacked awareness about its yield and profitability.

**Table. Choice of Tree Species by Land holders of Different Categories**

| S. No. | Name of the Species | Common Name | *Use | Total Responses |
|---|---|---|---|---|
| 1. | Eucalyptus spp. | Eucalyptus | T | 143 |
| 2. | Mangifera indica | Mango | F | 129 |
| 3 | Tectona grandis | Teak | T | 109 |
| 4. | Annona squamosa | Custard apple | F | 102 |
| 5. | Zizyphus mauritiana | Jujubee | F | 71 |
| 6. | Melia azedarach | Chinaberry | T | 52 |
| 7. | Tamarindus indica | Tamarind | F | 41 |
| 8. | Psidium guajava | Guava | F | 41 |
| 9. | Leucaena leucocephala | Leucaena, Subabul | T,Fo | 39 |
| 10. | Punica granatum | Pomegranate | F | 28 |
| 11. | Syzygium cumini | Jambolina | F | 22 |
| 12. | Moringa oleifera | Drumstick | F | 19 |
| 13. | Thespesia populnea | Portia | T | 18 |
| 14. | Azadirachta indica | Neem | Oi | 15 |
| 15. | Artocarpus heterophyllus | Jackfruit | F | 13 |
| 16. | Acacia nilotica var.telia | Gum acacia | Fu | 11 |
| 17. | Cocos nucifera | Coconut | F | 11 |
| 18. | Manilkara zapota | Sapota | F | 11 |
| 19. | Citrus medica | Sweet lime | F | 9 |

| | | | | |
|---|---|---|---|---|
| 20. | Casuarina equisetifolia | Casuarina | T | 6 |
| 21. | Anacardium occidentale | Cashew | F | 5 |
| 22. | Dendrocalamus strictus | Bamboo | T | 5 |
| 23. | Acacia auriculiformis | Australian acacia | Fu | 3 |
| 24. | Citrus reticulata | Mandarin | F | 3 |
| 25. | Agave sisalana | Agave | Fi | 3 |
| 26. | Polyalthia pendula | Ashok | O | 2 |
| 27. | Delonix regia | Gulmohar | O | 2 |
| 28. | Citrus sinensis | Sweet orange | F | 2 |
| 29. | Acacia nilotica var cupressiformis | Ramakathi acacia | Fu | 2 |
| 30. | Sesbania sesban | Sesbania | Fo | 2 |
| 31. | Emblica officinalis | Indian gooseberry | F | 3 |
| 32. | Annona reticulate | Ramphal | F | 1 |
| 33. | Dalbergia sissoo | Shishum | T | 1 |
| 34. | Santalum album | Sandalwood | Oi | 1 |
| 35 | Feronia limonia | Wood apple | F | 1 |

*T -Timber, F-Food, Fo-Fodder, Fu-Fuel, Fi-Fibre, Oi-Oil, O-Ornamental

The small holders had shown preference for fruit species, while the medium and large holders preferred timber species. Inspite of its popularity among farmers, it was surprising to observe that eucalyptus was not the most profitable species promoted under social forestry in India. This indicated that with wider publicity and market linkages and in the absence of knowledge on better alternatives, farmers are often influenced in making wrong judgments (Hegde, 1991). Profitability of Tree Species: The benefit-cost analysis of 14 important fruit and timber species based on the data collected from farmers is presented in Table 2. While pole timber such as melia, eucalyptus, leucaena, bamboo and portia start generating income from the third year, sesbania starts generating income through fodder during the first year itself and completes its economic life in 2-3 years. Melia, leucaena and eucalyptus coppice well and thus, the plantations can be maintained to harvest 3-4 crops. Portia trees are pollarded at an interval of three years and maintained for 20-25 years. Harvesting of bamboo starts in the third year and continues every year for about 20-25 years. While leucaena and eucalyptus have good demand as pulpwood, melia and portia are used as poles for housing and agricultural implements with limited demand.

Drumstick starts fruiting from the second year and continues to provide income for 10-15 years. Fruit trees like jujubee, custard apple, mango and cashew start fruiting from third year while tamarind starts producing fruits after 7-8 years. Neem starts fruiting after 7-8 years and continues for 75-100 years, yielding 50-100 kg seeds every year. As these species have different gestation period and various uses, it is extremely difficult for common farmers to take a quick decision about planting them. However, as all these species except neem, mango, cashew and tamarind, can be planted on field bunds without affecting arable crops, farmers do not mind planting these species if some support is given in the form of free seedlings and inputs. If they have to establish a sole plantation on good lands using their own resources, then they will certainly explore more about investment and profitability before taking a final decision. Among the above species, portia was the most profitable (Rs.52,000), followed by teak, drumstick, leucaena, melia, sesbania, eucalyptus, bamboo, custard apple, mango and neem. It is interesting to observe that except in certain districts of Maharashtra, by and large, farmers are not aware about the management of portia to induce poles and the use of poles as light wood for agricultural implements.

All the 14 tree species were most popular because of easy marketability of the produce and higher return.

**Table. Analysis of Income (in Rs.) from different Species**

| Name of the Species | Common Name | Duration | No. of trees/ Ha | Net/Tree /Year | Net/ha /year |
|---|---|---|---|---|---|
| Sesbania sesban | Sesbania | 2 | 5000 | 4.80 | 24000 |
| Melia azedarach | Chinaberry | 9. | 974 | 24350 | 2500 |
| Leucaena leucocephala | Subabul | 9 | 2500 | 13.88 | 34575 |
| Eucalyptus Hybrid | Eucalyptus | 9 | 2500 | 9.24 | 23100 |
| Dendrocalamus strictus | Bamboo | 10 | 625 | 23.33 | 14581 |
| Thespesia populnea | Portia | 10 | 625 | 83.93 | 52456 |
| Tectona grandis | Teak | 20 | 625 | 80.00 | 50000 |
| Azadirachta indica | Neem | 75 | 200 | 50.00 | 10000 * |
| Moringa oleifera | Drumstick | 10 | 400 | 124.00 | 49600 * |
| Annona squamosa | Custard apple | 10 | 400 | 29.69 | 11876 * |
| Zizyphus mauritiana | Jujubee | 10 | 400 | 48.52 | 19568 * |
| Mangifera indica | Mango | 50 | 100 | 100.00 | 10000 * |
| Anacardium occidentale | Cashew | 50 | 156 | 125.00 | 19500 * |
| Tamarindus indica | Tamarind | 50 | 45 | 463.00 | 20835 * |

* Income from wood not included

** According to prices of 1989-90

The popularity among these species varied from district to district based on local use, availability of planting material and extension efforts. There are many other species which can produce timber (White siris, siris, shishum, White teak), fruits (citrus, guava, sapota, coconut, jackfruit, jamun, kokum, oil seeds (pongamia and mahua) and other non-wood produce (Indian gooseberry, myrobalan, and soap nut) which can be promoted for cultivation, if cost-benefit analysis is carried out and silvicultural practices are standardised. Even neem can be profitable, if plants of elite genotypes, multiplied vegetatively are used for planting and the seeds are processed for bio-pesticide production. Likewise, many non-wood product species having different uses such as edible products, oil, gum, resin, wax, pesticides, tan, dyes, fibre, soap and medicines can be profitable, if plants produced through vegetative propagation are used for establishment and the produce is processed for value addition.

As compared to the economics of fruit and timber species, production of fuelwood is least attractive because the net annual income per ha is only about Rs.347/-. Thus, it is not attractive for farmers to grow fuelwood species inspite of intensive programme promotion and heavy incentives. If a ton of wood is sold for fuel, it would fetch only Rs.1000/-. The same wood when sold as pulpwood would fetch 50% more and as round timber, 200% more. When the wood is used as timber either for construction or furniture, it would fetch 400-500% higher price. In such a situation, naturally farmers would prefer species with higher returns. Under Social Forestry Programme, the poor farmers were persuaded to plant fuelwood and fodder, while large farmers had the option to grow wood for round timber, paper and pulp. Thus, unknowingly, there was discrimination and the poor were left out of an excellent opportunity to earn more from the programme. This was the major reason for lack of people's participation and failure of many projects, which were intended for the benefit of the poor.

Preference for different Tree Species: While calculating the profitability of different tree species, it is necessary to take their entire life cycle and convert into annual returns. For instance, teak and many timber trees mature after 60-100 years, while the round timber species are ready for harvest at the age of 15 to 30 years. Pulpwood will be ready in 4-6 years and fuelwood can be harvested in 2-5 years. In case of fruit trees, tamarind has a productive life of over 80 years, while mango and cashew have a productive life of 40-50 years. However, fruit trees start generating income from an early age and contribute to profit every year. In case of timber species, income is generated after a long gestation and only when trees are cut. Thus, fruit and non-wood tree species deserve to be promoted on a wider scale.

Even for expansion of various fruit crops, there are limitations of labour, resources and market beyond certain scales of operation. For instance, the area under fruit crops such as ber and amla could be expanded well during the last two decades. However, with larger volume of these fruits arriving in the market, which is more than the existing demand, the prices have started falling down. Unless efforts are made to process these fruits for value addition and preservation, farmers are not likely to cultivate these species on a large scale in the future. For crops like mango, in the absence of cold storage and processing, glut during a particular period in the year may affect the price realisation. Similarly, for crops like grapes which are highly labour intensive, farmers may not expand the area due to shortage and inefficiency of labour. In such a situation, farmers are likely to select the next best crops for cultivation. Looking to the present status of tree planting on private lands, it can be concluded that private land owners opt for different types of tree species in the following order of priority:

Preference for tree species in the order of priority:

- Fruits and nuts
- Round wood species and plywood
- Non-timber forest products and oil seeds
- Paper and pulpwood
- Forage and fuelwood

The above preference is based on current profitability and subject to availability of good soil, assured soil moisture and easy availability of inputs. The priority may change for different sites, based on adaptability of the species to local agro-climatic conditions, infrastructure for backward and forward integration, investment capabilities, etc. In areas prone to natural calamities, it is better to select hardier species even if the returns are low instead of growing sensitive crops capable of higher returns. There are many useful and valuable species like sandalwood, teak and red shishum, which are highly priced but the gestation period is very long. Farmers may plant these species on a small scale but not for income generation in the short run. The species covered in this paper are suitable for tropical regions and there are different species suited for sub-tropical and temperate regions.

To ensure selection of suitable species, it is better to prepare a land use plan, based on the soil productivity of the site earmarked for tree planting. Fertile soil with assured soil moisture is highly productive, where fruit trees grow well and give high returns. Hence, such lands can be reserved for establishing fruit orchards, if farmers are not intending to grow arable crops of high value. Medium quality soils with moisture stress, not suitable for fruit crops can be used for growing round wood, soft wood or ply wood. Soils of slightly inferior quality can be used for pulp and paper wood. Soils of low fertility with moisture stress, not suitable for above types of species can be used for establishing fuelwood plantation. There are shallow soils with moisture stress, where it is extremely difficult for tree species to survive. Such soils can be used for growing fodder shrubs and grasses. Thus, soil productivity and profitability of different tree species should be taken into consideration, while making final selection of tree species for growing on private lands.

## Strategy for Solving Fuelwood Crisis

From various studies, it is clear that establishment of tree plantations for fuelwood and fodder is neither economically viable nor attractive to farmers for cultivation, particularly when they have other options. Further, there is no scope for selling fuelwood at a higher price, firstly, because most of the poor who are dependent on fuelwood for cooking, try to fetch it from public properties, free of cost. Secondly, they do not have the purchasing power to buy at higher prices. Therefore, in the absence of easy supply of fuelwood at an affordable price, pressure on community lands and forests will further increase, resulting in further denudation of the natural resources. To reduce this problem, the following alternatives need to be considered.

Promotion of Commercial Plantation: In a forestry plantation for industrial raw materials and round timber, only about 40-50% wood is used for timber or industrial raw material and the rest is used as fuelwood. If the community forestry programme can promote commercial plantations to meet the annual demand of 65-70 m3 of wood, these plantations can also meet 25% of the demand for fuelwood. As the returns from commercial wood are very attractive, tree growers can afford to sell the leftover fuelwood as by-product at a lower price. Thus, the poor can be benefitted. Local grasses which grow in abundance in a well managed plantation, can be cut and carried to feed the stall-fed livestock.

Mixed Plantations: Introduction of fuelwood species in a mixed stand with fruit, timber and commercial tree species is feasible. Species like teak, mango, cashew, neem, etc. need wider spacing but the interspace remains idle for about 8-10 years, till trees attain normal size. It is possible to establish fuelwood species of short gestation between these trees and harvest them in 3-5 years. Selection of nitrogen-fixing tree species which are known for high calorific value, can further benefit farmers by nursing the main tree species through soil enrichment. This strategy can further boost the fuelwood production. Fodder cum fuelwood species like leucaenea, gliricidia, sesbania, acacia and albizia are ideal for establishing a mixed stand. Non-browsing tree species such as Australian acacia, kassod and casuarina are also useful as fuelwood species for planting on bunds and borders in fruit orchards.

Simultaneously, wood saving devices and alternate energy sources such as biogas plants, improved wood stoves, processing of biomass to improve the burning efficiency and solar cookers should be promoted.

## Trees under Agroforestry System

With depletion of agricultural lands and lack of irrigation facilities, agriculture in arid and semiarid regions is becoming uneconomical. Agroforestry provides a viable solution for such problems. Under this system, woody perennials are introduced in the agricultural field without hampering arable crop production. Trees serve as wind breaks, source of organic matter, shade and soil binder to prevent soil erosion while generating additional income. Depending on the fertility and depth of soil and moisture availability, different tree species can be introduced. In areas receiving more than 800 mm annual rainfall, it is possible to introduce various fruit crops while planting multipurpose tree species on field bunds and borders. The interspace can be used for cultivation of food crops for 8-10 years, till the trees spread widely in the field.

Establishing shelterbelts by planting tall growing trees on field bunds is very popular in India. Popular species used under shelterbelt planting are eucalyptus, poplars, casuarina, bamboo, acacia, dalbergia, leucaena, Silver oak, sesbania, gliricidia, melia, etc. To avoid adverse effects of these trees on agricultural crops, regular pruning of side branches and lateral roots will be helpful. These trees will be ready for harvest as poles, while contributing foliage and twigs for fodder, fuel and green manure. Shelterbelt plantations are profitable where farmers have fertile lands with irrigation facilities, with 200-300% cropping intensity like in Punjab, Haryana and Terai regions of Northern India.

Many farmers, particularly large land holders and absentee landlords, have been cultivating eucalyptus as a monocrop in non-irrigated areas, where very little care is needed after establishment. In a few districts of Andhra Pradesh, leucaena is cultivated as a sole crop and harvested at an interval of 3-5 years for pulp wood. This system is becoming popular as the local paper mills are offering a remunerative price, apart from arranging to harvest and transport wood from the field. Such buy back support is needed to expand tree plantations on a large scale.

The Wadi programme promoted by BAIF Development Research Foundation is another good model for promoting trees on degraded private lands particularly in hilly terrains. Under this programme, 0.4 ha land owned by each family is being developed under agri-horti-forestry system. The agricultural crops grown as intercrops between the fruit trees start generating income from the first year itself, while fruit trees start bearing fruits after 4-6 years. Large numbers of less known species of fruits, nuts and multipurpose plants are planted on the boundary and bunds to meet various household needs, while protecting the orchard. These orchards provide gainful employment all round the year, while improving the ecosystem of the location and income of the land owners.

Bio-diesel Plantation is a wave to promote non-edible oil tree plantations in the country. Major oil seed trees in the country are neem, mahua (Madhuca indica and Madhuca longifolia), pongamia (Derris indica), undi (Calophyllum inophyllum) and jatropha (Jatropha curcus). Among these, jatropha and pongamia have received wider publicity. The Government of India has launched a massive extension programme to promote jatropha cultivation, by projecting very high returns and providing partial financial support for establishing the plantation. The programme with good publicity, was launched to cover a larger area. However, the programme faded away as the farmers realised that they were not receiving the anticipated returns. This is an excellent example of how the programme can receive a severe setback if the anticipated benefits are not accrued. Pongamia cultivation is also not well accepted by farmers due to lack of precise information about the yield and incomes, although it has several other benefits such as tolerance to drought, ability to prevent soil erosion, source of woody biomass, green manure, bio-pesticides and better micro-climate. Species such as mahua, neem and pongamia are excellent for planting on community lands, village forest lands and along roads.

Trees with Religious Sentiments: Trees have religious and sentimental values. In ancient Hindu scriptures, uses of many trees for different purposes and their placement in home gardens have been very well described. Establishment of tree groves around the community temples with a wide range of tree species is also a traditional custom. These groves known as sacred groves, with a wide range of naturally grown and introduced trees and shrubs, are protected with respect by the community. Many species of ficus and acacia are also considered holy trees and people generally do not cut them. However, they do not want to plant a large number of such species unless they find some tangible benefits.

Trees for Beautification: Economics and tangible benefits are the primary considerations for selecting tree species by farmers. Trees are also planted for beautification, to improve the micro-climate, arrest soil erosion and many other functions. Trees provide an excellent ambience to the site, either residential or work areas. Trees absorb carbon dioxide, thereby reducing the harmful effects of air pollution. Thus, trees act as lungs of our cities to purify air and keep the surroundings cool, healthy and beautiful. Selection of suitable tree species will enhance the aesthetic value and beauty of the surroundings. The selection of species for various locations depends on the size, height and root system. A strong and deep root system ensures better establishment and prevents uprooting by wind.

Tall growing trees with wide branches to provide shade, such as mahogany, rain tree, ficus and rubber tree can be planted to bring the open area under tree groves. For the outer border of the campus, tall growing tree species may be planted. Along the internal roads on the campus, trees of small size, preferably with colourful and fragrant flowers may be planted. There are smaller flowering herbs which do not cause any damage to the buildings when planted close to the buildings. Plants like bamboo, bottle brush and weeping willow can be planted along lakes and canals. Apart from plants of small and large size trees, a wide range of creepers with colourful and fragrant flowers can also be introduced. Thorny hedges may be avoided, except for fencing as they require regular pruning.

Different tree species which can be selected for planting in Tropical gardens are presented in Annexure I.

**Conclusion**

While promoting tree planting on private lands, the preference of farmers should be considered. Tree species to be selected should be based on the quality of land, availability of moisture, suitability of climate, growth rate, gestation period, profitability and for fulfilling other specific objectives. While most of the farmers consider profitability as the primary consideration, beautification, conservation and improving micro-climatic are the other considerations. For the success of any afforestation programme on private lands, income being the primary consideration, arrangements should be made for backward and forward linkages. The extension programme to promote afforestation, should be based on well tested technical and economic data to guide the farmers and general public in the right direction.

# 7

# Cacao Trees: Growth and Production

## INTRODUCTION

Cacao is a preferentially alogamous tropical woody species formerly in the Sterculiaceae family (Cuatrecasas, 1964) and reclassified in the Malvaceae family (Alverson et al., 1999). Under natural conditions the tree can reach 20 to 25 m in height (Lachenaud et al., 1997), whereas under cultivation height varies from 3 to 5 m. The geographical origin of cacao is South America (Motamayor et al., 2002), where several wild populations can be found in the Amazon and Guyanian regions. It is considered one of the most important perennial crops on the planet, with an estimated world output of 3.46 million tons in 2008/9 (ICCO, 2009). Cacao is predominantly grown in the tropical areas of Central and South America, Asia and Africa (Marita et al., 2001) and is commercially exploited for seed output that is mainly destined for chocolate manufacturing. However, derivatives and byproducts of cacao are also transformed into cosmetics, fine beverages, jellies, ice creams and juices.

Traditionally cacao varieties in cropping systems belong to three main groups named Criollo, Forastero and Trinitario, which differ according to morphology and genetic characteristics and geographical origins (Bartley, 2005). Cacao Criollo was the first domesticated cacao and has been cultivated for at least two millennia in Central and South America (Sounigo et al., 2003). This group comprises varieties that produce fruits (pods) containing white or pinkish seeds that yield seeds highly prized for their flavor and mainly used in fine chocolates manufacture (Marita et al., 2001). Cacao Criollo is, however, infrequently cultivated because of its high susceptibility to diseases (Soria, 1970). Varieties of the Forastero group are widely cropped due to their high yields and resistance to diseases. They produce around 80% of the world output of cacao seeds (Marita et al., 2001). This group is subdivided into Lower and Upper Amazonian Forasteros; the former are cultivated in the Amazon Basin and their cultivars were the first to be introduced in Africa, while the latter are considered more genetically diversified and frequently used in breeding programs due to their vigor, precocity and disease resistance (Iwaro et al., 2001). The first Forastero varieties originated from the lower Amazon Basin and were cultivated mainly in Brazil and Venezuela (Sounigo et al., 2003). The Trinitario group is considered a recent hybrid that originated from crosses between Criollos and Lower Amazonian Forastero genotypes or intermediate types (Motamayor, 2001).

Traditionally, cacao is cultivated under the shade of a selectively thinned forest (Lobão et al., 2007) and represents one of the first agroforestry systems in tropical America. Cacao agroforestry has been known since pre-Colombians times by the Mayas (Bergman, 1969). In the Atlantic coastal forests of the states of Bahia and Espirito Santo, Brazil, around 4% of the world and 75% of the Brazilian cacao output is obtained using a system locally called Cabruca (Lobão et al., 2007). This system is a special kind of agroforestry in which the understorey is drastically suppressed to introduce cacao, and the density of upperstorey trees is reduced (Lobão et al., 2007). Cacao cultivation in this form shows much of the sustainability attributes of the natural heterogeneous forest and is considered an efficient agroforestry system (Figure 1) for protection of tropical soils against degradation agents (Alvim, 1989b).

Cacao is also intercropped around the world in planned systems with other species of economic value like Areca catechu, Cocos nucifera (Liyanage, 1985; Alvim and Nair, 1986; Daswir and Dja'far, 1988; Abbas and Dja'far, 1989), Hevea brasiliensis, Syzigium aromaticum, Cinnamomum zeylanicum, Erythrina fusca (Alvim, 1989a; 1989b), Bactris gasipaes (Almeida et al., 2002a) and other Amazonian species (Brito et al., 2002). Several agroforestry systems use more than three species in planned associations (Müller and Gama Rodrigues, 2007).

**Fig.** Some systems of cacao cropping: Cabruca, rubber x cacao and cacao without shade tree (from left to right). Photos by the authors.

In contrast, cacao is also cropped under unshaded conditions (Figure 1). In Ghana and Ivory Coast, for example, 50% of the total cacao farm area is under mild shade whilst an average of 10% in Ghana and 35% in Ivory Coast is managed under no shade (Padi and Owusu, 1998). In a shade and fertilizer trial conducted with Amazon cacao over a 20-year period in Ghana, yield of heavily shaded plots was about half that under the non-shade treatment (Ahenkorah et al., 1987). Despite this, the authors inferred that the economic life of an unshaded Amelonado cacao farm in Ghana might not last for more than 15 years of intensive cropping. Nonetheless, cacao can be produced economically with no shade if adequate management practices including water and nutrient replenishment are utilized.

We provide information about the effects of biotic and abiotic factors on growth and development of the cacao plant that could be useful in the design of strategies for new cropping systems. This is particularly important when also considering that the cacao planting frontiers are being expanded towards marginal lands in tropical and subtropical areas.

## NATURAL HISTORY

Cacao grows in the forest understory to a height of 6–12 metres (20–40 feet), usually remaining at the lower end of this range. Its oblong, leathery leaves measure up to 30 cm (12 inches) in length, and periodically the tree sheds them to grow a new set that is strikingly red when young. The foul-smelling or odourless flowers can be present at all times but appear in abundance twice a year. Growing from "flower cushions" on the trunk and limbs, the flowers are about 1 cm (0.4 inch) in height and breadth. They can be white, rosy, pink, yellow, or bright red, depending on the species. In many areas cacao relies on tiny flies called midges for pollination.

After four years the mature cacao tree produces fruit in the form of elongated pods; it may yield up to 70 such fruits annually. The pods, or cherelles, range in colour from bright yellow to deep purple. They ripen in less than six months to a length up to 35 cm (14 inches) and a width at the centre of 12 cm (4.7 inches). Each pod is divided into 10 sections by ridges running along its length, and 20 to 60 seeds, or cocoa beans, within are arranged around the long axis of the pod. The beans are dicotyledonous (having two seed leaves), oval shaped, and about 2.5 cm (1 inch) long and are covered with a sweet, sticky white pulp. After fermentation and drying, 20 to 30 cocoa beans will weigh 28 grams (1 ounce).

Cacao thrives at altitudes of 30 to 300 metres (100 to 1,000 feet) above sea level in areas where temperatures do not range much below 20 °C (68 °F) or above 28 °C (82 °F). Rainfall requirements depend upon the frequency and distribution of rain and the degree of water retention by the soil; minimum rainfall is about 1,000 mm (39 inches) evenly distributed throughout the year, but 1,500–2,000 mm (59–79 inches) is optimal. Successful cultivation also requires deep, well-drained soil that is porous and rich in humus. Protection against strong winds is necessary because of the tree's shallow root system. Insect and disease control is crucial, as over 25 percent of the world's crop is lost to damage each year.

### Cultivation

Because of the hazards of disease and pests, only 20 percent of the world's crop is grown on large plantations. Most of the world's cocoa is grown on small farms of less than two hectares (five acres). Cacao can also be grown in pristine rainforests at low densities, providing an economic use for protected land.

In cultivating cacao, plants are first grown from seeds or cuttings, then transplanted, being spaced at intervals of 1.5 to 4.5 metres (5 to 15 feet). Other tree crops such as banana, palm, or rubber are often planted with the cacao to provide shade for the young trees. Floral buds are removed from the trees until they are five years old. Commercial cocoa bean crop yields may vary from under 100 to over 3,000 kg per hectare (110 to 2,700 pounds per acre), with the world average being between 340 and 450 kg per hectare (300 and 400 pounds per acre).

Cacao is likely to be of hybrid origin, and many varieties exist. In Central America two additional species of cacao (T. bicolor and T. angustifolium) are grown to produce cocoa. In parts of Brazil and Colombia T. grandiflorum is cultivated for its tasty pulp. Cacao is related to the kola nut tree (genus Cola), and both are members of the plant family Sterculiaceae.

### Pests and diseases

The most commonly destructive diseases of the cacao tree are pod rots. A pod rot called black pod is caused by a fungus (Phytophthora) that spreads rapidly on the pods under conditions of excessive rain and humidity, insufficient sunshine, and temperatures below 21 °C (70 °F). Control requires timely treatment with copper-containing fungicides and constant removal of infected pods. Fungal diseases found in the Americas and West Indies include witches' broom and monilia pod rot. Asian cacao trees are affected by a fungus that causes the tree to dry out, starting from the branch tips—a condition called dieback. Swollen shoot is a viral disease transmitted to the plant by mealybugs that has devastated Ghanaian and Nigerian cocoa crops.

Some common diseases such as cherelle (young pod) wilt, cushion galls, and dieback are not thoroughly understood and may result from a combination of physiological, viral, nutritional, and fungal conditions. Many different insects cause vegetative and crop damage to cocoa, especially mealybugs, true bugs (heteropterans), thrips, and scale insects. In Southeast Asia the cocoa pod borer, the larva of mosquito-like insect, is a common pest.

## EFFECTS OF ABIOTIC FACTORS ON SEEDLING GROWTH AND DEVELOPMENT

Cacao exhibits a considerable genetic variability in morphological and physiological traits (Daymond et al., 2002a, b). When it is multiplied via seeds, the seedling initially displays an orthotropic growth pattern and exhibits repeated cycles of leaf flushes with spiral phyllotaxy (Vogel, 1975). Furthermore, leaf emission occurs in a rhythmic way relatively independent of climate, suggesting that growth rhythm is under endogenous control (Vogel,

1975). However, after achieving approximately 1.0 to 1.2 m in height, the orthotropic growth ceases and the plant emits plagiotropic branches (Garcia and Nicolella, 1985). The number of plagiotropic branches varies from three to five, forming what is generally named the cup or crown of the cacao tree (Cuatrecasas, 1964). Also, there are inter- and intra-specific variations for the orthotropic growth pattern (Batista and Alvim, 1981; Garcia and Nicolella, 1985). In contrast to Soria (1964) and Garcia and Nicolella (1985), Batista and Alvim (1981) demonstrated that the genetic background of orthotropic growth is influenced by environmental factors. Environmental factors exert an influence of approximately 70% on the development of the cacao plant at this stage (Garcia, 1973).

According to Greathouse et al. (1971) and Orchard et al. (1980), the specific stages of leaf flushing in the orthotropic growth phase are: F-1 - bud swelling: leaf initiation and unfolding; F-2 - leaf expansion: thin leaves, with strong anthocyanin pigmentation and apical bud dormant; I-1 - leaf expansion complete: rapid greening and apical bud dormant; I-2 - entirely expanded dark green leaves: apical bud dormant. Immediately after the emergence of the next leaf, leaves in the initial F-1 stage are red and are positioned almost vertically upwards. However, with expansion (F-2 stage) the stem:petiole angle decreases to approximately 90° such that the petiole remains horizontal and, with an angle of approximately 90° between the leaf blade and petiole, the blade hangs vertically downwards (Abo-Hamed et al., 1983). During the F-2 stage the leaves are very thin and flexible. This condition is maintained during leaf expansion until they enter the period of fast greening (beginning of stage I-1), when the stem:petiole and leaf:petiole angles increase. Up to the moment that the leaf blade becomes completely green, it thickens and becomes rigid with a pronounced cuticle, and both the stem:petiole and leaf blade:petiole angles increase resulting in the petiole and blade being positioned almost vertically. Subsequently the petiole:leaf blade angle decreases approximately 90° and the blade is positioned horizontally (lower leaves). This position is maintained until leaf senescence (Abo-Hamed et al., 1983).

Although the duration of the several stages of the flush cycle may vary depending on the growth conditions [cf., for example, Greathouse et al. (1971) and Vogel (1975)] the basic development pattern is relatively fixed, that is, the beginning of the cycle is marked by swelling of the apical bud followed by unfolding of all leaves in the flush (Abo-Hamed et al., 1983). The end of leaf production and the growth limitation of subsequent leaves within a flush, which are smaller in size, are probably consequences of carbohydrate limitations, as noted by Machado and Hardwick (1988b). They proposed that a new flush occurs following an interflush period in which the carbohydrate supply is replenished; with the return of a favorable carbohydrate balance a new leaf flush takes place. According to Baker et al. (1975), (i) plastid length, breadth, number of grana and lamella index per plastid, as well as the number of lamella per granum, increase with chlorophyll (Chl) content; (ii) the number of chloroplasts per cell remains constant (three) during leaf development; (iii) lutein, neoxanthin, violaxanthin, a- and ß-carotenes are the carotenoids found in measurable quantities in leaves; (iii) a-carotene is present in slightly higher quantities than ß-carotene; (iv) both xanthophylls and carotenes increase linearly with chl content during leaf development; and (v) the phase of maximum chloroplast development occurs after the end of leaf expansion. Baker and Hardwick (1973) did not observe changes in the molar relation of chl a:b.

However, Baker and Hardwick (1974) and Baker et al. (1975) observed that (i) the photosynthetic capacity and the chl content increased in parallel during flushing; (ii) the maximum photosynthetic capacity per chl unit remains constant during chloroplast development; (iii) the saturating irradiance (Is) increases with increasing chl content; (iv) the total number of photosynthetic units increases one by one during leaf growth; (iv) the photosynthetic efficiency increases per chl unit during leaf development; (v) the thylakoid formation occurs in parallel with chl synthesis during chloroplast development; (vi) the maximum phase of chloroplast development may not occur until after the end of leaf expansion; (vii) the phase of maximum synthesis of chl coincides with the end of the linear phase of leaf expansion; (viii) Rubisco activity increases with the content of chl and can be a limiting factor for photosynthesis; and (ix) the chl content can be used as a general indicator of chloroplast development.

The apparent absence of green pigmentation in young cacao leaves is not related to the delay in chloroplast development, but is due to the fact that chloroplasts are initially very small and usually few (Whatley, 1992). The delay in greening of leaves is associated with a late increase in the total extension of the thylakoid system inside each cell that accompanies the subsequent increase in plastid size (Whatley, 1992). Lee et al. (1987) reported that the changes in color shown by cacao leaves during its ontogeny resulted from varying levels of different pigments, like anthocyanins and/or phenols, which initially mask the chl content (Figure 2). Content of these pigments fluctuates during leaf expansion (total content of phenols remains high and that of anthocyanins decreases with maturity).

On the hypostomatous cacao leaf surface (Hardwick et al., 1981) four main kinds of hairs are found, the most common of which are thick and short, with a multicellular rounded head (Abo-Hamed et al., 1983). In 4-d-old leaves, stomata are found only in the midribs, primary lateral and minor veins; however, only those of the midribs are completely developed and can open. Stomata of the secondary and lateral ribs open within 2-4 d, and those in inter-veinal regions begin to develop and can open towards the end of the leaf expansion phase (Abo-Hamed et al., 1983). According to Abo-Hamed et al. (1983) the maturation of stomata is correlated with the presence of chl. The cuticle thickness increases at a constant rate during leaf expansion and achieves maximum thickness during the I-1 stage. Additionally, stomata and cuticle development follow a pattern closely correlated with the growth stages of the leaf.

**Fig.** Seedling flushing in the field and details of a flush. Note the color variability in the seedling leaves. Photos by the authors.

## SOLAR RADIATION

Müller et al. (1992), evaluating photosynthesis in cacao leaves of different ages grown under two irradiances, found a strong effect of light intensity during leaf development on photosynthetic capacity of mature leaves. In cacao, shade leaves often exhibit greater total chl concentrations per unit mass than sun leaves (Merkel et al., 1994). Ontogenetic changes in the leaf chl a:b ratio is a controversial issue, as it has been shown to increase continuously (Merkel et al., 1994) or to be held in check (Baker and Hardwick, 1973, 1974) during leaf development. Costa et al. (1998), evaluating the growth of cacao seedlings under different light levels and N concentrations, verified increases in leaf lamina thickness positively correlated with the increase of light intensity, independent of N concentration. Under shade conditions and high N concentrations, they observed that the spongy parenchyma had a reduced number of small cells distributed over vast intercellular spaces, while short cells constituted the palisade parenchyma, interspersed by quite large spaces.

Seedlings grow slowly under full sunlight and some degree of shade is beneficial for their establishment (Okali and Owusu, 1975). Thus, shelter trees have been invariably recommended for the establishment of cacao seedlings (Alvim, 1977), which would be gradually removed with increasing self-shade (Byrne, 1972). The slow growth of seedlings under full sunlight is due to constraints on leaf expansion, which is most likely caused by excessive transpiration that would induce leaf water stress (Okali and Owusu, 1975). In contrast, shaded leaves show higher relative water content and fewer stomata per unit leaf area than unshaded leaves. The perpendicular positioning of young leaves in the flush, the presence of superficial hairs, the rapid synthesis of cuticle and the development of stomata, which are all limited to the initial stages of leaf expansion, act to reduce the water loss of developing leaves (Abo-Hamed et al., 1983).

The saturating irradiance for photosynthesis of a mature cacao leaf varies from 6% (Baker and Hardwick, 1973) to 30% (Okali and Owusu, 1975) of full sunlight, while for a developing leaf Is increases with increasing chl content (Baker and Hardwick, 1976). At Is, younger leaves (PF1) in the flush showed photosynthetic rates significantly higher than those from both the oldest leaves of PF2 (middle age leaves) and PF3 (old leaves) (Machado and Hardwick, 1988a). Under field conditions, the total potential contribution of photosynthesis of the oldest leaves would be remarkably reduced as a consequence of mutual shading in the crown interior (Machado and Hardwick, 1988b).

Cacao seedlings show increasing rates of net CO2 assimilation (A) as the photosynthetically active radiation (PAR) increases to values in the range from near 400 (Joly, 1988) to 750 (Da Matta et al., 2001) µmol m-2 s-1, which corresponds to 20 to 35% of PAR at full sunlight, when Is is reached. About 95% of the maximum A is obtained with half Is. At full sunlight, A declines at light exposures above 30% of the global radiation (Okali and Owusu, 1975; Hutcheon, 1976), in which A of many plant species usually peaks (Alvim, 1965). Baligar et al. (2008), measuring A in three cacao clones, showed that increasing PAR from 50 to 400 µmol m–2 s–1 led to increases in A by about 50%, but further short-term increases in PAR up to 1500 µmol m–2 s–1 had no effect on A.

For leaves developed under shade conditions, a decrease in A occurs below 20% of full sunlight (Owusu, 1978), whereas Okali and Owusu (1975) found in shade leaves that Is occurred at around 3-4% of full sunlight. Therefore, the varying Is should be a reflection of differences in growth and measurement conditions. Cunningham and Burridge (1960) by submitting cacao seedlings to full sunlight and heavy shading (85%) showed that water and mineral nutrients were the most crucial factors for growth promotion at full sunlight. According to Okali and Owusu (1975) nutrients may be more important than shading for seedling establishment. Costa et al. (1998), by growing cacao seedlings under different light levels and N doses, observed a significant increase in chl content of mature and more shaded leaves, especially at higher N doses. The increase at all irradiances tested occurred in the presence of up to 10 mM of N, but above that concentration chl content leveled off. Okali and Owusu (1975) and Merkel et al. (1994) found similar results regardless of chl content and PAR for other cacao genotypes.

For cacao plantation systems it is hypothesized that once the nutritional demands are met, the yield of the understorey crop is dependent mainly on the accessibility to solar radiation (Cunningham and Arnold, 1962). In any case, Hutcheon (1973) evaluating the growth of the cacao clone Scavina-6 and two other Amelonado genotypes under full sunlight, verified that the former is more tolerant to excessive radiation than the Amelonados. Scavina-6 kept more leaves and grew well at full exposure, whereas the leaf area of Amelonado was halved, showing that seedling capacity to grow under full sunlight is also related to genetic factors.

**Temperature**

Cacao growth and development, as also occur with other tropical woody species, is highly dependent on temperature (Raja Harun and Hardwick, 1988b; Hadley et al., 1994). This can be observed through the reduction of photosynthetic rates at temperatures above or below the range considered optimal (Raja Harun and Hardwick, 1988b). Low temperatures affect mainly vegetative growth (Alvim et al., 1977), flowering (Alvim, 1988) and fruit development (Sale, 1969).

Daymond and Hadley (2004), evaluating the effect of temperature on the initial vegetative growth, emission of chl fluorescence and leaf chl content in seedlings of four cacao genotypes, showed the existence of genetic variability in response to temperature stress. In a study by Amorim and Valle (1990), in which gas exchanges were measured on 8-month-old cacao seedlings submitted to different root temperatures (10, 20, 30 and 40°C), it was found that at root temperatures between 20 and 30°C the water fluxes and stomatal conductance were greater than at any other soil temperature. Consequently, the resistance to water movement was low and A rates were higher.

Daymond and Hadley (2004) calculated cacao base temperatures (temperatures below the point at which development ceases) that varied from 18.6 to 20.8°C. These values are considerably higher than those obtained by Alvim (1977), who calculated a base temperature of 9°C for Catongo based in its pod growth ability. The higher base temperatures of Daymond and Hadley (2004) were probably due to the fact that their studies were conducted under conditions of low irradiance, while the studies by Alvim (1977) were carried out under field conditions.

Okali and Owusu (1975) observed that leaf temperatures of unshaded cacao seedlings are higher than that for shaded seedlings, and the transpiration rate (E) is usually at its maximum. The vertical orientation of developing cacao leaves decreases light interception, and, thus, leaves can be kept cooler and water loss may be limited in comparison to mature horizontally positioned leaves (Abo-Hamed et al., 1983). Low chl content and high anthocyanin pigmentation of expanding leaves also contribute to maintain relatively low leaf temperature and, together with the presence of relatively dense hair cover on the leaf surface, may contribute to further decreased E (Abo-Hamed et al., 1983).

## Relative Humidity

Stomatal opening in cacao leaves is related to the air relative humidity (RH) – stomata are kept more open at higher than lower RH (Sena Gomes et al., 1987). However, due probably to a high cuticular transpiration, stomatal closure does not always efficiently control water loss. In fact, some cacao genotypes do not show high stomatal resistance under water deficit and low RH (Raja Harun and Hardwick, 1988). This is in contrast to other genotypes where an efficient mechanism of stomatal regulation may strongly limit E under water deficit conditions. This may be an important acclimation strategy against drought (Balasimha, 1988). In fact, the extreme sensitivity of some cacao genotypes to low RH can be a limiting growth factor in areas where RH is low (Sena Gomes et al., 1987). In such areas, growth may be adversely affected as a result of stomatal closure and, consequently, a reduction of A is to be expected. The inefficient use of water (low water-use efficiency, WUE= A/E) at low RH probably would cause water deficit in shoots under limited soil water supply (Sena Gomes et al., 1987). According to Baligar et al. (2008), increasing VPD leads to decreases in A with parallel increases in E, thus ultimately lowering WUE. Furthermore, under the conditions of their measurements, Baligar and co-workers showed an unusually poor response of gs to VPD, which could limit the ability of cacao to grow where VPD is high. Nonetheless, high A and high WUE of cacao in a high RH regime are consistent with many species of the humid tropics (Alvim, 1977; Wood, 1985).

## Water Supply

Both water excess (flooding) and water deficit impair cacao growth and development. Flooding is an important barrier for the initial growth and establishment of cacao in places subjected to periodic waterlogging. Waterlogging occurs in some cacao producing areas in Brazil, Ghana, Nigeria and Ivory Coast, where total rainfall often largely exceeds evapotranspiration, which creates hypoxic conditions in the soil (Sena Gomes and Kozlowski, 1986). In the state of Bahia, Brazil, the hypoxic condition occurs after heavy rainfalls in locations with shallow soils, as well as areas with hydromorphic soils and margins of rivers after periodic floods. Decreases in growth, leaf area, gs and A, as well as epinasty, leaf fall, formation of hypertrophic lenticels and adventitious roots in submerged stems have been found during the flooding period. The extent of these changes is genotype dependent (Rehem, 2006). Sena Gomes and Kozlowski (1986) reported similar results for cacao seedlings of Catongo. Bertolde (2007) studying 35 cacao clones in the young phase found some tolerant genotypes and demonstrated that there are no significant relationships among physiological variables in response to flooding and heterozygosis patterns. In any case, growth inhibitions in plants submitted to soil flooding reduce pod production due to the delay of flowering and suppression of vegetative growth (Sena Gomes and Kozlowski, 1986). A similar result was also found by Mariano and Monteiro (1980), who showed an approximately 60% reduction in pod production after a flooding of some weeks. Water deficit, in addition to impairing crop yield of adult plants, also negatively affects growth of cacao seedlings. Almeida et al. (2002b) evaluated the effects of water stress in 5-month-old seedlings of eight cacao genotypes grafted on the variety Cacau Comum. They demonstrated that (i) drought resistance occurred, at least partially, through osmotic adjustment in most genotypes; (ii) the genotypes maintained values of relative water content at around 90% with a leaf water potential (?w) of approximately -1.0 MPa, gradually reaching 55% at -3.5 MPa; (iii) a significant increase occurred in leaf concentration of K and P during the dehydration process of some genotypes at ?w of -1.5 MPa.

In cacao, leaf carbon export rate is sharply reduced with the decline of ?w between -0.8 and -2.0 MPa. This reduction is strictly associated with both A and the export capacity, which is strongly reduced when A approaches zero (Deng et al., 1989). Deng et al. (1990) found values of A close to zero at ?w below -1.6 MPa, while average values of A of about 2.2 μmol CO2 m-2 s-1 were found at approximately -1.5 MPa. This was also observed by Gama-Rodrigues et al. (1995), who attributed to K fertilization a role in the mechanisms to reduce the negative effects of severe water stress in cacao, independently of the K source.

## NUTRIENTS – K

Cacao requires large amounts of K, and approximately 700 kg ha-1 are necessary to produce 1000 kg ha-1 of seeds per year (Thong and Ng, 1980). This element corresponds to about 70% of the nutrients in the sap of the cacao xylem (Martins, 1976). Orchard (1978) showed in cacao seedlings that high doses of K (5 mmol L-1) promoted increased leaf area without affecting dry biomass production. He also showed that there is an inverse relationship between leaf transpiration and K availability. Under field conditions cacao trees well supplied with K may be more tolerant to the adverse effects of water stress (Bosshart and Uexkhull, 1987). According to Gama-Rodrigues et al. (1995), KCl was the K source that induced the lowest gs, E and A, which differed significantly from the sources K2SO4 and Kalsilite. There were no significant effects among K sources on WUE, ?w and internal CO2 concentration. The largest dose of K added to the soil, however, resulted in smaller values of A and larger values of E, thus decreasing WUE but without affecting shoot dry biomass.

## CARE OF COCOA TREES IN THE HOME LANDSCAPE

Frequent applications of small amounts of fertilizer are best for continuous cocoa growth and fruit production. Fertilizer mixtures containing 6 to 10% nitrogen, 6 to 10% available phosphoric acid, 6 to 10% potash, and 4 to 6% magnesium give satisfactory results with young trees. Examples of commonly available fertilizer mixes include 6-6-6-2 [6 (N)-6 (P2O5)-6 (K2O)-2 (Mg)] and 8-3-9-2 [8 (N)-3 (P2O5)-6 (K2O)-3 (Mg)]. Young plants should be fertilized every 14 to 21 days with 1/8 to 1/4 lb (69-118 g) of a complete fertilizer (6-6-6, 8-3-9 or similar materials), with the amounts increasing as trees become larger. Once trees become about 12 to 18 months old, they should be fertilized with up to 1 to 2 lbs (474-948 g) every other month. Fertilizer rates may be reduced from November through February due to cool temperatures reducing plant growth. Minor elements should be applied every other month but are most effective during the warm summer months. Minor elements including manganese and zinc may be applied to the ground in soils with a pH less than 7 and foliarly applied for plants growing in high-pH soils. Similarly, iron sulfate may be applied to the ground for plants growing in low-pH soils. However, for plants growing in high-pH soils, chelated iron should be mixed in water and applied as a soil drench.

### Irrigation (Watering)

Watering is essential for best cocoa plant growth and fruit production. Cocoa plants that lack water (drought stress) may drop leaves, flowers and young fruit, and they may produce small fruit. Plants growing in sandy or rocky soils that are well drained and do not hold much water should be watered every other day or every day during hot, dry conditions and less often during cool parts of the year (late fall, winter). Plants growing in soil with a capacity to hold water (loams, sandy loams) should not be overwatered and therefore watered at 3- to 4-day intervals, especially during hot weather.

### Cacao Trees and Lawn Care

Cocoa trees in the home landscape are susceptible to trunk injury caused by lawn mowers and weed eaters. Maintain a grass-free area 2 to 5 or more feet (0.6–1.5 m) away from the trunk of the tree. Never hit the tree trunk with lawn mowing equipment and never use a weed eater near the tree trunk. Mechanical damage to the trunk of the tree will weaken the tree, and if severe enough can cause dieback or kill the tree. Roots of mature cocoa trees spread beyond the drip-line of the tree canopy, and heavy fertilization of the lawn next to cocoa trees is not recommended because it may reduce fruiting and or fruit quality. The use of lawn sprinkler systems on a timer may result in over watering and cause cocoa trees to decline. This is because too much water too often applied causes root rot.

### Mulch

Mulching cocoa trees in the home landscape is recommended. Mulch helps retain soil moisture, reduces weed problems next to the tree trunk, and improves the soil near the surface. Mulch with a 2- to 6-inch (5- to 15-cm) layer of bark, wood chips, or similar mulch material. Keep mulch 8 to 12 inches (20–30 cm) from the trunk.

**Pollination and Thinning**

Self-incompatibility of cocoa flowers may result in little to no pod production. Therefore either a self-compatible type (variety) should be grown or 2 or more compatible types (varieties), should be grown near each other. Since cocoa is normally pollinated by specific midge species (Forcipomyia spp.) that may not be present in Florida, hand pollination is one way to increase the chance of pod formation. Flowers usually open in the early morning, and hand pollination during the morning hours is best. This may be accomplished by using a small artist's paint brush that is first placed in contact with the anthers of an open flower and then placed in contact with the stigma of another flower. Cocoa plants may set a large number of fruit, which may lead to plant decline. In general, only 1 to 2 pods should be allowed to develop at any one flowering cushion on a limb. In other words, if more than 1 of the flowers from a cushion sets a fruit, leave only 1 or 2 to develop. Removing an excessive number of pods will result in faster development and larger pods of remaining fruit.

**Pruning**

Initial Pruning. Newly planted trees should be allowed to grow to 1 to 2 ft (0.3–0.6 m). If no branching occurs, then the top should be cut back to initiate branching. Three to 4 main branches should be allowed to develop. Remove all others. Maintenance Pruning. Plant height should be limited to 6 to 8 ft (1.8–2.4 m) to facilitate care and make protection from winds, excessive light, and cold temperatures easier. Periodically, selected branches should be removed to allow an increase in light and air movement inside the plant canopy. This will enhance pod production and reduce fungal disease problems. Damaged or diseased branches should also be removed regularly.

**Insect Pests**

Numerous insects are reported to attack cocoa plants, however, since cocoa is rarely grown in south Florida, insect problems may be reduced although not eliminated. Contact your local county extension agent for current control recommendations.

**Diseases**

Numerous diseases are reported to attack cocoa plants. Because cocoa is rarely grown in south Florida, however, disease problems may be reduced, although not eliminated. Several Phytophthora species common to Florida are reported to attack cocoa shoots, leaves, roots, and pods. Several pod rotting fungi may be present in Florida. Never propagate new cocoa plants from mother plants showing symptoms of distorted growth. The distorted growth may indicate a viral infection.

### Harvest

Pods should be harvested soon after developing their characteristic yellow or red and yellow peel color. The pods should be clipped off carefully. Pulling them off may damage the cushion and reduce pod production. The pod may be opened with a knife and the mucilaginous seeds removed.

## HABITAT AND RANGE OF THE CACAO TREE

Naturally Cacao grows under heavy rainforest canopy, it is cultivated underneath Banana or Casaca (Tapioca) or other large leaf, tree-like, grasses. It has unusually deep roots for a rainforest tree because it naturally tends to grow in the riparian zone. It requires a deep, slightly acidic, moist, well drained soil. In poorer soils, the low shade of the banana is ineffective and the high overhead shade of the canopy is required.

The Cacao Tree grows in lowland tropical forests with little seasonality. It needs a consistent climate: temperatures of 21 to 32 degrees Celsius year round -- never lower than 15 C, and 100 to 250 cm of rainfall, well distributed throughout the year with no month less than 10 cm. It grows only below 1000 meters of elevation, and usually below 300 meters. All of this means that it grows only in the tropics -- almost exclusively within 10 degrees latitude of the Equator and only in places that are not too mountainous and do not have monsoons or droughts. The largest number of species are found in northwestern South America, where the tree is native. However over half of the world supply of commercial Cacao comes from two East African countries: Cote D'Ivorie (Ivory coast) exports 41%, and it's neighbor, Ghana 13% of the world's supply.

Indonesia is third in world exports at 11%. Brazil, Cameroon, Ecuador, Madagascar, Nigeria, Sri Lanka, and Venezuela export significant amounts. And Cacao is also cultivated for export in Columbia, Congo/Zaire, Costa Rica, Cuba, Dominican Republic, Fiji, Gabon, Grenada, Haiti, Jamaica, Malaysia, south central Mexico, Panama, Papua New Guinea, Peru, Philippines, Sau Tome, Sierra Leone, Togo, Trinidad and Western Samoa. Unlike the two other popular drugs that grow in the tropics, coffee and coca, Cacao is not horribly damaging to the rainforest -- it doesn't require open land and in fact, requires the shade of the jungle to grow, although commercial plantations a usually clear of some or all of the forest to make it easier to harvest the pods. As with Coffee, Cacao is a crop that is grown in extremely poor areas with hot climates and manufactured into a product that is generally consumed in very wealthy areas with cool climates.

## Reproduction: fruit, flower & pollination

The flowers (and the fruit) are on the trunk of the tree, and it flowers (and fruits) all year long. This means that Cacao has the very unusual quality of having flowers and fruit on the tree at the same time!! It takes a long time, 5 to 8 months, to progress from blossom bud to ripe fruit. On cultivated Cacao plantations, only 3 out of 1000 flowers are pollinated, fertilized and progress to fruit!

There are a great many flowers, often waves of flowers covering the main stem of the tree! The white flowers are odorless. There are more flowers at the end of season than at the beginning. Cacao is pollinated by midges (gnat-like insects) and occasionally by bats. Pollination usually occurs in the morning and the flowers die in 24 hrs if not pollinated!

Although hermaphroditic, Cacao Flowers are self-incompatible, they cannot fertilize themselves. This creates a much healthier stock in the wild, but is terribly frustrating to plantation owners who cultivate Cacao. It turns out that the Cacao plantations themselves are the reason for the extremely low fertilization rate of the Cacao Flowers, and the potential Chocolate shortage that Chocolate companies are always warning against.

The insects that pollinate Cacao live in the rainforest. They require humid shade with a wide range of species and decaying matter on the ground; the natural habitat of Cacao. The midges have no reason to venture far from home into the sunny, dry neatly kept cultivated groves of Cacao trees. "The bigger a Cacao plantation, the more it frustrates the midges in their efforts to pollinate individual Cacao Flowers." -- Allen Young, biologist Cacao has over 400 distinct smells (compare that to 14 in the rose and 7 in the onion) however cultivated Cacao has only a small percentage of those, leaving the midges even more confused! Luckily some Fair Trade and organic Cacao farmers are working with the trees instead of against them. Smaller and more wild plantations are better for the trees, better for the Chocolate, and better for the workers as well.

Like most high production food plants, Cacao was almost certainly engineered to produce large and plentiful fruit by the natives of the area many hundreds (or thousands) of years ago. The seeds in these fruit were valued highly by the people living in that region, as well as their northern neighbors. They continue to be highly valued today. The seeds are encased in a large colorful pod which grows close to the tree after a flower. The large pod is green while maturing and and turns yellow, orange, red or purple when ripe (some varieties are still green when ripe). The pods vary significantly in size, shape and texture. They range from about 10 cm to greater than 40 cm in length! They have 5 to 10 veins or longitudinal ridges and are spherical to oblong, shaped roughly like an American football. Fruits are produced throughout the year, simultaneous with more flowering. It takes take 4 to 5 months to achieve the pod size, and then yet another month to ripen! A ripe pod can be left on the tree for 2 or 3 weeks without spoiling. It is important for the flavor that it is harvested only when ripe, although it will not open and lose it's seeds when overripe. If separated from the pod the seeds soon become infertile, but retain their fertility for a long time within the pod.

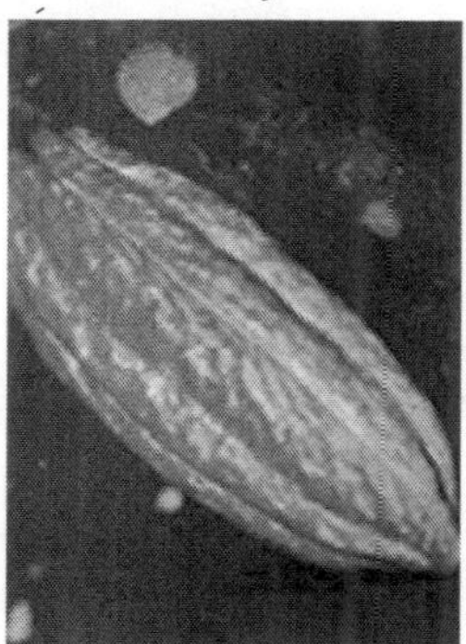

Cacao Pod from chocovic.es

The pulp of the fruit is edible, but it is NOTHING like Chocolate. It is yellow, slippery and sweet and a bit less dense than an apple. I have seen it described as vaguely lemony, although some have suggested it tastes a bit like mango. I have never had it, but i'd love to try it! Seeds are dispersed by monkeys and other small mammals which break through the pod wall to eat the pulp.

# THE CACAO TREE AND ITS FRUIT

## THE CACAO TREE

Cacao comes from the cacao tree, which is very difficult to grow. It only bears fruit inside the band of 20 degrees north and 20 degrees south of the equator. But even within this band of the tropics, the tree will not grow if the altitude is so high that the resulting temperature falls below 15° Celsius (60°F). Cacao also needs year-round moisture. Poor growing conditions make it even more susceptible to a multitude of diseases. On the other hand, when conditions are perfect for the tree, the seeds will sprout in a few days and the tree will start bearing fruit after four years.

Photo courtesy of Forest & Kim Starr

Some interesting facts about the cacao tree:

- it grows to a height of 12 to 15 metres
- it bears fruit continuously
- it stops bearing fruit after 25 years, although it can live much longer
- in the wild, it always grows beneath the much taller rainforest canopy to be protected from direct sunlight and wind
- it requires ground cover to maintain soil moisture; its dead leaves serve this purpose
- it has flowers directly on its trunk and major branches
- it has flowers pollinated by little midges and not bees
- it has no way to release its seeds by itself; it needs either animals or humans to open its pods

### Cacao leaves

When cacao leaves fall to the forest floor, they mix with the leaves of other plants and decay. Fungi and other organisms decompose this debris, which are going to feed the soil with essential nutrients, thus fertilizing the tree. In addition, decaying leaves provide the perfect breeding ground for midges, the tiny insects that pollinate cacao flowers.

## Cacao roots

Cacao roots soak up rainfall and nutrients from the soil and leaf litter. Stretching across the thin forest floor, these roots also anchor the cacao tree and help prevent soil erosion. Cultivated cacao's root system is shallow, however, and relies heavily on the decaying cover of leaf litter to remain healthy. Most of the nutrients in rainforest soils can be found in the topmost layer of decaying vegetation. When grown naturally from seed the tree has a 2 metre deep taproot. However in cultivation, most plantations use vegetative reproduction (cuttings) that results in a tree without taproot.

## Cacao flowers

Once they reach maturity, cacao trees flower continuously during the entire year. The flowers of the cacao tree are tiny pink and/or white five-petaled blossoms. They are found on the trunks and lower branches of the tree, while in general, trees produce their flowers and fruit only on the smallest branches. Botanists refer to this phenomenon as cauliflory. Fewer than 5% of cacao flowers are pollinated. These flowers can only be pollinated by small, gnat-like midges that can work their way through a cacao blossom's complicated parts.

**Cacao cherelles**

"Cherelles" are small pods that die on the tree before they mature. Even though fewer than 5% of flowers are pollinated, the tree will still produce more fruit than it can healthfully support. In consequence, the tree naturally "weeds out" some of these energy-draining youngsters (up to 90%), which blacken and shrivel during their early stages of growth. These sticky cherelles contribute to leaf litter and provide nice, juicy homes for the midge population.

**Cacao pods**

Cacao pods are the fruits of the cacao tree. Successfully pollinated flowers mature into this ribbed and oval fruit. The ripening process takes about five months. It is quite common to see both flowers and pods together on the same tree throughout the year. The thick shelled cacao pod contains "mucilage", a sweet white pulp that surrounds the bitter cocoa beans. Each tree produces about 20 pods and each pod contains between 20 and 60 seeds. In order to produce one kilogramme of cocoa paste about 10 pods are required. Healthy mature pods will eventually rot on the tree unless picked by an animal or farmer. When the pods ripen they turn from green or yellow to orange or red.

**Cocoa beans**

Cocoa beans, the base for making chocolate, are the seeds of the cacao tree. They are found inside the cacao pods, surrounded by a sweet white pulp. Each pod contains between 20 and 60 cocoa beans. A variety of chemicals, including theobromine which is very similar to caffeine, give the seeds a bitter flavour. Beans can only germinate within 2 weeks of being harvested. When monkeys, birds, human or other animals break open the pods to reach the delicious sweet pulp, they spit out the bitter-tasting seeds. This is the clever adaptation that the cacao tree evolved so that its seeds hit the forest floor and sprout into new trees.

## SCIENTIFIC CLASSIFICATION OF THE CACAO TREE

Theobroma cacao (literal translation: "cacao, food of the gods")

| | |
|---|---|
| Kingdom | Plantae – plants |
| Subkingdom | Tracheobionta – vascular plants |
| Superdivision | Spermatophyta – seed plants |
| Division | Magnoliophyta – flowering plants |
| Class | Magnoliopsida – dicotyledons (e.g. like coffee) |
| Subclass | Dilleniidae (e.g. like tea) |
| Order | Malvales |
| Family | Sterculiaceae – cacao family |
| Genus | Theobroma L. – theobroma |
| Species | Theobroma cacao L. – cacao |

Dried cocoa beans

### The three principal types of cacao

Criollo (nicknamed the "Queen of Cacao")

- History: criollo is the earliest known variety, first cultivated 2,000 years ago by the Maya
- Physiology: it's very difficult to grow, it's very susceptible to diseases, it produces fewer seeds than the other varieties and it is classified as "Fine grade" (it's considered a delicacy)
- Appearance: criollo pods are yellow or red; they are deeply grooved and warty (it remains, however, extremely difficult to determine the type of cacao based on the exterior appearance of the pod; experts will always split the pod open and look at the beans in order to establish the type)

- Seeds: criollo cocoa beans are large and white
- Flavour: it has a sweet aroma; there is almost no bitterness and it has a delicate, well-balanced, exquisite taste
- Geography: it's mainly grown in Mexico, Central America, Venezuela, Madagascar and Indonesia
- Production: only about 5 to 10% of the world's production

**Forastero**

- History: native of the upper Amazon region (Ecuador and Venezuela)
- Physiology: it has a high yield and is very resistant to disease
- Appearance: forastero pods are usually yellow and superficially grooved (it remains, however, extremely difficult to determine the type of cacao based on the exterior appearance of the pod; experts will always split the pod open and look at the beans in order to establish the type)

- Seeds: forastero cocoa beans have a flat shape
- Flavour: it has a perfumed aroma with a fruity and bitter taste; because of its full flavour, the forastero cocoa bean is chiefly used for blending

- Geography: the most widely cultivated variety of cacao, particularly in Africa (Ghana, São Tomé and Príncipe, Ivory Coast)
- Production: more than 85% the world's production

**Trinitario**

- History: a cross between forastero and criollo
- Physiology: it combines some of the best aspects of both types: it has a greater resistance to disease and damage, and is classified as "Fine grade"
- Appearance: trinitario pods can be red or yellow (it remains, however, extremely difficult to determine the type of cacao based on the exterior appearance of the pod; experts will always split the pod open and look at the beans in order to establish the type)
- Seeds: trinitario cocoa beans can be round or flat
- Flavour: it has a fruity and slightly acid aroma; it is spicy and sharp
- Geography: it is mainly grown in Haiti, Jamaica, Trinidad and Tobago, Grenada, Venezuela, Cameroon and Papua New Guinea
- Production: 10 to 15% of the world's production

## FINE ROOTS IN CACAO TREES

The capacity of a tree to absorb water and nutrients depends, among other factors, on fine root dynamics (mortality rates and regrowth of fine roots) as well as its variation in time. The turnover of fine roots can contribute a significant proportion of recycled nutrients in agroforestry systems, where the competition between forest trees and crops for water and nutrient depends on the temporal pattern of fine root regrowth (Muñoz and Beer, 2001). Studies done under greenhouse conditions have shown that cacao seedlings, as with many tropical woody species, have alternate phases of root and shoot growth (Taylor and Hadley 1987). Field evaluations of fine root growth in cacao trees have been very much limited to static inventory data. For example Beer et al. (1990) assessed biomass only at a single harvest. Although Kummerow et al. (1981, 1982) accomplished a pioneer study with cacao root dynamics, their methodology involved sequential coring, which can lead to significant over or underestimation of root productivity (Anderson and Ingram, 1993).

Biomass of active fine roots (= 2 mm at 0-10 cm soil depth) of 11-yr-old cacao plants shaded by Erythrina glauca was reported as 0.8 Mg ha-1 (Kummerow et al., 1981) and 0.4 Mg ha-1 for roots = 1 mm (Kummerow et al., 1982). In cacao plantations shaded by E. poeppigiana or Cordia alliodora the biomass of cacao fine roots, on average, was approximately 1.0 Mg ha-1 and changed little during the year (Muñoz and Beer, 2001). The maximum value found at the beginning of the rainy season was 1.85 Mg ha-1 in the cacao-C. alliodora system compared to 1.20 Mg ha-1 for cacao-E. poeppigiana. According to Muñoz and Beer (2001), if the annual turnover of fine roots were constant and close to 1.0 Mg ha-1 in both systems, the input of nutrients would be 23-24 (N), 2 (P), 14-16 (K), 7-11 (Ca) and 3-10 kg ha-1 yr-1, which would represent 6-13% and 3-6% respectively of the total input of nutrients in the organic matter for the C. alliodora and E. poeppigiana systems.

Growth and longevity measurements of fine roots (diameter < 1 mm) performed by Silva and Kummerow (1998) directly in the soil-litter interface, in a Brazilian cacao plantation shaded with E. fusca, revealed that the elongation rates of fine roots were, on average, 3.7 and 1.8 mm d-1 for ramifications of first and second order, respectively. They also found that the average period of functional life of these roots was 3 d, with a range of 1 to 10 d, and that growth of fine roots was closely correlated with rain frequency. Data reported by Silva and Kummerow (1998), Kummerow et al. (1982) and Medeiros et al. (1987) suggest that life spans of fine roots are short, but, in compensation, they have high turnover rates.

According to Muñoz and Beer (2001), the low fine root production between October and January in Costa Rica is probably the result of internal distribution of carbon to pod production during the six months that it takes for fruits, harvested in December-January, to develop and mature. Furthermore, these authors reported that partial pruning of cacao in December might also reduce cacao fine root production during the following months due to reduction in photoassimilate production and internal competition between flushing and root renewal (Muñoz and Beer, 2001). A negative correlation between number of new leaves per flush and number of growing root tips was reported by Kummerow et al. (1982). The observations of Muñoz and Beer (2001) in Costa Rica are consistent with observations in Brazil that the number of active cacao root tips (0-10 cm soil depth) was lowest in the driest month and highest in the wettest month (Kummerow et al., 1982). Therefore, this temporary increase in fine root production seems to be part of the normal phenological cycle of cacao (Muñoz and Beer, 2001) since formation of new root tips in cacao grown in nutrient solution was reported only during the rainy month of July in Costa Rica (Rodriguez et al., 1963). Additionally, the results regarding fine root dynamics of shaded cacao plantations suggest that a greater proportion of nutrients in organic or inorganic fertilizers is absorbed by cacao at the beginning, rather than at the end, of the rainy season (Muñoz and Beer, 2001).

Experiments (B.V. Leite, 2007; unpublished results) conducted with 35-month-old cacao plants in a semi-arid region of Bahia, Brazil, in which irrigation was performed by placing one or two plastic pipes at 0.30 m from the trunk with drip emitters 0.50 m apart, revealed that in plants grown in plots with two irrigation pipes, the longitudinal root length (along the planting line) was about 1.70 m and the perpendicular (to the planting line) root length was about 1.0 m for both sides of the trunk. For the cacao plots with only one pipe the longitudinal root length (along the planting line) was, on average, 2.0 m. Perpendicularly to the planting line a clear root displacement towards the dripping pipe was observed. At the opposite side, without a pipe, the roots were finer, fewer and smaller in length. Clearly, the greater concentration of radicels was observed near the irrigation pipe. In the two-pipe plots the tap roots reached a depth of 1.3 m against 1.0 m for the one pipe irrigation plot, indicating that the growth limitation was due to the size of the wetted soil area (Figure 3).

**Fig.** Root system of cacao plants grown in plots with one (left) and two (right) irrigation pipe lines. Note the skewed distribution of roots in the plant grown in plots irrigated with one pipe line. Photos by the authors.

The wetted area, after 2 h of irrigation, reached 0.7 m in depth and 0.6 m diameter. In this system 3% of the roots were concentrated between 0-0.1 m, 80% between 0.1 to 0.6 m, and 17% below 0.6 m depth. For the one-pipe plot the wetted area was 0.4 m in depth and 0.5 m diameter after 2 h of irrigation. In this system 9% of the roots were distributed between 0-0.1 m, 56% between 0.1-0.4 m, and 35% below 0.5 m depth, and these were skewed to the side of the irrigation pipe. The distribution of fine, medium and coarse roots followed the same patterns, that is, in the two-pipe plots, they were distributed along the two sides of the pipes. The fine roots were at a higher percentage while the medium and coarse roots presented about the same percentage. In the one-pipe plots the percentage of fine roots was higher than medium and coarse roots. However, fine roots were generally low in number as well as distributed preferentially on the side of the irrigation pipe. From these preliminary results (B.V. Leite, 2007, unpublished data) it can be concluded that the behavior of the plant as a whole, and the root system in particular, is different from the traditional cropping areas in which soil moisture depends on rainfall. In the traditional cacao growing areas the fine root system is concentrated in the first 0.2 m soil depth, and most fine roots are located in the first 0.10 m (Leite and Cadima Zevallos, 1991).

## THE FLUSHING CYCLE

Formation of a cacao branch is characterized by alternate periods of growth and quiescence, i.e. the flush cycle (Greathouse et al., 1971). During flush activity a variable number of leaves is produced in fast succession, after which the leaves rapidly expand (Baker and Hardwick, 1973). Under field conditions, the flushing cycle is activated usually by environmental changes (Almeida et al., 1987). However, the cycle persists under constant conditions (Orchard et al., 1980) that indicates there is an endogenous form of control along the flush cycle. Studies reported by Alvim et al. (1974b), Orchard et al. (1981) and Abo-Hamed et al. (1981) showed that the alternating growth and quiescence periods of the apical bud of the cacao branch (the flush cycle) is controlled by a regular cycle of accumulation and depletion of growth promoter and inhibitor compounds in the bud. Instead of in situ synthesis, transport of abscisic acid (ABA) (Alvim et al., 1974b) and cytokinins (Orchard et al., 1981) towards the apical bud has been proposed.

**Growth Promoters and Inhibitors**

The highest levels of auxin and cytokinin are found in young, expanding cacao leaves of a new flush (Orchard et al., 1981). This suggests that these leaves could be acting either as sinks, importing auxin and cytokinins from other parts of the plant, or as auxin and cytokinin sources, from where these promoters are exported to the apical bud and other plant parts. The auxin and cytokinin levels in young leaves close to the apex could also reflect the increase in transport of these compounds to the higher parts of the flush, including the apex (Abo-Hamed et al., 1984).

Analyses of new and previous flushes of cacao plants under controlled environment (Orchard et al., 1980) or under water deficit conditions (Alvim et al., 1974b) showed high leaf levels of ABA when the apex is in the quiescence phase of the cycle. In agreement with the distribution of [14C]ABA in cacao, it is suggested that ABA is exported from the leaf, via phloem, to areas of low ABA concentrations such as buds and young leaves, which is consistent with source/sink relationships. ABA induces branch quiescence or, alternatively, helps to maintain the branch in the non-active state. Appearance of radioactivity in mature leaves can be due to the lateral movement of [14C]ABA in the xylem due to transpiration flow (Abo-Hamed et al., 1981).

Cacao seedlings under optimal irrigation conditions also show flush growth, which is related with changes in the flush ABA levels (Orchard et al., 1980). These observations suggest that, despite adequate water supply, the seedlings suffer from an internal water deficit when the water uptake rate by the root system is lower than the rate of losses by leaf transpiration. This would cause the accumulation of ABA in the apex and, consequently, the imposition of apical bud inactivity at least in part of the interflush period. This hypothesis was confirmed by Abo-Hamed et al. (1983) after relating physical parameters (leaf blade position, population of hairs, cuticle thickness and number of stomata) of leaves of a new flush with transpiration and, therefore, with the water balance of cocoa seedlings in different stages of the flush cycle. The decrease in diffusive resistance and increase of water loss at the I-1 stage coincide with the period of higher ABA levels in leaves of the new as well as previous older flushes (Orchard et al., 1980). This formation of ABA after stage F-2 could be the result of water stress created by leaf transpiration exceeding the capacity of water uptake. The rate of maximum water loss of a new flush is only reached when the leaves are fully expanded, stomata completely developed and the leaves horizontally positioned (Abo-Hamed et al., 1983).

Iserentant (1976), comparing the effects of removal of all leaves of cacao seedlings in stages equivalent to F-1 and I-2 of the flushing cycle, verified the break of apical dormancy after defoliation in both stages. Iserentant (1976) concluded that quiescence in seedlings was due to correlative inhibition instead of a true dormancy. Similar results were found by Abo-Hamed et al. (1981) when only leaves in stages F-2 and I-2 of a new flush were removed. This suggests that there are no reversible or irreversible stages in cacao seedlings (Abo-Hamed et al., 1981), as described by Vogel (1975) for cacao trees grown under field conditions. However, even in plantlets, it seems that two physiological stages exist in the period of bud growth inhibition (Abo-Hamed et al., 1981).

**Nutrient X ABA Interactions**

Vogel (1975) described the initial phase of quiescence as a reversible phase that occurs during F-2 (Abo-Hamed et al., 1981). It appears that this phase in cacao seedlings can be initiated by the competition for nutrients between the apex and the developing leaves of the flush. According to Orchard et al. (1980), F-2 leaves of a new flush contain low levels of ABA and, consequently, are not the source of this compound. Since the removal of leaves in the flush stimulates the apical bud to produce new leaves, this stimulus could result in the removal of a powerful drain for nutrients instead of the removal of an ABA source (Abo-Hamed et al., 1981). Only the second phase, previously known as the irreversible phase (dormancy in I-1 and I-2), is controlled by the transport of ABA from mature leaves to the apex. In this case, the role of ABA seems to be in maintaining rather than initiating apical bud quiescence (Abo-Hamed et al., 1981).

Removal of new flush leaves at the I-2 stage causes reduction in the quiescence period of the apex (Abo-Hamed et al., 1981). Since in this period the new flush leaves contain high ABA levels (Orchard et al., 1980), probably the inhibitory effect of new flush leaves in I-2 is due to ABA and its transport to the apex (Abo-Hamed et al., 1981). The removal of previous flush leaves at the I-1 stage also promotes reduction in the apex dormancy period. These mature leaves are sinks for nutrients (Abo-Hamed et al., 1981) and also contain high ABA levels (Orchard et al., 1980) and, therefore, they could serve as an

ABA source for the apex, like mature leaves of new flushes (Abo-Hamed et al., 1981).

Application of ABA always extends the inactive period of growth. Alternatively, the application of growth promoters (GA3 or zeatin) reduces the duration of the subsequent quiescence stage of the apex. Probably, endogenous inhibitors in leaves are overcome by the effects of growth promoters that shorten the quiescence period. On the other hand, the fact that GA3 applied to new flush leaves in the F-2 stage does not promote a reduction effect on the duration of the apex dormancy period reinforces the hypothesis that the apical dominance, at this stage, is due to competition by nutrients instead of the presence of hormonal inhibitors in the leaves (Abo-Hamed et al., 1981).

## Source–Sink Relationships

Developing leaves display a low photosynthetic capacity (Baker and Hardwick, 1973) and are considered sinks for photoassimilates during most of the expansion phase. During this period, leaf contents of fructose, glucose and sucrose increase, presumably due to imports from older leaves. However, when leaf expansion ends, glucose and fructose pools decline and sucrose concentration increases, which reflects the end of photoassimilate import and the beginning of endogenous sucrose synthesis (Baker and Hardwick, 1975). The simultaneous development of several leaves within a flush constitutes, however, a very large carbohydrate consumption event during a period of 10 to 15 d that exceeds the available current photosynthate (Machado and Hardwick, 1988b). Although there is little competitive influence among developing leaves within the flush, leaves that develop at the end of a normal flush, are smaller in size (Machado and Hardwick, 1988a). This strongly suggests that the subsequent leaves in a normal flush grow under the increasing deficit of a presumable factor that supports growth, such as carbohydrate, which may negatively influence the potential for full size growth. In the interflush the stock of carbohydrates is reestablished, and, with the return of a favorable carbohydrate balance, the appearance of a new flush can occur (Machado and Hardwick, 1988b).

F-2 leaves of a new flush also represent a powerful sink for nutrients, which would be in competition with the shoot meristem (Abo-Hamed et al., 1981). Chemical composition of cacao leaves showed significant decrease in total N levels of mature leaves during the development of a new flush (Santana and Igue, 1979). This decline was attributed to an increase in N mobilization from mature to expanding leaves. At this stage in the flushing cycle, there is a large quantity of endogenous auxin (Orchard et al., 1981) and a high rate of 14C translocated out of the labeled leaf as compared with other stages (Sleigh, 1981). Additionally, at the beginning of F-2, the amount of carbohydrate translocated from mature leaves to the developing flush exceeds (positive balance) that required for leaf growth. This excess can be stored in the newly formed stem and used for its expansion, mainly in mid F-2, when the leaves of the flush constitute a very strong carbohydrate drain, and the amount required for leaf growth considerably increases (Machado and Hardwick, 1988b).

In the I-2 stage, new flush leaves are completely expanded, show maximum photosynthetic capacity, and are capable of photoassimilate export (Baker and Hardwick, 1975). At the end of this stage, cytokinin, imported from the root, accumulates in the apical bud and helps to activate the breaking of bud quiescence in the transition to stage F-1 (Alvim et al., 1974b; Orchard et al., 1981).

Observations reported by Sleigh (1981) and later confirmed by Machado (1986) showed that, at the F-2 stage, expanding leaves import considerable amounts of 14C that is assimilated by the mature leaves. Unexpectedly, if a high proportion of photoassimilates is removed from mature leaves during the F-2 stage, a corresponding increase of photosynthesis does not occur.

This suggests that the requirement of carbon for growth, at least for leaves, does not directly control photosynthesis of mature cacao leaves (Machado and Hardwick, 1988b). According to Baker and Hardwick (1973), A of older leaves, with 90% more chl, is only 10-20% higher than that of leaves at the end of flushing. Consequently, the photosynthetic efficiency of older leaves is significantly smaller. Therefore, in leaves of cacao, the decline in photosynthetic efficiency occurs during the period between the end of flushing and onset of senescence.

## Climate

Information exists about the periodicity of cacao leaf production in relation to changes of climatic elements, especially the effect of temperature (Couprie, 1972), solar radiation (Snoeck, 1979), and rainfall (Alvim and Alvim, 1978). Almeida et al. (1987), carried out a study in Ilhéus, Bahia state, Brazil, that investigated the relationship between actual evapotranspiration and reference evapotranspiration to quantify the effects of water deficit on cacao flushing. The actual-to-reference evapotranspiration ratio was used in the original expression of degree-day in order to evaluate the concomitant effects of temperature and water on flushing. Nominating this weighted function penalized degree-days (PDD), Almeida and co-workers verified that the weighted function represented approximately 80% of the existing interactions between flushing and weather factors. The high degree of significance of this function is due to the fact that PDD represents the joint effects of energy and water. For the conditions of Bahia the PDD that best explained flushing was that calculated three weeks before leaf emission.

In Bahia, the main flushing period, characterized by relatively short periods of vegetative rest, is from September to February. During those months, the energy curve (averages day-1) is in ascension and coincides with the period of more intense flushing (Almeida et al., 1987). Furthermore, in that area, the mean air temperature at the 8th week, combined with the solar radiation and insolation of the 9th week, exert great influence on flushing. In contrast, the period of decreasing energy (solar radiation, temperature and PDD) coincides with those months in which no flushing occurs (Almeida et al., 1987).

The seasonal influence of temperature on phenological events is evident in the flowering, fruiting and pod growth profiles found by Cazorla et al. (1989), who also observed low flushing during June to September. This period is characterized by relatively low temperatures. Therefore, the absence of flushing and flowering in south Bahia is intimately related to the decrease of fruiting, which culminates with an almost total absence of pods from January to March of the following year (Leite and Valle, 2000). In Bahia, cacao usually stops flushing from March to July and remains in a relative vegetative rest afterwards (Alvim, 1977). Therefore, the period of minimum flushing varies from 90 to 120 d – June to September (Alvim et al., 1974a).

Alvim et al. (1974a), Alvim (1977) and Almeida et al. (1987) studied the influence of rainfall and water balance on flushing within defined months of the year. According to Alvim (1977), in most cacao producing areas, the beginning of leaf emission follows the first rains after a dry period. For Sale (1968), rainfall or irrigation, after a dry period, is a necessary condition to elicit flushing. However, under field conditions, flushing begins in response to an increase of plant ?w, but not necessarily to rainfall, after a period of humidity stress (Almeida et al., 1987). Once apical bud growth resumes, rainfall plays a decisive role on leaf expansion. This may be one of the possible reasons to best explain the vigorous flushing associated with rainfall that follows a period of humidity stress (Alvim, 1977).

Almeida et al. (1987) showed that during the vegetative inactive period, the changes in soil moisture availability and, consequently, the sequence of water deficits and excesses, do not seem to play a direct role in the phenomenon of bud dormancy break. However, once the apical bud dormancy is broken, water availability seems to exert great influence on the intensity of leaf emission. Therefore, flushing becomes relatively intense when the period of water deficit (storage of water in the soil below 50 mm) is followed by rains that replenish soil water content (Almeida et al., 1987). Similar results were found by Sale (1968), Alvim (1977) and Machado and Alvim (1981).

Physiological responses of cacao are affected when the soil water content is below 60-70% of the maximum available soil water capacity (Alvim, 1960). Therefore, the development of water deficit during the dry season is considered an important factor in the control of flushing in field-grown cacao plants (Alvim et al., 1974a). At the beginning of a new flush cycle the transition from inactivity to apical growth of a branch occurs immediately following the decrease of water deficit. This is exhibited through leaf shedding at the beginning of the rainy season and the consequent reduction of the high shoot ABA levels in the preceding period of water deficit (Alvim et al., 1974b). In contrast, the effects on the apical bud activity, triggered by dry periods followed by rain, seem to occur through modifications of the ABA:cytokinin balance. Increase in the ABA:cytokynin ratio inhibits flushing (Alvim et al., 1974b). Removal of mature leaves produces modifications in that balance as well. Therefore, after defoliation, ABA content in the apical bud decreases, while cytokinins content increases considerably (Alvim and Alvim, 1976).

In tropical climates temperature oscillations are relatively small compared to temperate climates. Nonetheless, trees grow intermittently within seasonal climatic sequences, and flushing is considered the beginning of a growth cycle (Alvim, 1977). For cacao grown in southeastern Bahia, the flowering, fruiting and pod growth patterns show low flushing in June to September (Cazorla et al., 1989), a period characterized by relatively lower temperatures (Leite and Valle, 2000). According to Almeida et al. (1987), temperatures above a reference value (around 23°C) coincide with periods of increased flushing. Similar responses were observed under controlled conditions by Sale (1968), when cacao plants submitted to 23.3, 26.6 and 30°C emitted leaves respectively at 95, 36 and 20 d intervals. Therefore, it seems that high temperatures accelerate flushing initiation.

## ROLE OF ENDOGENOUS FACTORS ON CACAO FLOWERING

Cacao produces caulescent flowers that arise from meristematic tissues above leaf scars on the woody stems that are at least 2-yr-old (Figure 4). The floral meristem or cushion produces flowers throughout the life span of the tree (Aneja et al., 1999). Cacao flowers contain different polyphenolic compounds such as: (i) hydroxycinnamic acids, found in in the periphery of the organs, except in the ovary where parenchyma cells are present and the ovules occur; (ii) tannins; and (iii) anthocyanins. Both tannins and anthocyanins are located in the epidermis of different floral parts. Anthocyanins are mainly restricted to ornamentation in petals and staminoids (Alemanno et al., 2003). These external locations of phenolic compounds may function as chemical protective barriers against damages caused by pests and diseases (Alemanno et al., 2003).

In each flower the sepals begin to dehisce in late afternoon, and the flower is completely open at the beginning of the following morning. During this period the anthers liberate pollen and the stigma is receptive (Aneja et al., 1999). Although cacao flowers are hermaphroditic, the anthers are covered by the lower half of the petals. Without the intervention of insects, mainly Forcipomyia sp. (Diptera: Ceratopogonidae), pollination is impaired, since the viscous pollen does not come into contact spontaneously with the stigma (Dias et al., 1997). In Bahia, Brazil, high populations of this insect are found from March to July, which coincides with the increased pollination that results in the main harvest.

**Fig.** Counterclockwise: flowering on the trunk and branches of an adult tree, flowering in a 9-month-old seedling in the field and details of a flower cushion. Photos by the authors.

The fly populations are low during the high temperature months. This is reflected in the low and sporadic productivity that follows this fruit setting period (K. Nakayama, 2007; unpublished results). If the flowers are not pollinated, they abscise 24-36 h after anthesis, without visible signals of senescence (Aneja et al., 1999; Hasenstein and Zavada, 2001). If pollination is accomplished and fertilization occurs, the ovary increases in size, the pedicel enlarges, and the corolla wilts and deteriorates (Aneja et al., 1999).

The genus Theobroma represents an atypical example of an ovarian auto-incompatibility system that is distinct from most vegetal incompatibility systems (de Nettancourt, 1977). Incompatibility in cacao was first reported by Pound (1932), who noted the occurrence of changes from self-incompatibility to self-compatibility during certain periods of the year. The incompatibility system in cacao is complex, since it includes different degrees of cross and self-incompatibility of some genotypes as well as a high level of self-pollination (Lanaud et al., 1987).

Although there is no causal link between expression of enzymatic activity and incompatibility, Warren and Sunnite (1995) argued that enzymes like isocitrate dehydrogenase (EC 1.1.1.41 and 1.1.1.42), malate dehydrogenase (EC 1.1.1.37) and acid phosphatase (EC 3.1.3.2) are indicative of incompatibility systems. In contrast, Aneja et al. (1992, 1994) showed that the pollen grains of the self-incompatible clone 'IMC-30' did not germinate at low CO2 partial pressure. However, 'IMC-30' pollen grains germinated under high CO2 concentration, which helped overcoming incompatibility in self-pollinated flowers. The main effect of overcoming incompatibility systems in self-incompatible genotypes occurs when self-incompatible and compatible (foreign) pollen are mixed, which permits self-fertilization in self-incompatible trees (Glendinning, 1960; Bartley, 1969; Lanaud et al. 1987). These results imply that, in Theobroma, a substance other than CO2 may condition the maternal plant self-fertilization (Hasenstein and Zavada, 2001).

The percentage of flowers that sets pods in cacao is usually very low and varies from 0.5 to 5% (Aneja et al., 1999). This is partly due to the fact that the effective rate of self-fertilization in auto-incompatible trees is low, whereas in self-compatible trees self-fertilization can reach up to 43% (Yamada and Guries, 1998). The success of fruit (pod) set can depend on two factors, the degree of pollen compatibility and the number of pollen grains deposited on the stigma (Lanaud et al., 1987). There is some evidence that a high number of pollen grains is beneficial for overall pod set (Hasenstein and Zavada, 2001). Moreover, a high proportion of pollinations results from the visit of a single pollinator (Yamada and Guries, 1998). At anthesis floral auxin concentration is low and the retention of flowers, as well as fertilization and possibly pod development, depend on an external source or signal for satisfactory pollen germination or fertilization, as found by Hasenstein and Zavada (2001). They also noted an inverse correlation between the number of flowers per plant and floral auxin content and suggested that the genetic control of self-incompatibility in cacao may be modulated by the hormonal content of the flower. Hasenstien and Zavada (2001) concluded that this variation in floral hormonal content could explain the great number of pods set in some instances, even after incompatible pollinations.

A better understanding of the factors involved in floral abscission can be of great horticultural value (Aneja et al., 1999). Abscission of flowers in cacao occurs within 2-3 d after anthesis in the absence of pollination or fertilization. Even after manual pollination, only a small percentage of pollinated flowers develops and produces pods (Hasenstein and Zavada, 2001). Although there is no recognizable abscission area in the flower pedicel at anthesis, within 24 h starch grains appear in the pedicel cortex in the area of the incipient abscission zone, and cell separation occurs in the epidermis at 2-3 mm acropetally from the point of pedicel attachment to the flower pad (Aneja et al., 1999). Flower abscission due to incompatible pollination begins before the pollinic tube reaches the ovules (Aneja et al., 1994; Baker et al., 1997).

It has been observed that auxin concentration of developing floral buds is high and depends on the number of flowers formed. However, when the buds mature, the auxin level decreases during the initiation of the abscission process (Hasenstein and Zavada, 2001), while ABA levels of the flower increase significantly before abscission (Aneja et al., 1999). In contrast, most of the increase in ethylene production occurs simultaneously with abscission (Aneja et al., 1999). There is no abscission in the absence of ABA even when ethylene is present. Therefore, in cacao flowers, ABA seems to be the primary inductor of abscission, while ethylene, although accelerating abscission does not seem to be required for the process (Aneja et al., 1999). The increase of indoleacetic acid (IAA) levels in the cacao ovary, that follows compatible pollination and fertilization (Baker et al., 1997), or application of naphthalene acetic acid, can overcome the effect of ABA in the formation of the abscission area and reduce floral senescence (Aneja et al., 1999).

Hormonal changes after compatible and incompatible pollination indicates a strong increase of endogenous auxin after compatible pollination and a strong increase of ethylene after incompatible pollination (Baker et al., 1997). These hormonal responses occur before the interaction of the pollin tube with the ovule and suggest that the incompatibility system may be influenced by auxin (Hasenstein and Zavada, 2001). According to Aneja et al. (1994), the ability of CO2 to overcome abscission and the subsequent self-compatibility can be the result of minimization of ethylene effects. However, it is conceivable that high auxin levels also override the abscission signal, affecting or controlling the auto-incompatibility response in Theobroma (Hasenstein and Zavada, 2001).

Since flower production represents a considerable plant investment (Valle et al., 1990), the number of flowers can affect the auxin partition to each flower. This may explain the low concentration of endogenous auxin in clones with high flower productions (Hasenstein and Zavada, (2001). Auxin concentration is significantly higher in cacao clones with low production of flowers. Producers of a high number of flowers can show a 20-fold improvement in pod set, in spite of the existence of a poor relationship between pod and flower number. In addition, there is a strong negative correlation between the number of flowers produced and the endogenous content of IAA, while the content of ABA did not differ between the two sets (Hasenstein and Zavada, 2001).

## EFFECTS OF ABIOTIC FACTORS ON YIELD

The adult stage of cacao is reached when the tree enters reproductive growth. Pod production is small during the first years and increases as the tree matures. In general, most of the pod yield is found on the trunk and branches (Figure 5). Bartley (1970), studying the yield of cacao per tree, proposed to initiate selection of potentially high-yielding cacao trees in the 8th yr of planting. Since with young trees fruit bearing begins approximately at the 4th yr, it has been suggested that only at the 8th or 10th year after planting do the trees express their productivity potential (Dias and Kageyama, 1997).

Also, measurements of diameter and stem height, average crown diameter and leaf area of cacao trees between 1- to 3-yr-old have been used as variables to estimate the yield capacity of cacao (Batista and Alvim, 1981; Garcia and Nicolella, 1985). However, the same abiotic stresses that reduce pod production can also indirectly decrease yield by increasing the duration of the juvenile period, which varies widely within cacao genotypes. Flowering, for instance, in some cacao hybrids can begin 18 months after planting in the field, while in other genotypes flowering may be initiated after 3 to 5 yr (Sena

Gomes and Kozlowski, 1986). The most important determining biotic parameters of cacao yield are related to (i) light interception, photosynthesis and capacity of distributing photoassimilate, (ii) maintenance respiration, and (iii) pod morphology and seed formation (Zuidema et al., 2005). These three items are crucial to seed yield and were examined in initial studies concerning cacao production (Yapp and Hadley, 1994). On the other hand, these biotic parameters can be modified by abiotic factors, such as those inhibiting, for instance, initiation of floral buds (Alvim, 1977) or those that, after flowering and pod set, influence pod development (Wood, 1985). Climatic changes exert outstanding effects in the flush cycle, an important determinant factor of yield (Almeida et al., 1987).

In the definition of an appropriate climate for cacao cultivation in forest understorey conditions, it is well known that their microclimatic variables are completely different from the standard weather conditions outside that environment (Bonaparte and Ampofo, 1977).

**Fig.** Most cacao pods are beared in the trunk. Cacao pods of two different genotypes. Photos by the authors.

Also, there are indications that the structure and dynamics of the cacao agroecosystem exert some influence in the microclimate (Beer, 1987). Air temperature, RH and rainfall are correlated and affect the seasonality of microclimatic conditions inside a cacao plantation (Miranda et al., 1994). The magnitude of the meteorological parameters varies less in a hot week than in a cooler week. The predominantly clear summer (average of 9 h daily sunshine) and the cloudy winter (average of 5 h daily sunshine) regulate, to a great extent, the behavior of microclimatic parameters above and inside a cacao plantation.

## Temperature

According to Daymond and Hadley (2004), temperature is one of the main limiting factors for cacao production, since temperature stress affects the seasonal variation in seed yield. In a cacao plantation, the difference in temperature measured outside and inside canopy layers was around 2 ± 0.5°C, although higher values have been registered during summer weeks. The undercanopy maximum temperatures are characterized by drops varying from 1 to 3°C and temperature differences inside the canopy vary from 2.5 to 4°C, while the above canopy maximum temperature decreased from 1.5 to 3.5°C (Miranda et al., 1994). When measuring leaf temperature in the upper canopy layer of a non-shaded cacao tree on a sunny day, Valle et al. (1987) verified that it was always 2°C above the leaf temperatures of the middle and lower canopy layers. Leaves of the upper canopy layer showed higher differences between leaf and air temperature and, therefore, water vapor deficit, in comparison to lower layers.

**Water and Light**

Several factors influence leaf dynamics, among them water and light availability, which largely vary between locations and cropping systems (Zuidema et al., 2005). Additionally, wind conditions during the dry season reduce cacao yield due to water loss and reduction of leaf area (Alvim, 1977; Leite et al., 1980). Southeastern Bahia is a region with an average annual temperature of 23.5°C, annual total rainfall of approximately 1700 mm, no pronounced dry season, and trees are shaded with E. fusca and/or E. poeppigiana. Here the relative light intensity received by cacao trees cultivated varies between 30 and 100% of full sunlight and 4 to 10% radiation at soil level. The average leaf area index and the extinction coefficient are about 3.9 and 0.61, respectively (Miyaji et al., 1997a). Leaf longevity varies with position inside the canopy and/or incident irradiance, time of emergence and height in relation to the soil level. According to Miyaji et al. (1997b) for Catongo trees of about 7-yr-old the leaf duration at heights greater than 220 cm and above 60% full sunlight is 181 d, which is half the duration of leaves at heights between 0-150 cm and 5 to 20% sunlight. Factors like soil water deficit and leaf respiration, due to high air temperature and high radiation can be causes for the intensive leaf fall in the middle and upper canopy layers of the cacao crown (Alvim and Alvim, 1978).

The cacao yield is strongly related to rainfall in the dry season. It also depends on soil type and consequently water retention capacity (Zuidema et al., 2005). According to Leite and Cadima Zevallos (1991), the elevation and residence of the water table lead to increase and maintenance of humidity levels above the subsurface of the soil, a zone in which the cacao root system concentrates in traditional cacao cropping areas. To reach high levels of cacao yield in south Bahia, the monthly rainfall totals should be equal or greater than 200 mm, while the permanence of the water table should be more than 30 d (Leite and Valle, 2000). The fluctuation of the water table can be considered a climatic variable (derived from rainfall) that can influence cacao production as a form of natural sub-irrigation (Leite and Valle, 2000).

Using a simulation model, Zuidema et al (2005) showed that annual solar radiation and rainfall during the dry season explained 70% of the variations in annual seed yields obtained for 30 sites along the tropical cacao producing areas of the world. Drought promotes decreases in leaf area (Orchard and Saltos, 1988), A (Hutcheon, 1977), flowering (Sale, 1970) and yield (Khan et al., 1988). Changes in carbon distribution and export will depend on the severity and duration of the stress, as well as its occurrence in relation to flowering, pod set and pod maturation (Deng et al., 1989). There are positive correlations between yield and rainfall during months that precede harvest (Alvim, 1978), where rain distribution is more important than its magnitude (Atanda, 1972).

Alvim (1988) considered that harvest can be influenced by the rain distribution that occurred six months prior. This relationship seems valid for countries such as Brazil (southeastern Bahia), Ghana, Ivory Costa and Malaysia. In Bahia, the agricultural cacao calendar year begins in April and extends to March. The yield is divided into two distinct periods, minor harvest (temporão) from April to August and main harvest (safra) from September to March. The relative importance of each harvest depends on rain distribution (Alvim, 1973). Furthermore, cacao pods take approximately from six to seven months to complete their development (Almeida and Valle, 1995). This explains the importance of higher rainfall and longer phreatic residence of the water table at the soil subsurface in one year for high production in the following year (Leite and Valle, 1990; Dias and Kageyama, 1997).

According to Valle et al. (1987), data modeled for transpiration showed that, on a cloudy day, a non-shaded cacao tree could transpire about 45 L (1.2 L m-2 d-1). This assumes that all leaves are transpiring at the same rate. A shaded cacao tree, in contrast, could transpire about 26 L (0.8 L m-2 d-1). However, on a sunny day both trees showed a substantial difference in transpiration when compared with a cloudy day. The daily E was estimated to be 90 L tree-1 (2.4 L m-2 d-1) for the unshaded tree and 40 L tree-1 (1.2 L m-2 d-1) for the shaded one when 100% of its leaf area was exposed to incident irradiance (Valle et al., 1987). Estimates of daily E on a cloudy day for both tree conditions and for the shaded canopy layers on a sunny day were of the same order of magnitude of those calculated by Cadima and Alvim (1983). However, for unshaded trees, on a sunny day, Valle et al. (1987) showed an average E more than twice the estimated by Cadima and Alvim (1983). From the estimations of LAI (Miyaji et al., 1997a) and radiation (Valle et al., 1987), it is concluded that more than 60% of the water can be transpired by the upper canopy layers of unshaded cacao trees. In the shaded trees less than 5% of the absorbed water might be transpired by the upper layers, which is caused by the greater irradiance received in the lower layers than in the top layers due to the shade tree that limits light penetration to the upper layer even at lower solar angles.

## Shade vs Unshaded Cropping

In its natural habitat wild cacao grows under shade trees, but with a low pod production (Murray and Nichols, 1966), since the shade trees compete with cacao for water, nutrients and light (Bonaparte, 1975). In fact, many people believe that shade is indispensable for cacao cultivation (Owusu, 1978) and, therefore, earlier cacao plantations were conducted under shade regardless of yield (Urquhart, 1961).

Evans and Murray (1953) in Trinidad, Alvim (1959) in Costa Rica and Cunningham and Lamb (1959) in Ghana were the pioneers to critically examine the role of shade and light in cacao cultivation. In a shade and manurial experiment in Ghana strong reductions in yield were observed due to shading (Ahenkorah et al., 1974). In fact, cacao grown at full sunlight produced more pods than when grown under shade (Owusu, 1978). Control of both growth and density of cacao trees, and the increase in aeration and light penetration throughout the canopy are fundamental cultural practices necessary to promote increased seed production (Vernon and Sunderam, 1972). More recently the comprehensive work of Ahenkorah et al. (1987) further elucidates the importance of appropriate fertilization in unshaded cacao plantations.

Studies on light interception and photosynthesis show that seed production is light limited, probably due to external and internal shade (self-shading) in the cacao stand and light extinction inside the canopy (Yapp and Hadley, 1994). In fact, light intensities below 1800 h yr-1 suppress flower production with a significant negative effect on pod yield (Asomaning et al., 1971). Heavy shade not only reduces seed yield (Zuidema et al., 2005), because of low photosynthate production (Alvim, 1977; Ng, 1982), but also increases the incidence of diseases (Alvim, 1977). On the other hand, cacao is a shade tolerant species (Guers, 1985), in which appropriate shading could lead to adequate photosynthetic rates, growth and seed yield. Shading also alleviates effects of unfavorable ecological factors, such as wind velocity and excessive evapotranspiration (Miyaji et al., 1997a; Beer et al., 1998) and, consequently, decreases in both humidity stresses and extreme air and soil temperatures during the dry season are to be expected (Anim-Kwapong, 2003). This is essential for survival and establishment of cacao seedlings in dry and seasonally humid environments (Beer, 1987), since the seedlings are highly susceptible to dry periods (or soil moisture stress) (Leite et al., 1980). Furthermore, shelter trees used for shade greatly contribute to the formation of soil organic matter, carbon sequestration, nutrient recycling and maintenance of biodiversity (Lobão et al., 2007; Müller and Gama Rodrigues, 2007). Specifically in regions with low access to inorganic fertilizers, the multistrata plantation is used to maintain soil fertility with the subsequent increase in nutrient availability for cacao (Isaac et al., 2007).

When nutrient availability is not a growth-limiting factor, there is a positive correlation between cacao yields and light (Bonaparte, 1975; Ahenkorah et al., 1987). Vernon (1967) concluded that the relationship between cacao yield and available light was approximately linear from 30 to 60% full sunlight. However, when modeled from 0 to 100%, a quadratic model showed better adjustment than the simple linear model, suggesting that some degree of shading is desirable.

The higher production of non-shaded cacao results in a shorter productive lifespan, a larger demand for fertilizers (Owusu, 1978), a decrease in the incidence of cherelle wilt (Asomaning et al., 1971) and larger investments (Ahenkorah et al., 1987). The subsequent decline in productivity (Bonaparte, 1973) is attributed to high losses of exchangeable bases in the soil, attack of insects and diseases, excessive leaf transpiration and increase in soil evaporation (Ahenkorah et al., 1974). The inherent high risk of unshaded cacao cultivation was illustrated by the economic analyses of Cunningham (1963), the extra expenditure and work associated with clear-felling and growing unshaded cacao with great amounts of fertilizers would probably be justifiable only if yields of more than three tons are obtained. However, most of the cacao experiments involving shade and fertilizers show that shade reduces the responses to fertilizer applications (Alvim, 1977), and such conditions are rarely economically justifiable (Beer, 1987). This shows that cacao production is positively related to light intensity (Vernon, 1967), but this relationship depends primarily on the availability of nutrients in the soil (Figure 6).

In commercial plantations it is hard to understand and detail the potential productivity in relation to pod-set and yield, since the cacao canopies are under several shade levels (Alvim, 1977). With shade management practices cacao production can increase during the first decade and thereafter stabilizes (Rosenberg and Marcotte, 2005). Cacao trees can produce during 30 to 80 yr (Stevenson, 1987); however, yield subsequently decreases (Rosenberg and Marcotte, 2005). Furthermore, it is probable that the physiology of trees under heavy shade would be modified, because shaded trees can be more efficient in photosynthesis and leaf dynamics (rates of production and abscission), as, for example, the increase in leaf lifespan (Miyaji et al., 1997a). According to Owusu (1978), there is a close relationship between these physiological processes and carbohydrates in cacao. Improved growth, increase of flushing, enhanced flowering and increased yield under unshaded conditions can, however, be explained in terms of increase in carbohydrate production (Asomaning et al., 1971). This is exemplified by the internal competition between flowers and pods (Alvim, 1954), in spite of the small total energy expenditure for flowering (Valle et al., 1990). Furthermore, Alvim (1977) considers that besides the competition within pods, the annual variation of pod set is also subject to competition with other vegetative events, such as flushing and root and cambium growth.

## Conclusion

The expectation for the next 10 yr, without considering India and China, is a consumption increase of nearly one millions tons of cacao seeds. The big question is from where would the extra demand come? A strong possibility is an increase in productivity in West African countries, but this would be at the expense of significant resource inputs and occur over a long period. Actually, the chocolate industry foresees difficulties regarding future supplies. Today the world faces a new situation with the search for new renewable energy sources. As an example, Malaysia has replaced cacao plantations with oil palm, and Indonesia has become less and less interested in producing cacao.

According to the World Cocoa Foundation (www. worldcocoa foundation.org) there is a tendency towards an increase in the consumption of chocolate with high cacao content (43%) in Europe and USA. Also, there is a world trend in the proliferation of small companies that process cacao and make chocolate. Their search for cacao is remarkable. There is also a consensus that in countries with a better level of development, such as Brazil, the future will be the investment in quality and production of differentiated cacao. Therefore, the efforts in Bahia to produce fine cacao is in reality a world tendency for other locations with similar environments, thus, constituting a great opportunity for expansion in this market. Furthermore, the markets under greatest expansion are those of the emerging nations. Therefore, expectations are strong for the inclusion of China and India in the chocolate consumption market. However, if the markets show a promising tendency, more and more of the cacao farmers will feel the burden of high costs of resource inputs, mainly for control of diseases and insects that are steadily increasing. Several pests and diseases attack cacao, with some estimates putting losses as high as 30 to 40% of global production (www.icco.org/about/pest.aspx). In this sense farmers seek new lands that offer an escape from known diseases, and in this respect we may highlight the importance of studying the behavior of the crop under semi-arid conditions as mentioned earlier. The knowledge of the effects of factors that affect the cacao growth and development under such conditions should even make it possible to engineer new practices for the traditional cropping areas.

## ENVIRONMENTAL EFFECTS OF COCOA PRODUCTION

The environmental effects of cocoa production, such as herbicide resistance and deforestation are becoming much more severe as demand for cocoa increases in present day.

### BACKGROUND

Cocoa beans are a high demand consumer item all over the world. They are used in products such as chocolate, candy bars, beverages and cocoa powder. However, cocoa farming and the production of cocoa beans are extremely fragile and labour intensive processes. The process begins with a Cacao plant, or Theobroma Cacao, in which the beans are extracted from pods that grow directly on branches. Each pod contains roughly 30 to 50 beans.After the beans are extracted they must go through a time-consuming process of natural fermenting and drying. The farming process of cocoa can damage the environment depending on the practices of the farmer, as well as be limited by the environment itself. Global Climate Change, for example, causes longer drought seasons making it more difficult for farmers to plant and sustain new Cacao trees. Most of the environmental impact comes from Co2.

## PROCESS OF FARMING

Cocoa farming can only occur 15 degrees north or south of the equator. It can take approximately three years after planting for the trees to be fruitful enough to harvest the pods. Ripe cocoa pods, which are yellow in colour, are cut down from the trees using a machete. They can be very low on branches and easily accessible or higher up on thick branches. Once they are gathered, they are sliced open and the cocoa beans are extracted from the pods. The beans are then spread out, usually between banana leaves, for a number of days to ferment. Next, the seeds are placed in the sun to dry for several more days. After drying, they are gathered, placed into bags and taken to collection offices. From there, they are shipped around the world to be processed into end products.

## ENVIRONMENTAL IMPACT

### Full sun cocoa

Cocoa farms are generally small, family owned and operated businesses. There are approximately 4.5 million cocoa farms around the world. The majority of cocoa farms are located in Côte d'Ivoire and Ghana. In Ghana, Cocoa contributes 64% of all exports. Traditional cocoa farms are planted in the shade among other crops and trees. They are especially found in the Tropical Rainforest areas.Farming cocoa beans is a long process with many factors that can affect the farm's yearly output.

Farms' cocoa crop outputs struggle to match the increasing demand for chocolate. It is estimated that the demand for chocolate will increase twofold by the year 2050.Farmers have shifted towards unsustainable, less environmentally conscious practices to meet these demands.

Some farmers have shifted their crops out of the shade and into direct sunlight.This practice yields a greater quantity in a short period and at lower quality. Cacao trees with no shade tend to accumulate more weeds as well as be more susceptible to diseases such as Witches Broom and Frosty Pod Rot. If the crops begin to accumulate pests, farmers use large amounts of herbicides to rid the crops of these pests. The herbicides used damage the land and the health of the sprayers applying the herbicide. Excessive spraying of pesticides can also cause the weeds and insects to build up a resistance which will eventually create more harm to the crops.

### Deforestation

Cocoa farming also contributes to rainforest and old growth forest deforestation. By clearing land into these forests, farmers decrease the biodiversity and interactions between the organisms that naturally live in this area. Many wildlife habitats are destroyed and the plant species diversity is drastically reduced. Nutrients begin to leach out of the soil due to poor irrigation and inadequate soil protection, which can increase the erosion of the soil.The more intense the farming practices are, the more damaging they are to the ecosystem. Cocoa farming becomes a destructive circle as farmers wear out the soils and cut further into the forest to obtain fresh land. All of these processes stress the Cacao trees and result in lower yields, giving the opposite effect to what the farmers expect from these practices.

Some of the forests in Ghana and other Cocoa producing countries have been declared protected by the government after observing the Tropical Rainforest destruction. However, with a shortage of fresh land to plant Cacao trees, some farmers are beginning to illegally cut down parts of these protected forests. It has been estimated that approximately 50% of these protected forests have been cut down.

## SOLUTIONS

### Education

Through groups and programs such as the World Cocoa Foundation, Rainforest Alliance and Roundtable for a Sustainable Cocoa Economy, cocoa farming can return to its sustainable roots through education programs and help in finding ecologically and economically sound resources to further their farming. As a last resort, some programs will help farmers to access pest control products such as biocides as an alternative to the harmful pesticides being used. Other programs promote proper irrigation, composting, suitable soil management, and intercropping, meaning planting other trees and fruit crops in the surrounding land of the Cacao trees. Some farmers will burn old, fermented pods and place them back on the soil as a form of composting and fertilizer. To stop the process of deforestation, it is suggested that farmers replant on their current land while using the practices previously mentioned.

### Shade cacao

It has been suggested that Cocoa farmers go back to the original and natural ways of farming, by planting within the natural tree-cover and without cutting down existing trees. When an area has already been clearcut another possibility exists. Planting trees, especially fruit trees around and within the plantation, helps with growth of Cacao plants. These trees can provide shade to the Cacao plants and be a source of oxygen replenishment to the environment. If the shade trees are fruit-bearing, this can also provide additional income to the farmer. Shade trees return organic matter to the soil through falling leaf litter and decaying branches. The shade provided by these trees also helps to keep soil moist in dry seasons which results in less damaging irrigation practices. Shade trees will raise the amount of infiltration and slow erosion of the soil. Since shade inhibits the growth of weeds, farmers are able to use less or perhaps no pesticides which can decrease the occurrences of Witches Broom in these crops. Cacao plants that grow in the shade provide the environment with more biodiversity, allowing natural populations and habitats to flourish. Finally, shade can be extremely helpful in keeping and lengthening the productivity of old growth Cacao plants.

# 8

# Planting the Tree

Plant the tree out in the open. It cannot compete with mature trees for your attention, light, water and fertilizer. Remove and replace entire rows. Three 20-foot spaced rows can be replaced by four 16-foot spaced rows. This will allow you to rework the soil and irrigation system in the gap, and the new trees can be treated as a unit separate from the rest of the orchard.

Take good care of the trees once you take them out of nursery storage. Nurseries keep the trees cool and the roots moist. You should try to do the same until shortly before you plant them. Tree roots will dry in a few minutes if placed in the sun. Keep the unplanted trees' roots moist in a water barrel, or under wet burlap before planting. Dry roots die, which leads to poor growth. Trim broken or obviously damaged roots, but leave as much root as possible. Plant trees into warm, moist soil. Tree roots benefit from oxygen, temperatures above 45F and moderate moisture. You must make an effort not to put them into dry soil, followed by early irrigation, or cold, muddy soil.

If you must water them in a planting hole, avoid firming the wet soil by pressing it with your feet soon after watering, wet soil compacts and turns to brick under pressure. Wait a day or three for the wet soil to drain, then firm it and finish refilling the hole. If you rip the soil, which as previously discussed is the best option under almost every circumstance, you can plant trees by shovel, sometimes faster than by any method other than machine planting. Mark the desired tree spacing on a wire stretched between temporary posts you have placed to mark the row. The planting crew can start anywhere they wish along the marked portion of the row and plant two or three trees each, without getting into each others' way.

If you plant by shovel, you may be able to install much of the irrigation system prior to planting. If you are unable to rip the soil before planting, dig a hole worthy of the tree. Dig the holes a very short time ahead of tree planting, so you will have moist soil going back into the hole. If the hole is augured, widen the hole at least a shovel full in all directions.

Just spade the surface soil into the deep augered hole until you have broken the steep sides all the way around. The tree is helped more by a wide hole than a deep one. You may have to wait until your tree is in place before you can finish the last few shovels full around the hole. In difficult to auger soils, you may dig holes with a back-hoe. If you are using new soil in the planting hole, first fill the hole with your orchard soil to within 6 inches of the level where you want to place the young tree roots.

Then add a six-inch layer of moist new soil, then the tree (on a mound of new soil, all roots sloping down) then about 6 inches more moist new soil. Fill the remainder of the hole with orchard soil. When you are finished, the soil level should be at least four to six inches higher than the general soil level around the tree. Otherwise, as the soil in the hole compacts when wetted, it will slump. If your soil is not mounded at planting, the young tree will spend its' early years in a hole, prime for collar rot, root rot, and scion rooting.

The graft union must be placed at least a hand-width (4-6 inches) above the soil mound so that it will be above the soil level when the planting hole soil is finished slumping. If the bud union comes in contact with the soil, "scion rooting" may occur, which will cause your tree to develop the vigour similar to a tree on "seedling" root.

## PLANTING AND CARE OF A YOUNG ORCHARD

If you honestly analyse the economic returns from individual blocks on your orchard, (and you should) it may have become painfully obvious that some older blocks are not even covering the "out of pocket" costs of production.

As production of apples has increased in Washington, our standard of excellence has risen. Much of what once qualified as Washington extra fancy grade is sold now as a lesser grade. Many of those high quality strains of Red Delicious that you planted 30 years ago, or those Galas and Fujis that you planted in the early 1990's are not considered red enough to pack in a premium grade. They need special fertilizing and pruning to produce the grade that newer strains do without apparent effort.

You may have already started planting a few blocks of the newer varieties, or your neighbour has, so you know that returns can be double or triple what your older block receives. You may have to do something positive to stay in business. Move up, move on, or move out.

However, you may have a problem renovating your orchard. When you plant trees on your property, they do not grow well. They grew much better when you first planted the block. You hear about high early yields that others receive in young orchards, and know that you must get early yields to cover your planting investment. Is it possible that your site does not qualify for replanting under these new production standards? maybe it would be easier to retire, rather than replant. Maybe you should sell and move to an area where there is "new soil" where trees grow well. When the soil became unproductive, nomads once struck their tents, packed their belongings and moved on to better pastures. You may not consider relocation a good idea. For most people, an orchard is more than a business, it's a home. You live on your site, and prefer to stay there. The soil may seem a bit "tired", and newly planted trees may grow poorly, but the site is otherwise world-class.

The summer and winter temperatures are ideal for apple and pear production, frost is controllable, and late spring and summer rains are infrequent, so cherries are an option. Hail is so infrequent that insurance is not a good option. Where else will you find a site that is so well suited for fruit production? They are not making any more sites like yours. Your site gives you a natural, long-range competitive advantage; and besides all that, your family already lives there and doesn't want to move. Some growers around the world must spend extra time and money to try to change their orchard environments to try to match what you receive on your site for free. Getting young trees to grow properly may be the only horticultural problem holding your orchard out of its world-class potential.

If you could be sure that new trees would grow and produce well, you could start rotating out of your blocks that have unacceptable quality and production, replacing them with improved strains or varieties.

These new fruit trees could provide better support your family and make your orchard a better long-term investment. If you wish to pass the orchard on to your children, they may be able to support themselves (and you?) on a thriving, modern orchard. If you decide to sell it to someone else, the trees could be a valuable part of the sale, rather than something that will be removed and replaced. Can trees grow properly in old orchards? What are people doing that makes trees thrive in some areas, while trees on similar sites sit and sulk for years? There must be something that makes the difference, as too many people are succeeding under too many situations for luck to play an important role. It's time to investigate what others have done to ensure healthy trees and high production on "old" orchard sites

## TREE GROWING BASICS

High quality growth of young trees on a replanted site is usually the reward for well planned and carried out site preparation. Taking care of only three-fourths of the vital steps will not often lead to success. Taking care of these interacting factors falls under the heading of "management."

Below are listed some topics that may increase your chances that the replanted orchard will thrive. If we treated young trees as we do vegetables, they would thrive. The payoff, or the economic impact of failure is apparent in a season of growing carrots or corn. We have to wait a few years to see if everything went right when we plant fruit trees.

Few row crops would grow without careful yearly soil preparation, disease control, insect and weed management, balanced fertilization, and careful irrigation. Should we offer less to a crop we will count on to support us for the next 20 years? You should never plan to plant more trees than you can properly plant or tend. You will never regret having planted trees that have the potential to grow and produce well for as long as you wish to leave them in the ground.

### ORCHARD-PLANTING PROBLEM

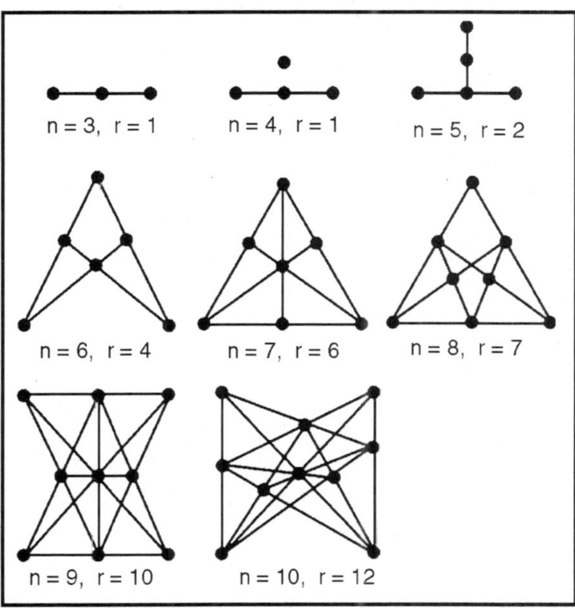

The orchard-planting problem (also known as the orchard problem or tree-planting problem) asks that n trees be planted so that there will be r(n, k) straight rows with k trees in each row. The problem of finding the maximum number of lines of three points for n points is due to Sylvester. The following table gives maxk (r(n, k)) for various k and n.

| **n/k** | **3** | **4** | **5** |
|---|---|---|---|
| **Sloane** | **A006065** | **A003035** | **A008997** |
| 3 | 1 | -- | -- |
| 4 | 1 | 1 | -- |
| 5 | 2 | 1 | 1 |
| 6 | 4 | 1 | 1 |
| 7 | 6 | 2 | 1 |
| 8 | 7 | 2 | 1 |
| 9 | 10 | 3 | 2 |
| 10 | 12 | 5 | 2 |
| 11 | 16 | 6 | 2 |
| 12 | 19 | 7 | 3 |
| 13 | [22, 24] | ≥9 | 3 |
| 14 | [26, 27] | ≥10 | 4 |
| 15 | [31, 32] | ≥12 | ≥6 |
| 16 | 37 | ≥15 | ≥6 |
| 17 | [40, 42] | ≥15 | ≥7 |
| 18 | [46, 48] | ≥18 | ≥9 |
| 19 | [52, 54] | ≥19 | ≥10 |
| 20 | [57, 60] | ≥23 | ≥11 |
| 21 | [64, 67] | | |
| 22 | [70, 73] | | |
| 23 | [77, 81] | | |
| 24 | [85, 88] | | |
| 25 | [92, 96] | | |

*Sylvester showed that*:

$$r(k=3) \geq \frac{1}{6}[(n-1)(n-2),] \tag{1}$$

where [x] is the floor function. Burr *et al*. have shown using cubic curves that

$$r(k=3) \geq 1\left[\frac{1}{6}n(n-3)\right], \tag{2}$$

except for n = 7, 11, 16, and 19, and conjecture that the inequality is an equality with the exception of the preceding cases. For n≥4,

$$r(k=3) \leq \left[\frac{1}{2}\left[\frac{1}{2}n(n-1)-\left[\frac{3}{7}n\right]\right]\right] \tag{3}$$

where [x] is the ceiling function

# PLANTING THE CITRUS ORCHARD

## Planting Density

Tree spacing is affected by factors such as the species of citrus concerned, the cultivar, the type of rootstock, the environment, the size of the orchard, and the manage-ment practices which the grower will be using. For example, if he will be using machinery, he must leave enough space between the rows for the machines to pass when the trees are mature.

## Close Spacing

Small or dwarf citrus species should be planted at a higher density.

## Wide Spacing

Tall citrus trees, and those with rapid growth, should be more widely spaced.

## Planned Close Spacing

For maximum benefit per unit area, spacing can be as close as possible when the trees are planted. Later, as the trees grow in size, they can be thinned.

## The Layout

Most orchards on level ground can be planted in a square or a rectangle. For close spacing or planned close spacing plantings, a quincunx system can be used. (In a quincunx arrangement, four trees are planted at each corner of a square, with a fifth tree in the middle). In orchards on sloping land, trees should be planted along the contours.

## Time of Planting

The optimum time for planting citrus trees varies according to the species. It also depends on the condition of the roots and shoots of the seedling. Transplanting the seedlings out in the field is best done at the beginning of the wet season. Growers should avoid transplanting seedlings during hot summer weather.

## Handling the Seedlings

Seedlings with bare roots should be planted out as soon as possible after they have been removed from the soil or growing medium. Seedlings grown in containers have generally become root-bound by the time they are ready for transplanting. Roots which are severely bent and twisted should be pruned before the tree is transplanted. Overcrowded leaves and shoots should also be pruned.

## Planting

In orchards on sloping land, or in orchards where the trees are widely spaced, seedlings are usually planted by digging a hole for each seedling. If the orchard is to have closely planted trees, it is preferable to dig a furrow along which the row of seedlings will be planted. Covering the soil around the planting hole with mulch will help conserve soil moisture and keep down weeds. An orchard planted at a high density can make a good profit in the early years after planting. However, after the trees become mature, they are too crowded and ventilation is poor. This increases the relative humidity, and the incidence of pests and diseases. At this stage, closely planted trees should be thinned or trimmed, to open up space between them.

## IMAGES

**Fig**. Newly Planted Citrus Orchard

**Fig**. Planting Holes for New Orchard

**Fig**. Seedling with Straw Mulch to Conserve Soil Moisture and Keep down Weeds

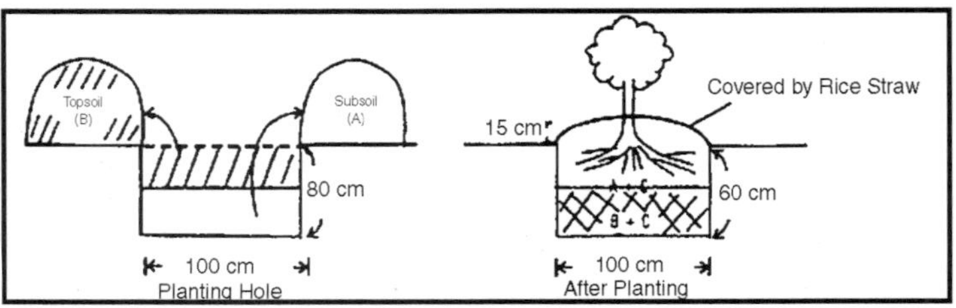


**Fig**. Diagram of Planting Hole for Young Seedling

## DOUBLE PLANTING OF ORCHARDS

The wide spaces - thirty to forty feet - between the small trees of apple or pear have tempted many to plant peaches, dwarf pears, or plums and cherries between the trees both ways, with the intention of taking them out when the permanent trees needed the whole space. But this is not satisfactory, as the double planting interferes with proper culture, the use of cover-crops, spraying, and all needed orchard care, and too often robs the permanent trees before the interspaces are cleared.

A far better plan is to plant only one way with early bearing varieties of apple or dwarf pear, leaving open spaces north and south for air-circulation, culture, cover-crops, and spraying. At the north where the open spaces are narrowed the strawberry can be profitably grown in newly planted orchards. The partial protection from wind-sweep will benefit the plants, and the strawberry rows, with the winter covering of straw raked between the rows, make a good cover-crop in summer and a fair protection of the surface-roots in winter. After picking the second crop the turning under of the rows and the mulching adds needed humus to the soil and benefits the succeeding crop of strawberries as well as the orchard trees.

### Distance Apart of Peach, Plum, and Cherry

In peach-growing centres where the trees are properly pruned the usual distance apart is only fifteen feet each way. But farther north, where less cutting back and pruning is done, twenty feet apart is the usual distance. Where orchards of the duke and heart cherries are planted for profit in rows running north and south, twenty feet apart, with thirty-foot spaces between the rows, gives ample room. The Morello varieties are usually planted twenty feet apart both ways in relatively mild climates.

In the prairie States the preferred plan is that of planting the Morello varieties and the native plums only twelve feet apart in rows running north and south, with spaces between the rows of not less than twenty feet. This plan gives the needed circulation between the rows and shades the stems and main limbs from the noonday sun in summer and to a great extent in winter. The spacing of other orchard fruits and the small fruits are given in connection with their discussion.

### Planning and Staking the Orchard Site

A little advance work in the way of setting a small stake where each tree is to stand will save much time and give straight rows in all directions. Prepare at a leisure time as many small stakes as there are trees to plant. The small stakes or pins can be split from sections of inch-board about fifteen inches long. Also secure, by purchase or borrowing, a surveyor's steel tape four rods long. With this chain set stakes that can be seen four rods apart across the orchard in a straight line by sighting. These sight-stakes can be set for all the rows at the same time. By stretching the steel tape from one stake to another the small stakes where the trees are to be set are stuck at proper distances apart as indicated by figures on the tape. To make quick work it is best to tie red yarn on the tape at the figures, giving equal spacing. When all the small stakes representing the trees are stuck, what is known as the "planting-board," as represented by Fig. 56, comes into use. The centre notch in the board is placed around the little stake and wooden pins are stuck through the holes at the ends. These pins remain in place as the board is taken forward until the work is completed. The board should not be less than nine feet long and the central notch should be exactly equidistant from the holes in the ends. When the holes are dug the stakes are taken up, but the pins remain as a guide in replacing the board on the same side of the tree or stake. If reversed on alternate rows it will make a crook in the rows.

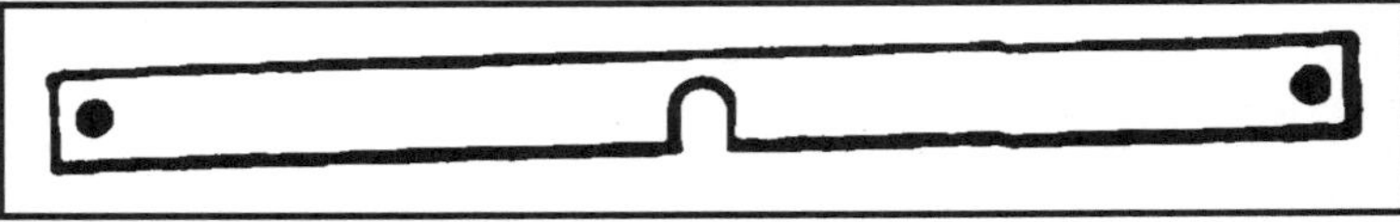

**Fig.** Planting-board.

## FRUIT TREE PRUNING

Pruning fruit trees is a technique that is employed by gardeners to control growth, remove dead or diseased wood or stimulate the formation of flowers and fruit buds. The most economical pruning is done early in the season, when buds begin to break, and one can pinch off the soft tissue with one's fingers (hence the expression "nipped in the bud"). Many home fruit growers make the mistake of planting a tree, then neglecting it until it begins to bear. But careful attention to pruning and training young trees will ultimately determine their productivity and longevity. Good pruning and training will also prevent later injury from weak crotches that break under snow or fruit load.

### OVERVIEW

To obtain a better understanding of how to prune plants properly, it is useful to have some underlying knowledge of how pruning works, and how it affects the way in which plants grow. Plants form new tissue in an area called the meristem, located near the tips of roots and shoots, where active cell division takes place. Meristem growth is aimed at ensuring that leaves are quickly elevated into sunlight, and that roots are able to penetrate deeply into the soil. Once adequate height and length is achieved by the stems and roots, they will begin to thicken to give support to the plant. On the shoots, these growing tips of the plant are known as apical buds.

The apical meristem (or tip) produces the growth hormone auxin, which not only promotes cell division, but also diffuses downwards and inhibits the development of lateral bud growth which would otherwise compete with the apical tip for light and nutrients. Removing the apical tip and its suppressive hormone allows the lower dormant lateral buds to develop, and the buds between the leaf stalk and stem produce new shoots which compete to become the lead growth. Manipulating this natural response to damage (known as the principle of apical dominance) by processes such as pruning (as well as coppicing and pollarding) allows the horticulturist to determine the shape, size and productivity of many fruiting trees and bushes. The main aim when pruning fruit trees is usually to obtain a decent crop of fruit rather than a tree with an abundance of lush yet unproductive foliage. Unpruned trees tend to produce large crops of small, worthless fruit often damaged by pests and diseases, and much of the crop is out of reach at the top of the tree.

Branches can become broken by the weight of the crop, and the cropping may become biennial (that is, only bearing fruit every other year). Overpruned trees on the other hand tend to produce light crops of large, flavourless fruit that does not store well. Pruning is therefore carried out to achieve a balance between shoot growth and fruit production.

**Formative Pruning of Bush Trees**

Formative pruning of apple and pear trees (the pome fruits; the stone fruits such as cherries, plums, gages, etc., have different requirements and should not be pruned during the dormant months) should be carried out during the dormant winter months between November and March (or June and September in the southern hemisphere) and during the early years of the tree's life to develop a strong framework capable of bearing the weight of the crops that will be borne in later years. This involves hard pruning, although in later years pruning will be lighter and carried out to encourage fruiting.

One year old tree

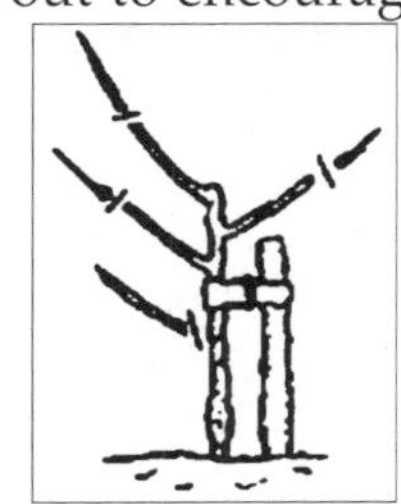

Two year old tree

Three year old tree

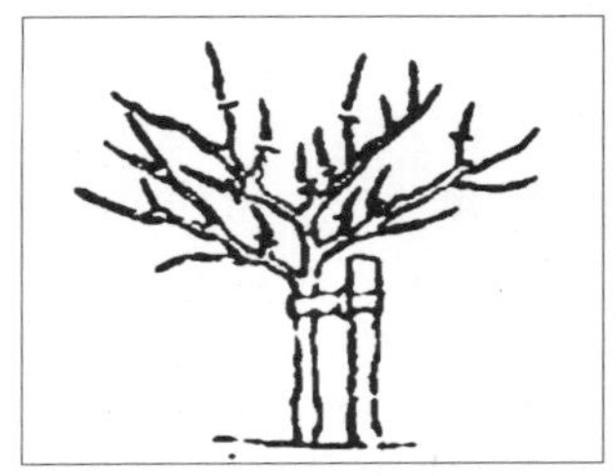

Four year old tree

### Maiden Tree

A maiden whip (that is, a one year old tree with no side shoots) should be pruned to a bud with two buds below it at about 80 cm from the ground immediately after planting to produce primary branches during the first growing season. A feathered maiden (that is, a one year old tree with several side branches) should have its main stem pruned back to three or four strong shoots at 80 cm from the ground. Side shoots should be shortened by two thirds of their length to an upward or outward facing bud. Lower shoots should be removed flush with the stem.

### Two Year

Remove any lower shoots and prune between three and five of the best placed shoots by half to an upwards or outwards facing bud to form what will become the tree's main structural branches. Remove any inwards facing shoots.

### Three Year

Prune the leading shoots of branches selected to extend the framework by half to a bud facing in the desired direction. Select four good laterals to fill the framework and shorten these by a half. Prune any remaining laterals to four buds to form fruiting spurs.

### Four Year

The tree will have begun to fruit and only limited formative pruning is now required. Shorten leaders by one third and prune laterals not required to extend the framework to four buds.

### Five year and Onwards

The tree is considered to be established and should be annually pruned as described in

### Pruning the Cropping Tree

Before pruning it is important to distinguish between spur bearing and tip bearing varieties. The former, which is the most common type, bear most of their fruit on older wood, and include apples such as *Cox's Orange Pippin, James Grieve* and *Sunset,* and pears such as *Conference, Doyenne du Commice* and *Williams Bon Chretien*. Tip bearers on the other hand produce most of their fruit buds at the tips of slender shoots grown the previous summer, and include the apples *Worcester Pearmain* and *Irish Peach,* and the pears such as *Jargonelle* and *Josephine de Malines*. There are basically three types of pruning that are applied once the main shape of the tree has been established.

## Renewal Pruning

- Spur pruning:Spur bearing varieties form spurs naturally, but spur growth can also be induced.
- Renewal pruning: This also depends on the tendency of many apple and pear trees to form flower buds on unpruned two year old laterals. It is a technique best utilised for the strong laterals on the outer part of the tree where there is room for such growth. Pruning long neglected fruit trees is a task that should be undertaken over a lengthy period, with not more than one third of the branches that require removal being taken each year.
- Regulatory pruning: This is carried out on the tree as a whole, and is aimed at keeping the tree and its environment healthy, eg, by keeping the centre open so that air can circulate, removing dead or diseased wood, preventing branches from becoming over crowded (branches should be roughly 50 cm apart and spurs not less than 25 cm apart along the branch framework), and preventing any branches from crossing.

## Pruning of tip Bearers

Tip bearers should be pruned lightly in winter using the regulatory system. any maiden shoots less than 25 cm in length should be left untouched as they have fruit buds at their tips. Longer shoots are spur pruned to prevent over-crowding and to stimulate the production of more short tip bearing shoots the following year. Branch leaders are 'tipped', removing the top three or four buds to a bud facing in the desired direction to make them branch out and so produce more tip bearing shoots.

# PLANT GROWTH

Meristems are regions of cell division and there are two types of plant meristems. An apical meristem is located at the tip of every shoot and root. As cells divide in these apical meristems, the shoots and roots elongate as cells are piled one on another. Behind the region of cell division is a region of cell differentiation, where cells enlarge and differentiate into various tissues. In the axil of each leaf is a small apical meristem called an axillary meristem that forms an axillary bud, which usually remains dormant until well after the subtending leaf is fully developed. An axillary bud may remain dormant or develop into a lateral branch or a flower.

There are two distinct layers of meristematic tissue within the stem or root responsible for secondary growth, the vascular cambium and the cork cambium. The vascular cambium is a cylinder of specialized cells, usually five to ten cells thick, running the length of the plant, including the roots, and is responsible for the radial growth of plant parts. Phloem cells are produced to the outside of the cambium and xylem cells are produced to the inside of the cambium. Downward transport of sugars, nutrients, and hormones from the top of the tree to the roots occurs in the phloem tissue. Xylem cells are tube shaped, become hollow and die to form a pipe-like system through which water, hormones and mineral nutrients move from the roots to the top of the tree. Most of the radial growth of woody plants is due to activity of the vascular cambium, but a small amount results from activity in another lateral meristem, the cork cambium, located outside the vascular cambium. The cork cambium together with the cork cells, constitute the periderm: a protective layer of suberized dead cork cells forming the bark. Suberization is the impregnation of cell walls of cork tissue with a fatty substance called suberin. Each season new layers of cells are produced and appear as growth rings when viewed in cross-section. Over time, the xylem cells at the centre of the trunk or limb are crushed and become nonfunctional as transport pipes, but they do provide structural support to hold the plant upright. While grafting it is important to line up the cambiums of the scion and the rootstock to ensure a successful graft union.

## BUDS

Buds are important to the vegetative and reproductive growth of trees. Fruit tree training and, to a lesser extent, pruning primarily involves bud manipulation. Buds are actually undeveloped shoots. When a vegetative bud is sliced longitudinally during the winter and viewed under magnification, the apical meristem at the tip, leaf primordial (developing leaves), axillary meristems, developing axillary buds, and procambial tissue (tissue that will develop into the cambium) are all visible.

Buds on fruit trees usually have about seven leaves and initial shoot elongation in the spring results from cell expansion. During late June and July some of the shoot apices will flatten out and develop into flower buds. Flower buds are actually modified shoots and the various flower tissues are actually modified leaves. Although the process of switching from vegetative to reproductive buds is not fully understood, hormones that can be influenced by environmental factors, stresses, and plant nutrition control the process. There are several things we can do to influence whether or not a bud becomes a flower bud or remains vegetative. In general, factors that favour rapid growth, such as high nitrogen levels in the shoot tissues, inhibit the development of flower buds. Applying growth-promoting plant growth regulators such as gibberellins usually inhibits flower-bud induction, whereas ethylene may promote flower-bud development. Mild stresses such as shoot bending and water stress may also promote flower-bud development.

Producing annual crops of high-quality fruit requires a balance between reproductive and vegetative growth. Fruit producers use various techniques, including pruning, branch bending, and plant growth-regulator sprays, to manipulate tree growth and flowering. Often these techniques affect bud dormancy, so knowledge of buds and bud dormancy is essential if we are to understand how pruning influences tree growth. It is also important to be able to identify the different types of buds on a tree, especially to distinguish between flower and vegetative buds. Buds may be classified as to location, contents, or activity. Classification by content Several types of buds commonly develop on fruit trees. Vegetative buds only develop into leafy vegetative shoots. Flower buds produce only flowers. Stone fruit trees (peach, nectarine, apricot, plum, and cherry) produce vegetative buds and flower buds. Apple and pear trees produce vegetative and mixed buds. Both leafy shoots and flowers emerge from mixed buds.

### Classification by Location

Terminal buds are located at the tip of a shoot. On stone fruit trees terminal buds are vegetative buds. Terminal buds on apple and pear trees are usually vegetative; however, some varieties such as Rome Beauty produce mixed buds terminally and are referred to as tip bearers or terminal bearers. Most mixed buds on apple and pear trees are formed terminally on short, less than sixinch, shoots that terminate with a rosette of leaves. These short shoots are called spurs. Lateral buds form in the axils of leaves and are often referred to as lateral buds or axillary buds. On stone-fruit trees, lateral buds may be either vegetative or flower. Nodes on one-year-old shoots may have one to three buds, some of which may be flower buds and others vegetative buds. Flower buds are larger with tips that are relatively round, whereas vegetative buds are small, narrow, and pointed. In the case of apple and pear trees, lateral buds on the previous season's growth are usually vegetative. However, lateral buds on some varieties, especially on the dwarfing rootstocks, may be mixed buds.

### Classification by Arrangement on the Stem

The bud arrangement influences the arrangement of a fruit tree's branches and thus the tree's shape and how easy it is to manage. A node is the joint on a stem where a leaf is or was attached. Axillary buds are located in the axis above where a leaf is attached to the stem. In apples there is usually only one leaf per node, whereas three leaves often arise from a node on peach shoots. When a leaf falls in the autumn, a leaf scar remains just below the axillary bud. Buds are opposite when there are two at the same node but on opposite sides of the stem. Forsythia is an example of a plant with opposite buds. Buds are alternate when there is only one from each node and no one bud is on the same side of the stem as the one next above or below it. Deciduous fruit trees have buds that spiral along a shoot. The spiraling three-dimensional arrangement of leaves around a stem is known as Phyllotaxy and is expressed as a fraction, where the numerator is the number of turns to get to a leaf directly above another and the denominator is the number of buds passed.

### Classification by Activity

Buds are dormant when they are not visibly growing. When shoots develop around large pruning cuts, they usually are sprouting from dormant buds. Adventitious buds form irregularly on older portions of a plant and not at the stem tips or in the leaf axils. They form on parts of the root or stem that have no connection to the apical meristems. They may originate from either deep or peripheral tissues. For example, shoots often arise from adventitious buds growing from callus tissue around wounds. Root suckers (vigorous upright shoots developing from the roots) develop from adventitious buds on the roots.

## FRUIT DEVELOPMENT

Ethylene production is largely pre-determined in both time and amount by the genetics of the fruit and, depending on when the fruit blossoms and the climate, will determine to a large degree the development of the respiratory climacteric. During fruit development, respiration (*i.e.*, the generation of carbon dioxide), which is a measure of metabolic activity, declines gradually throughout the season until several weeks before it ripens where it reaches what is known as a preclimacteric minimum. At this point, the metabolic functions of the fruit are in a near resting stage in preparation for a burst of metabolic activity signifying ripening. During ripening both carbon dioxide and ethylene increase significantly. The main developmental stage of the fruit is referred to as maturation, in which photosynthate is converted to starch. The ripening phase is when the starch is converted to sugar. Senescence is that stage in which the membrane functions break down due to degradation of lipid bilayers leading to cell damage and necrosis.

Optimum harvest is a subjective measurement defined as a fruit with good keeping quality and good eating quality. If picked too early during the maturation stage, insufficient starch will be converted to sugar and the fruit will keep well enough, but the eating quality will be poor. On the other hand, if the fruit is picked too late, there will be insufficient starch and acid reserves for metabolic maintenance in storage, but the eating quality will be good. Therefore, the optimum timing is critical to the proper storage (and marketing) of apples. Researchers throughout the world have spent decades defining the optimum time of harvest for long-term CA storage and yet, as new varieties enter the marketplace, optimums must be redefined for each individual cultivar.

### ETHYLENE PRODUCTION

After the respiratory preclimacteric minimum comes the respiratory climacteric. This is the point at which the fruit will generate a high, sustained respiration and ethylene biosynthesis. It occurs at different times for different cultivars. For example, a Yellow Transparent apple, which ripens in early July, has a very quick respiratory climacteric and may ripen within a few days. In fact, there are times when I have gone out one day and seen green fruit on the tree and two days later, these fruit are yellow and on the orchard floor. The respiratory climacteric of this apple is quick and peak is high. On the other hand, a Delicious apple, which ripens sometime in mid-September, has a somewhat longer climacteric period.

This fruit may develop a respiratory climacteric over a period of 10 days, for example. This is advantageous to a grower because it allows time to harvest all of the fruit without being so concerned that the fruit will ripen early and even drop from the tree. Further, there are cultivars that ripen later, such as Fuji, Braeburn, or Granny Smith. These cultivars in turn have an even broader climacteric and the peak of respiration (when similarly measured) is lower than in the previous scenarios. Consider, for example, ethylene from the cultivar Braeburn. In one of our trials, fruit was picked over a period of 10 weeks beginning around August 15. If we measure the ethylene from the core by sampling with a syringe 24 hours after the fruit has been harvested, we see a very low ethylene production within the fruit tissue—generally less than 2 ppm. On the other hand, if we wait for four days before sampling the core gases of the fruit, we see significantly higher levels of ethylene reaching about 10 ppm at proper harvest. Waiting for seven days after the fruit is harvested before sampling, results in ethylene values in the 60-70 ppm range at harvest time.

This suggests that delaying cooling (storage) of the fruit after it has been picked hastens the ripening process and, therefore, is deleterious to the condition of the fruit out of storage. This is important because often times a grower will harvest his fruit in bins and it may stay in his orchard or on the warehouse receiving dock before it is properly cooled. The longer the delay between harvest and storage the shorter will be the storage life and the poorer the condition of the fruit after it is removed from storage.

## SAMPLING FRUIT ETHYLENE

Clearly, the accurate measurement of ethylene in apples is vital to their proper storage and therefore their condition out of storage. But with so many fruit (100 million boxes of roughly 100 apples each in Washington State alone, and at least 4 times that throughout the world) the problem has become not the accuracy of sampling, but the unfathomable number of samples to assess. Early efforts to analyse gas samples used a bulk approach. A bushel of apples would be placed in a large glass container and a sample of the headspace gas would be withdrawn and sampled. If each apple were very similar, this approach would accurately depict the physiological state of the individual fruit and, therefore, the proper regime for optimum storage. This is hardly the case. Data obtained from this method was qualitative, however, because there always exists the possibility that one apple could be riper than the others, thereby generating most all the ethylene. Thus, although all of the other fruit could have extremely low ethylene levels, the single mature apple would indicate the "average" ethylene was quite high. Because there are differences due to size, colour, tree position, stress, variety, rootstock, soil, and a host of other factors, the best that can be said of this method is that it would always overestimate the average ethylene production. Individual apples were then evaluated for ethylene production. The most common method was to insert an 18-gauge needle into the apple through the calyx end extending through to the core. A 0.5 to 1.0 mL sample of gas could be withdrawn and subsequently analysed by gas chromatography. (Routine analysis of ethylene by megabore gas chromatography is sensitive to the ten parts per billion range.) By increasing the number of samples and examining the variability among samples, one could obtain a more accurate picture of the physiological state of the part of the orchard sampled.

Harvest decisions could then be based on both the amount of ethylene being produced as well as on the degree of variation of samples. High ethylene blocks (experimental sites within and orchard) with high variability might well indicate that very few fruit are producing significant amounts, whereas, high ethylene blocks with little variability might indicate the entire block is ripening more evenly. The concern with this method lies in not knowing the physiological meaning of a gas sample taken from the fruit centre. Again, there are differences due, for example, to size of seed cavity, amount of water in the fruit, amount of interstitial space, cell density and the amount of wounding from the needle insertion. Generally, any damage to the fruit tissues induces ethylene biosynthesis.

The method presently under design of assessing individual fruit ethylene is to place each apple in a small, 3-litre Plexiglas chamber with a low flow of scrubbed air. Micro-processors may sample the effluent gas as often as required 24 hours a day. This takes a single person about 1 hour to set up the system capable of analysing gas composition of hundreds of apples—a process that manually takes about 2 to 3 minutes per sample. The greater number of samples increases the likelihood that apples will reach the consumer with optimum quality. What is needed is a method to assess quickly and accurately the ethylene generated by the whole apple, or one that uses part of the fruit that is most directly correlated with the physiological stage of fruit maturity. As for sampling a portion of tissue, it almost seems as though Heisenberg's principal is at work here in that any invasive action to the fruit will change its physiological stage and response.

Perhaps the closest thing the apple industry has to assessing individual fruit maturity is in the fruit sorting process. In modern fruit packing warehouses individual fruit are mechanically sorted into weight categories and optically scanned for colour sorting. Not far away are sensors for inline determination of firmness, sugar content and internal disorders. Practically everything can be determined about an individual fruit without damaging it. So far, however, only ethylene gives a meaningful picture of the physiological stage of ripeness. Thus, the problem remains; how does the apple industry nondestructively assess ethylene of 10 billion apples.

## INHIBITING ETHYLENE BIOSYNTHESIS

Abbott Laboratories has recently developed a new chemical tool for use in apple production. AVG (aminoethoxyvinylglycine) was discovered in 1976 by researchers at Hoffmann-LaRoche, but it wasn't until the last few years that it was made commercially available. This compound is a substituted amino acid produced by a streptococcal mold and is specifically targeted to inhibit the enzymatic production of ACC, which is the precursor of ethylene. In 1980, Williams in Wenatchee showed that AVG could be used to inhibit internal ethylene production thereby reducing fruit abscission, but could not be used to compensate for external ethylene, no matter the source. In 1981, we did some experiments on Golden Delicious apples, in which we took fruit that was harvested at optimum harvest time and dipped them in solutions containing 0, 100, 200, or 400 ppm AVG (Curry, *et al.*). These fruit were stored in Plexiglas chambers and the ethylene and carbon dioxide were measured daily for a period of seven weeks. Ater about 10 days, the untreated fruit showed an increased rate of respiration that coincided with the climacteric rise.

For this treatment, ethylene exceeded 50 ppm for the remainder of the study. Fruit dipped in 100 ppm AVG had a respiratory climacteric delayed about six weeks. At seven weeks, ethylene levels were about 20 ppm and by the ninth week, ethylene had not exceeded 30 ppm. At 200 ppm, it took approximately nine weeks to see a slight increase in carbon dioxide and at that time, ethylene was approximately 4 ppm. Lastly, at 400 ppm, Golden Delicious apples dipped after harvest never showed an increase in respiration and the ethylene levels were near zero, even at nine weeks after harvest. At the end of the nine weeks, the control apples were yellow and shriveled whereas apples dipped in AVG were still green and noticeably firmer. In another trial, Williams (1980) treated Delicious apples with pre-harvest applications of AVG approximately three days before harvest. These fruit were stored in regular cold storage and after three weeks, showed no difference from those that were not treated with AVG. By 16 weeks, however, differences began to occur. Firmness of untreated apples was 13.9 pounds, whereas those treated with 450 ppm AVG three days before harvest showed an average firmness of 15.2. Thirty weeks after harvest control fruit had a firmness of 12.8 pounds, whereas those treated with AVG were a full pound firmer.

## PHYSIOLOGY OF PRUNING FRUIT TREES

Woody plants are pruned to maintain a desired size and shape and to promote a certain type of growth. Ornamental plants are pruned to improve the aesthetic quality of the plant, but fruit trees are pruned to improve fruit quality by encouraging an appropriate balance between vegetative (wood) and reproductive (fruiting) growth. Annual pruning of fruit trees always reduces yield, but enhances fruit quality.

Pruning increases fruit size because excess flower buds are removed and pruning encourages the growth of new shoots with high-quality flower buds. Pruning improves light penetration into the canopy, and light is required for flower-bud development, fruit set and growth, and red colour development. Pruning also makes the canopy more open and improves pest control by allowing better spray penetration into the tree; air movement throughout the canopy is increased, which improves drying conditions and reduces severity of many diseases.

This publication describes why plants respond to pruning and other forms of plant manipulation used to train trees. This information applies to all plants, but application to fruit trees is emphasized. Pruning fruit trees is somewhat of an art based on an understanding of plant physiology and development. In other words, if we understand how plants grow and how they will respond to different types of plant manipulations, we can alter vegetative growth and fruiting to obtain trees and fruit with desirable characteristics.

A basic understanding of certain aspects of plant physiology is a prerequisite to understanding pruning. Unlike animals, plants continue to increase in size throughout their lives. There are only two ways plants can grow. Primary growth is the increase in length of shoots and roots, and is responsible for increases in canopy height and width. Secondary growth is the increase in thickness of stems and roots. Both types of growth require cell division followed by cell enlargement and differentiation.

## FRUIT TREE PROPAGATION

Fruit tree propagation is usually carried out through asexual reproduction by grafting or budding the desired variety onto a suitable rootstock. Perennial plants can be propagated either by sexual or vegetative means. Sexual reproduction occurs when male pollen from one tree fertilizes the ovules (incipient seeds) of the flower of another, stimulating the development of fruit. In turn this fruit contains a seed or seeds which, when germinated, will become a new specimen. However, the new tree will inherit many of the characteristics of both its parents, and it will not grow 'true' to the variety from which it came. That is, it will be a fresh individual with many unpredictable characteristics of its own. Although this is desirable in terms of increasing biodiversity and the richness of the gene pool (such sexual recombination is the source of most new cultivars), only rarely will such fruit trees be directly useful or attractive to the tastes of humankind. A tendency to revert to a wild-like state is common. Therefore, from the orchard grower or gardener's point of view, it is preferable to propagate fruit cultivars vegetatively in order to ensure reliability. This involves taking a cutting (or scion) of wood from a desirable parent tree which is then grown on to produce a new plant or 'clone' of the original. In effect this means that the original Bramley apple tree, for example, was a successful variety grown from a pip, but that every Bramley since then has been propagated by taking cuttings of living matter from that tree, or one of its descendants.

**Methods**

The essentials of our present methods of propagating of fruit trees date from pre-Classical times. Grafting as a technique was first developed in China from where it was imported to Greece and Rome. Classical authors wrote extensively about the technical skills of fruit cultivation, including grafting techniques and rootstock selection. The oldest surviving named varieties of fruits date from classical times.

The simplest method of propagating a tree asexually is rooting. A cutting (a piece of the parent plant) is cut and stuck into soil. Artificial rooting hormones are sometimes used to assure success. If the cutting does not die of desiccation first, roots grow from the buried portion of the cutting to become a complete plant. Though this works well for some plants (such as figs and olives), most fruit trees are unsuited to this method. Root cuttings (pieces of root induced to grow a new trunk) are used with some kinds of plants. This method also is suitable only for some plants.

A refinement on rooting is layering. This is rooting a piece of a wood that is still attached to its parent and continues to receive nourishment from it. The new plant is severed only after it has successfully grown roots. Layering is the technique most used for propagation of clonal apple rootstocks. The most common method of propagating fruit trees, suitable for nearly all species, is grafting onto rootstocks. These are varieties selected for characteristics such as their vigour of growth, hardiness, soil tolerance, and compatibility with the desired variety that will form the aerial part of the plant (called the scion). For example, grape rootstocks descended from North American grapes allow European grapes to be grown in areas infested with *Phylloxera,* a soil-dwelling insect that attacks and kills European grapes when grown on their own roots. Grafting is the process of joining these two varieties, ensuring maximum contact between the cambium tissue (that is, the layer of growing plant material just below the bark) of each so that they grow together successfully. Two of the most common grafting techniques are 'whip and tongue', carried out in spring as the sap rises, and 'budding', which is performed around July and August.

## Apple Rootstocks

Another reason for grafting onto rootstocks is that this enables the grower to determine the tree's eventual size. Apple tree size classes number one to ten in increasing height and breadth. A "1" is a dwarf which can be productive and as short as three (3) feet with proper pruning. A "10" is the standard sized tree with no dwarfing and will grow to twenty(20) or more feet tall and wide, dependent upon the variety chosen. In general the class range is (1) 10-20 per cent of full size, (2) 20-30 per cent, (3) 30-40 per cent and so forth to size 10 which is 100 per cent of full size.

Apple tree rootstocks are referred to by numbers prefixed by letters indicating the developer of the rootstock. "M" designates Malling series developed stocks. East Malling Research is a pioneer in the development of dwarfing rootstocks. East Malling Research Station in Kent, England collected clones of the Paradise stocks from France in 1912 from which 24 "M" were designated with no particular order to the rootstock characteristics other than where they were located in the garden at the time the numbers were assigned. In other words, M.2 is larger tree than M.9 while M.27 is smaller than M.26.

"MM" designates Malling-Merton stocks developed from joint breeding programme by John Innes Institute, in Merton, England, and East Malling Research Station in the early 1950s. The "MM" series was developed primarily to provide resistance to Woolly Apple Aphid (Eriosomatinae) infestation.

"EMLA" designates East Malling/Long Ashton research stations who took the "M" stocks and developed virus free versions. *E.g.*, EMLA 7 is M 7 with a guaranteed virus free stock. EMLA characteristics are often different from the parent "M" rootstock. Note that nearly all the apple rootstocks in the industry are now virus free.

"CG" or "G" designates Cornell-Geneva stocks which are those developed via the Cornell and USDA collaboration at the New York Agricultural Experiment Station in Geneva, NY. The "G" is the old designation. All newer stocks are "CG" followed by numbers that actually provide some information about the stock. As one might surmise, this is a huge improvement in the classical naming scheme which has no identification method at all.

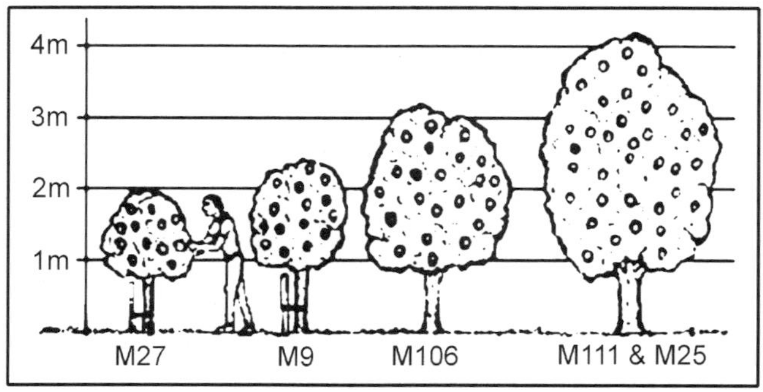

Diagram illustrating comparative sizes of apple trees depending on rootstock grafting

- M.27 Malling 27: A very dwarfing rootstock. Unless the central leader is supported, the tree will be very small. Often only used as an intermediate stem piece on MM.106 or MM.111. If handled and spaced properly, it can be a very productive stock for a vertical axe system. Trees can be grown three to four feet tall and produce about 45 fruit, roughly 2 pecks, depending on fruit cultivar.
- M.9: Very dwarfing-Reaches a height of 8 to 10 ft (2.4 to 3.0 m), coming into fruit after 3–4 years, reaching full capacity of 50 to 65 pounds (23 to 29 kg) after 5 to 6 years. It will grow under average soil conditions, but needs a good rich soil to thrive. A good choice where space is limited and fertility is high. Permanent staking is required, as is routine feeding and watering. Trees on this rootstock always require leader support. The rootstock is very susceptible to fire blight and can develop burr knots.
- G.41 Geneva 41, released in 2005, produces trees the size of M.9. The rootstock was developed from a cross between M.27 and Robusta 5 made in 1975. Resistant to Crown|Collar|Root rot(Phytopthora) and fire blight.
- M.26: Dwarfing-Similar to M9 in effect, although somewhat more vigorous and generally stronger, with a higher expected eventual yield of 65–75 pounds (29–34 kg) and height of 8 to 10 ft (2.4 to 3.0 m). A good choice where soil quality is average and compact growth is required. Comes into fruit after 3-4 years, reaching full cropping capacity after 5 to 6 years. Staking needed for first five years of its life. It is susceptible to collar rot and fire blight and should not be planted in a wet site. Certain varieties when grafted onto this rootstock may exhibit signs of graft union incompatibility(*i.e.,* the union breaks).
- G.11 Geneva 11 is the second release of the Cornell breeding programme similar in size to M.26 but more productive. Has the advantage of being resistant to fire blight and crown rot as well as only rarely producing suckers or burr knots.
- G.202 Geneva 202(CG 5202) is a semidwarfing rootstock that produces a tree in class 5 slightly larger than M.26 and is more productive than M.26. It was developed from a cross of M.27 and Robusta 5 to be fire blight and Phytopthora resistant as well as having resistance to woolly apple aphids. In a 9-year study with the scion cultivar of the "Liberty" apple, G.202 was about 50 percent smaller than M.7 but had much greater production efficiency.
- M.7 Malling 7 rootstock produces a semidwarf tree of Class 6 that is freestanding in deep well drained soils but in rocky, steep, or

shallow soils, it tends to lean. The rootstock may sucker profusely and is susceptible to collar rot(Phytopthora).

- MM.106: Semi-dwarfing-Sometimes referred to as semi-vigorous, this is the most widely used of rootstocks. It is probably the best choice for the average garden under average conditions, being tolerant of a wide range of soils, and producing a tree with an eventual size of 14 to 18 ft (4.3 to 5.5 m). Trees on this stock begin producing fruit within three to four years, and yield 90 to 110 pounds (41 to 50 kg) after some seven or eight years. MM106 is very suitable for use with weaker varieties that would produce under sized bushes with more dwarfing rootstocks. Can be trained as a half standard tree, but is rather too vigorous for cordons unless the soil is poor. Requires staking for the first five years or so of its life. Trees on MM.106 are highly susceptible to collar rot especially when planted in soils that remain wet(poor percolation).
- M.111 : Vigorous-Not generally suitable for garden scale growing, being both too large and spreading (18-25'), and too slow to come into cropping. They are however suitable for growing as specimen standards in the large garden, or for producing medium sized bushes on poorer soils. Begins to fruit after six or seven years, reaching full capacity of 160 to 360 lb (73 to 160 kg) after eight to nine years. It is the most cold-hardy rootstock readily available. Planting depth of this rootstock is critical. The union should be no higher than 1 to 2 inches above the final soil line.
- M.25: Very vigorous-Suitable for a grassed orchard, and to grow on as a full standard. Plant 20 ft (6.1 m) apart, makes a tree of 15 to 20 ft (4.6 to 6.1 m) or more height and spread, eventually yielding 200 to 400 lb (91 to 180 kg) per tree. This rootstock is primarily used in UK and is rarely seen in USA where M.111 is used for this size tree.
- Seedling: Very vigorous trees produced on a rootstock grown from seed. There is greater variability than with the vegetatively propagated rootstocks. Apples used for production of seedling rootstocks include 'Dolgo' and 'Antonovka', which are both extremely hardy and vigorous.

That is only a sample of some of the more important current apple rootstocks that are available. There are at least a hundred more that have been developed to either provide enhancement or prevent potential damage from one kind of pest or another. The problem with growing fruit trees, especially apple trees, is that they are subject to many different types of damage from bacteria, fungi and insects. The general approach of the commercial industry has been to use as many chemicals as necessary to insure attractive and marketable fruit. The attitude, still prevalent, has been "Who cares? Nobody eats a tree!" but as environmental problems increase and the general public pushes for low or no-spray fruit, there has become a commercial need for fruit that does not require such intensive spray programmes. This is being achieved, albeit slowly, by rootstocks and trees that are bred to have natural disease and pest resistance.

The Malling series and clones have been standard rootstocks for apples for many years and remain the standard "workhorses" for the commercial industry (in USA). However, since most of them are susceptible to disease some Malling rootstocks are being replaced by new breeds, including the Cornell-Geneva series. One of the newest rootstocks, only released commercially in 2004, is CG5202 (G.202) which adds resistance to the woolly apple aphid(WAA) for the "CG" series of stocks which already has resistance to the major problems preventing quality production of apples utilizing organic control systems. Combined with highly resistant trees such as "Liberty" it is showing great potential. *That leads to another characteristic of rootstocks that is or can be bred into them*: environmental adaptability. This may be tolerance to wet|dry soil conditions, acidity|alkalinity of soil or even hot|cold air temperature. Some new rootstocks based on Siberian Crab Apple are being used in colder areas for more cold tolerance.

The ability of rootstocks to modify or augment characteristics of fruit trees is limited and often disappointing in the final results. It takes ten years to get a full picture of the effects of any one rootstock so a rootstock that appears promising in the first five years of a trial may fail in the last five years. The Mark (apple) rootstock was such a stock and has now fallen mostly into disfavour. Another, the G.30, has proved to be an excellent stock for production but it was only after a number of years of trials that it was found to be somewhat incompatible with "Gala" apple(and possibly others) so that it is now recommended to be staked and wired.

To get a clear picture and push the industry forward, a consortium was founded and the so-called "NC-140" trials of rootstocks began. These test many pome rootstocks in many different sites across the USA and thereby provide growers, be they backyard or commercial, a clearer picture of what to expect when growing fruit trees on specific stock, in specific planting methods in their specific area of the USA. As one can imagine, this has the potential for a large economic benefit to both growers and consumers as well as going a long way to eliminating the need to spray pesticides as frequently as is currently required.

### Pear Rootstocks

Pears are usually grafted onto quince rootstocks, which produce small to medium sized trees. Some varieties however are not compatible with quince, and these require double working. This means that a piece of pear graft-work compatible with both the quince rootstock and the pear variety is used as an intermediate between the two. If this is not done the pear and the rootstock could eventually separate at the graft. Varieties that require double working include 'Bristol Cross', 'Dr Jules Guyot', 'Doyenné d' été' and 'Williams Bon Chrétien'.

- *Quince C:* Moderately vigorous- Makes a bush pear tree about 8 to 18 ft (2.4 to 5.5 m) tall, bearing fruit within four to eight years. Suitable for highly fertile soils and vigorous varieties, but not where conditions are poor. Used for bush, cordon and espalier growing. Old stocks of Quince C may be infected with a virus, so care should be taken to obtain certified virus free stock. If in doubt, use Quince An as there is not a great amount of difference in vigour between the two.
- *Quince A:* Medium vigour- Slightly more vigorous than Quince C, this is the most common variety upon which pears are grafted. Bears fruit between four to eight years, making a tree of some 10 to 20 ft (3.0 to 6.1 m) in height and spread. Suitable for all forms of pear trees except standards.

Pear stock: Very vigorous- Pears grafted onto pear rootstocks make very large standard trees, not suitable for most gardens.

### Cherries

Until the 1970s, cherries were grown of the vigorous Malling F12/1, Mazzard (*Prunus avium*), or Maheleb rootstocks, which required much space and time before cropping began, thus the growing of cherries was not a realistic option on a garden scale. The introduction of the rootstock 'Colt' enables trees reaching a maximum height of 12 to 15 ft (3.7 to 4.6 m) to be grown, and if trained as a pyramid it is possible to restrict growth to about 10 ft (3.0 m). The popular sweet variety 'Stella' could even be grafted onto a 'Colt' rootstock and successfully grown in a pot on the patio.

### Plums

*Plum rootstocks include;*

- Pixy- A dwarfing rootstock, suitable for bush trees planted 8-10 (3 m) apart.
- St. Julien A- A semi vigorous rootstock suitable for bush and half standards planted 12 to 15 ft (3.7 to 4.6 m) apart. Also suitable for peaches, nectarines and apricots.
- Brompton or Myrobalan B- Suitable for half standards planted 18 to 22 ft (5.5 to 6.7 m) apart. Also suitable for peaches, nectarines and apricots.
- Myro-29C- Semi-dwarf rootstock. Shallow, vigorous, good choice for hard soils. Somewhat drought tolerant.

- Citation- Semi-dwarf rootstock. Shallow, vigorous, good choice for hard soils. Prefers a wetter soil.

**Own-Root Fruit Trees**

Some species of fruit are commonly grown on their own roots; new plants are propagated by rooting, layering, or modern tissue-culture techniques. In these cases there are may be no great advantages to using a special rootstock or improved rootstocks are not available. Fig, filbert, olive, pomegranate, gooseberry, bramble, and other fruits are commonly grown without any special rootstock.

# 9

# Fruit Quality and Nutrition

## INTRODUCTION

Fruit quality goes back to tree stock, growing practices and weather conditions. Closer to the shipper and processor, however, are the degrees of maturity and ripeness when picked and the method of picking or harvesting. There is a distinction between maturity and ripeness of a fruit. Maturity is the condition when the fruit is ready to eat or if picked will become ready to eat after further ripening. Ripeness is that optimum condition when colour, flavour and texture have developed to their peak. Some fruit is picked when it are mature but not yet ripe. This is especially true of very soft fruit like cherries and peaches, which when fully ripe are so soft as to be damaged by the act of picking itself. Further, since many types of fruit continue to ripen off the tree, unless they were to be processed quickly, some would become overripe before they could be utilised if picked at peak ripeness. From a technological point of view, fruit characterisation by species and varieties is performed on the basis of physical as well chemical properties: shape, size, texture, flavour, colour/ pigmentation, dry matter content (soluble solids content), pectic substances, acidity, vitamins, etc.

These properties are directly correlated with fruit utilisation. The proper time to pick fruit depends upon several factors; these include variety, location, weather, ease of removal from the tree (which change with time), and purpose to which the fruit will be put. For example, oranges change with respect to both sugar and acid as they ripen on the tree; sugar increases and acid decreases. The ratio of sugar to acid determines the taste and acceptability of the fruit and the juice. For this reasons, in some countries there are laws that prohibit picking until a certain sugar-acid ratio has been reached. In the case of much fruit to be canned, on the other hand, fruit is picked before it is fully ripe for eating since canning will further soften the fruit.

## QUALITY MEASUREMENTS

Many quality measurements can be made before a fruit crop is picked in order to determine if proper maturity or degree of ripeness has developed. Colour may be measured with instruments or by comparing the colour of fruit on the tree with standard picture charts. Texture may be measured by compression by hand or by simple type of plungers. As fruit mature on the tree its concentration of juice solids, which are mostly sugars, changes. The concentration of soluble solids in the juice can be estimated with a refractometre or a hydrometre. The refractometre measures the ability of a solution to bend or refract a light beam which is proportional to the solution's concentration. A hydrometre is a weighted spindle with a graduated neck which floats in the juice at a height related to the juice density.

The acid content of fruit changes with maturity and affects flavour. Acid concentration can be measured by a simple chemical titration on the fruit juice. But for many fruits the tartness and flavour are really affected by the ratio of sugar to acid. Percentage of soluble solids, which are largely sugars, is generally expressed in degrees Brix, which relates specific gravity of a solution to an equivalent concentration of pure sucrose. In describing the taste of tartness of several fruits and fruit juices, the term "sugar to acid ratio" or "Brix to acid ratio" are commonly used. The higher the Brix the greater the sugar concentration in the juice; the higher the "Brix to acid ratio" the sweeter and lees tart is the juice.

# HARVESTING AND PREPROCESSING

## HARVESTING

The above and other measurements, plus experience, indicate when fruit is ready for harvesting and subsequent processing. A large amount of the harvesting of most fruit crops is still done by hand; this labour may represent about half of the cost of growing the fruit. Therefore, mechanical harvesting is currently one of the most active fields of research for the agricultural engineer, but also requires geneticists to breed fruit of nearly equal size, that matures uniformly and that is resistant to mechanical damage.

A correct manual harvesting includes some simple but essential rules:

- The fruit should be picked by hand and placed carefully in the harvesting basket; all future handling has to be performed carefully in order to avoid any mechanical damage;
- The harvesting basket and the hands of the harvester should be clean;
- The fruit should be picked when it is ready to be able to be processed into a quality product depending on the treatment which it will undergo.

It is worth emphasising the fact that the proximity of the processing centre to the source of supply for fresh raw materials presents major advantages; some are as follows:

- Possibility to pick at the best suitable moment;
- Reduction of losses by handling/transportation;
- Minimises raw material transport costs;
- Possibility to use simpler/cheaper receptacles for raw material transport.

Once it has left the tree, the organoleptic properties, nutritional value, safety and aesthetic appeal of the fruit deteriorates in varying degrees. The major causes of deterioration include the following:

- Growth and activity of micro-organisms;
- Activities of the natural food enzymes;
- Insects, parasites and rodents;
- Temperature, both heat and cold;
- Moisture and dryness;
- Air and in particular oxygen;
- Light and
- Time.

## Reception—Quality and Quantity

*Fruit reception at the processing centre is performed mainly for following purposes:*

- Checking of sanitary and freshness status;
- Control of varieties and fruit wholeness;
- Evaluation maturity degree;
- Collection of data about quantities received in connection to the source of supply: outside growers/farmers, own farm.

Variety control is needed in order to identify that the fruit belongs to an accepted variety as not all are suitable for different technological processes. Fruit maturity degree is significant as industrial maturity is required for some processing/preservation methods while for others there is the need for an edible maturity when the fruit has full taste and flavour. Special attention is given to size, appearance and uniformity of fruit to be processed, mainly in the form of fruit preserved with sugar using whole/half fruits ("with fruit pieces"). Some laboratory control is also needed, even if it not easy to precisely establish the technological qualities of fruit because of the absence of enough reliable rapid analytical methods able to show eventual deterioration. The only reliable method for evaluating the quality is the combination of data obtained through organoleptic/taste controls and by simple analytical checks which are possible to perform in a small laboratory: percentage of soluble solids by refractometre, consistency/texture measured with simple penetrometres, etc.

## Temporary Storage before Processing

This step has to be as short as possible in order to avoid flavour losses, texture modification, weight losses and other deterioration that can take place over this period.

*Some basic rules for this step are as follows:*

- Keep products in the shade, without any possible direct contact with sunlight;
- Avoid dust as much as possible;
- Avoid excessive heat;
- Avoid any possible contamination;
- Store in a place protected from possible attack by rodents, insects, etc.

Cold storage is always highly preferred to ambient temperature. For this reason a very good manufacturing practice is to use a cool room for each processing centre; this is very useful for small and medium processing units as well.

## Washing

Harvested fruit is washed to remove soil, micro-organisms and pesticide residues. Fruit washing is a mandatory processing step; it would be wise to eliminate spoiled fruit before washing in order to avoid the pollution of washing tools and/or equipment and the contamination of fruit during washing. Washing efficiency can me gauged by the total number of micro-organisms present on fruit surface before and after washing - best result are when there is a six fold reduction. The water from the final wash should be free from moulds and yeast; a small quantity of bacteria is acceptable. Fruit washing can be carried out by immersion, by spray/showers or by combination of these two processes which is generally the best solution: pre-washing and washing.

*Some usual practices in fruit washing are:*

- Addition of detergents or 1.5 per cent HCl solution in washing water to remove traces of insect-fungicides;
- Use of warm water (about 50°C) in the pre-washing phase;
- Higher water pressure in spray/shower washers.

Washing must be done before the fruit is cut in order to avoid losing high nutritive value soluble substances (vitamins, minerals, sugars, etc.).

## Sorting

*Fruit sorting covers two main separate processing operations:*

- Removal of damaged fruit and any foreign bodies (which might have been left behind after washing);
- Qualitative sorting based on organoleptic criteria and maturity stage.

Mechanical sorting for size is usually not done at the preliminary stage. The most important initial sorting is for variety and maturity. However, for some fruit and in special processing technologies it is advisable to proceed to a manual dimensional sorting (grading).

### *Trimming and Peeling (Skin Removal)*

This processing step aims at removing the parts of the fruit which are either not edible or difficult to digest especially the skin. Up to now the industrial peeling of fruit and vegetables was performed by three procedures:

- Mechanically;
- By using water steam;
- Chemically; this method consists in treating fruit and vegetables by dipping them in a caustic soda solution at a temperature of 90 to 100° C; the concentration of this solution as well as the dipping or immersion time varying according to each specific case.

### Cutting

This step is performed according to the specific requirements of the fruit processing technology.

### Heat Blanching

Fruit is not usually heat blanched because of the damage from the heat and the associated sogginess and juice loss after thawing. Instead, chemicals are commonly used without heat to inactivate the oxidative enzymes or to act as antioxidants and they are combined with other treatments.

### Ascorbic/Citric Acid Dip

Ascorbic acid or vitamin C minimises fruit oxidation primarily by acting as an antioxidant and itself becoming oxidised in preference to catechol-tannin compounds. Ascorbic acid is frequently used by being dissolved in water, sugar syrup or in citric acid solutions. It has been found that increased acidity also helps retard oxidative colour changes and so ascorbic acid plus citric acid may be used together. Citric acid further reacts with (chelates) metal ions thus removing these catalysts of oxidation from the system.

### Sulphur Dioxide Treatment

*Sulphur dioxide may function in several ways:*

- Sulphur dioxide is an enzyme poison against common oxidising enzymes;
- It also has antioxidant properties; *i.e.,* it is an oxygen acceptor (as is ascorbic acid);

- Further $SO_2$ minimises non enzymatic Maillard type browning by reacting with aldehyde groups of sugars so that they are no longer free to combine with amino acids;
- Sulphur dioxide also interferes with microbial growth.

In many fruit processing pre-treatments two factors must be considered:

- Sulphur dioxide must be given time to penetrate the fruit tissues;
- $SO_2$ must not be used in excess because it has a characteristic unpleasant taste and odour, and international food laws limit the $SO_2$ content of fruit products, especially of those which are consumer oriented (*e.g.* except semi-processed products oriented to further industrial utilisation).

Commonly a 0.25 per cent solution (except for semi-processed fruit products which are industry oriented and use a 6 per cent solution) of $SO_2$ or its $SO_2$ equivalent in the form of solutions of sodium sulphite, sodium bisulphite or sodium/potassium metabisulphite are used. Fruit slices are dipped in the solution for about two to three minutes and then removed so as not to absorb too much $SO_2$. Then the slices are allowed to stand for about one to two hours so that the $SO_2$ may penetrate throughout the tissues before processing. Sulphur dioxide is also used in fruit juice production to minimise oxidative changes where relatively low heat treatment is employed so as not to damage delicate juice flavour. Dry sulphuring is the technological step where fruit is exposed to fumes of $SO_2$ from burning sulphur or from compressed gas cylinders; this treatment could be used in the preparation of fruits (and some vegetables) prior to drying/dehydration.

## Sugar Syrup

Sugar syrup addition is one of the oldest methods of minimising oxidation. It was used long before the causative reactions were understood and remains today a common practice for this purpose. Sugar syrup minimises oxidation by coating the fruit and thereby preventing contact with atmospheric oxygen. Sugar syrup also offers some protection against loss of volatile fruit esters and it contributes sweet taste to otherwise tart fruits. It is common today to dissolve ascorbic acid and citric acid in the sugar syrup for added effect or to include sugar syrup after an $SO_2$ treatment.

## Fresh Fruit Storage

Some fruit species and specially apples and pears can be stored in fresh state during cold season in some countries' climatic conditions. Fruit for fresh storage have to be autumn or winter varieties and be harvested before they are fully mature. This fruit also has to be sound and without any bruising; control and sorting by quality are mandatory operations. Sorting has to be carried out according to size and weight and also by appearance; fruit which is not up to standard for storage will be used for semi-processed product manufacturing which will be submitted further to industrial processing. Harvested fruit has to be transported as soon as possible to storage areas. Leaving fruit in bulk in order to generate transpiration is a bad practice as this reduces storage time and accelerates maturation processes during storage. In order to store large quantities of fruit, silos have to be built.

## Processing of Fruit Bars

The fruit bar processing method developed for FAO only involves a single major operation, which it drying the fruit pulp after mixing it with suitable ingredients. It can be used to produce mango, banana, guava or mixed fruit bars. A dual-powered dryer, working by solar energy during the day and by electric or steam power at night and on rainy days, with cross-flow movement of air and controlled temperature (from 55° C at the beginning of processing to a high of 70° C), is well suited for dehydration of the pulp to the desired moisture level of 15 to 20 per cent.

*Mango fruit bar* — Fully ripe mangoes are selected and washed in water at room temperature. The peeled fruit is cut into slices and passed through a helicoidal pulper to extract the pulp. The required amount of sugar to adjust the Brix (the unit measure for total solids in fruits) of the mixed pulp to 25 degrees Brix is then added. Two grams of citric acid per kilogram of pulp (or 20 ml of lime or lemon juice) are added to inhibit possible growth of micro-organisms during drying. The mixture is then heated for two minutes at 80° C and partially cooled; the heat treatment serves to inactivate the enzymes and destroy the micro-organisms. Potassium or sodium metabisulphite is added (two grams per kg of prepared mixture), so that the concentration of SO2 is 1000 ppm. The mixture is then transferred to stainless steel trays which have been previously smeared with glycerine (40 ml/$m^2$). Each tray must be loaded with 12.5 kg of mixture per square metre. Drying could be carried out by a dual-powered dryer for a total of 26 hours:

- 10 hours by solar energy at about 55° C and
- 16 hours by electric or steam power at 70° C.

At the end of the drying operation, when moisture content is between 15 and 20 per cent, the pieces of suitable shape and size are wrapped in cellophane paper, packed in cartons and stored at ambient air temperature. Pieces of unsuitable shape and size are further cut into small pieces and used to prepare, along with peanuts and cashews, a variety of "cocktail mixtures". Banana fruit bar. - Banana varieties which give smooth pulp without serum separation must be used for this purpose. Ripe, suitable fruit is selected. The hand-peeled fruits are soaked in 0.3 per cent citric acid solution for about 10 minutes (lime or lemon juice can replace citric acid). The drained fruit are pulped to obtain smooth pulp. The rest of the procedure is the same as in the case of the mango bar.

### *Guava Fruit Bar*

A mixture of pink and yellow varieties is best suited for preparing the bar. The washed fruit is hand peeled and stem and blossom ends trimmed. The peeled fruit is cut into quarters which are passed through a helicoidal extractor to separate seeds and fibrous pieces (the holes in the stainless steel screen should be between 0.8 and 1.10 mm). To get the maximum yield of pulp, the material is passed through the extractor twice. After adjusting the refractometric solids to 25 degrees Brix, the fruit bar can be prepared by following the same procedure as for mango pulp.

### *Mixed Fruit Bar*

Mango and banana pulp, as well as papaya and banana pulp, can be mixed in a calculated ratio for preparing mixed fruit bars. The rest of the procedure is the same as in the case of pure mango pulp.

### *Packing and Storage*

The dried pulp is removed from the dryer and cut into square pieces of 5 x 5 cm at a thickness of about 0.3 cm. These pieces, arranged in three layers make up blocks of about 0.9 cm thickness weighing between 25 and 28 grams. An unit pack consist of two such blocks and weights between 50 and 56 grams. Each block is separately wrapped in cellophane and the unit pack is filled in a printed cellophane bag of size 15 x 8 cm. Two hundred unit packs are packed in a master carton of size 34 x 22 x 14 cm, with a net weight of about 10 kg. Shelf-life is about one year at room temperature.

### *Fruit Leathers*

Fruit leathers are manufactured by drying/dehydration of fruit purées into leathery sheets. The leathers are eaten as confections or cooked as a sauce. They are made from a wide variety of fruits, the more common being apple, apricot, banana, cherry, blackcurrant, grape, peach, pear, pineapple, plum, raspberry, strawberry, kiwi fruit, mango and papaya. A description of procedures for mango, banana, guava and mixed fruit bars is given in this document. Another product with good potential is ciku leather; ciku fruit is grown in Malaysia. A standard process is carried out using ripe fruits which are washed, peeled, diced and the seeds removed. The fruits are blanched for 1 minute at 80° C and blended into puree in a food processor. Ciku leather is prepared by mixing ciku puree with 10 per cent sugar, 10 per cent pre-gelatinous rice flour, 150 ppm sorbic acid an 500 ppm sodium metabisulphite (Na2H2SO4). The mixture is cooked on a water bath at 60° C and then made into sheets 1.8 mm thick on trays spread with glycerol to reduce stickiness. This is then further dried in a forced-air dehydrator at 45° C for 3.5 hr or until the surface no longer feels sticky when touched with the fingers. The dried and cooled leathers are cut into 12 x 12 cm squares and wrapped in polypropylene (PP) of 0.1 mm thickness.

## Osmotic Dehydration in Fruit and Vegetable Processing

Osmotic dehydration is a useful technique for the concentration of fruit and vegetables, realised by placing the solid food, whole or in pieces, in sugars or salts aqueous solutions of high osmotic pressure. It gives rise to at least two major simultaneous counter-current flows: a significant water flow out of the food into the solution and a transfer of solute from the solution into the food.

### *Process Variables*

*Main process variables are:*

- Pre-treatments;
- Temperature;
- Nature and concentration of the dehydration solutions;
- Agitation;
- Additives.

*In the light of the published literature, some general rules can be noted:*

- Water loss and solid gain are mainly controlled by the raw material characteristics and are certainly influenced by the possible pre-treatments;
- It is usually not worthwhile to use osmotic dehydration for more than a 50 per cent weight reduction because of the decrease in the osmosis rate over time. Water loss mainly occurs during the first 2 hr and the maximum solid gain within 30 min.;
- The rate of mass exchanges increases with temperature but above 45 ° C enzymatic browning and flavour deterioration begin to take place. High temperatures, *i.e.* over 60° C, modify the tissue characteristics so favouring impregnation phenomena and thus the solid gain;
- The best processing temperature depends on the food; mass exchanges are favoured by using high concentration solutions;
- Phenomena which modify the tissue permeability, such as over-ripeness, pre-treatments with chemicals ($SO_2$), blanching or freezing, favour the solid gain compared to water loss because impregnation phenomena are enhanced;
- The kind of sugars utilised as osmotic substances strongly affects the kinetics of water removal, the solid gain and the equilibrium water content. Low molar mass saccharides (glucose, fructose, sorbitol, etc.) favour the sugar uptake;
- Addition of NaCl to osmotic solutions increases the driving force for drying.

Synergistic effects between sugar and salt have also been observed.

***Applications***

The effects of osmotic dehydration as a pre-treatment are mainly related to the improvement of some nutritional, organoleptic and functional properties of the product. As osmotic dehydration is effective at ambient temperature, heat damage to colour and flavour is minimised and the high concentration of the sugar surrounding fruit and vegetable pieces prevents discolouration. Furthermore, through the selective enrichment in soluble solids high quality fruit and vegetables are obtained with functional properties "compatible" with different food systems. These effects are obtained with a reduced energy input over traditional drying process. The main energy-consuming step is the reconstitution of the diluted osmotic solution that could be obtained by concentration or by addition of sugar.

**Drying**

Air drying following osmotic dipping is commonly used in tropical countries for the production of so-called "semi-candied" dried fruits. The sugar uptake, owing to the protective action of the saccharides, limits or avoids the use of SO2 and increases the stability of pigments during processing and subsequent storage period.

The organoleptic qualities of the end product could also be improved because some of the acids are removed from the fruit during the osmotic bath, so a blander and sweeter product than ordinary dried fruits is obtained. Owing to weight and volume reduction, loading of the dryer can be increased 2-3 times. The combination of osmosis with solar drying has been put forward, mainly for tropical fruit. A 24 hour cycle has been suggested combining osmodehydration, performed during the night, with solar drying during the day. Two-three-fold increase in the throughput of typical solar dryers is feasible, while enhancing the nutritional and organoleptic quality of the fruits. A two-step drying process, OSMOVAC, for producing low moisture fruit products was described. The osmotic step is performed with sucrose syrup 65-75 Brix until the weight reduction reaches 30-50 per cent. By osmotic dehydration followed by vacuum drying puffy products with a crisp, honeycomb-like texture can be obtained at a cost comparatively lower than freeze-drying. Commercial feasibility of the process on bananas has been studied, based on the results of a semi-pilot scale operation; the process scheme. Osmotically dried bananas retained more puffiness and a crisper texture than simple vacuum dried ones, and the flavour lasted longer at ambient temperature. The combination of osmotic dehydration with freeze-drying has been proposed only at laboratory scale.

### Appertisation

A combination of osmotic dehydration with appertisation has been proposed to improve canned fruit preserves. The feasibility of a process, called osmo-appertisation, to obtain high quality fruit in syrup, has been assessed on a pilot scale. The key point of this technique is the pre-concentration of the fruit to about 20-40 Brix, that causes, together with the enhancement of the natural flavour, an increase of the resistance of the fruit to the following heat treatment, especially for colour and texture stability. The products obtained are stable up to 12 months at ambient temperature and show a higher organoleptic quality than canned preserved alternatives. Furthermore, because of their higher specific weight and diminished volume, the filling capacity of jars or pouches is increased.

### Freezing

The frozen fruit and vegetable industry uses much energy in order to freeze the large quantity of water present in fresh products. A reduction in moisture content of the material reduces refrigeration load during freezing. Other advantages of partially concentrating fruits and vegetables prior to freezing include savings in packaging and distribution costs and achieving higher product quality because of the marked reduction of structural collapse and dripping during thawing. The products obtained are termed "dehydro-frozen" and the concentration step is generally carried out through conventional air drying, the additional cost of which has to be taken into account. Osmotic dehydration could be used instead of air drying to obtain an energy saving or a quality improvement especially for fruit and vegetable sensitive to air drying.

### Extraction of Juices

An osmotic pre-step before juice extraction was reported to give highly aromatic fruit or vegetable juice concentrates.

### Further Developments

So far only applications on a pilot plant scale are reported in the literature. For further developments on a larger scale, theoretical and practical problems should be solved. The industrial application of the process faces engineering problems related to the movement of great volumes of concentrated sugar solutions and to equipment for continuous operations. The use of highly concentrated sugar solutions creates two major problems. The syrup's viscosity is so great that agitation is necessary in order to decrease the resistance to the mass transfer on the solution side. The difference in density between the solution (about 1.3 kg/litre) and fruit and vegetables (about 0.8 kg/litre), makes the product float. Another important aspect, so far not investigated, is the microbiological safety of the process, which should be studied thoroughly before further industrial development.

# TYPES OF FRUIT

## SEEDLESS FRUITS

Seedlessness is an important feature of some fruits of commerce. Commercial cultivars of bananas and pineapples are examples of seedless fruits. Some cultivars of citrus fruits (especially navel oranges), satsumas, mandarin oranges table grapes, grapefruit, and watermelons are valued for their seedlessness. In some species, seedlessness is the result of *parthenocarpy*, where fruits set without fertilization. Parthenocarpic fruit set may or may not require pollination. Most seedless citrus fruits require a pollination stimulus; bananas and pineapples do not. Seedlessness in table grapes results from the abortion of the embryonic plant that is produced by fertilization, a phenomenon known as *stenospermocarpy* which requires normal pollination and fertilization.

### Seed Dissemination

Variations in fruit structures largely depend on the mode of dispersal of the seeds they contain. This dispersal can be achieved by animals, wind, water, or explosive dehiscence. Some fruits have coats covered with spikes or hooked burrs, either to prevent themselves from being eaten by animals or to stick to the hairs, feathers or legs of animals, using them as dispersal agents. Examples include cocklebur and unicorn plant. The sweet flesh of many fruits is "deliberately" appealing to animals, so that the seeds held within are eaten and "unwittingly" carried away and deposited at a distance from the parent. Likewise, the nutritious, oily kernels of nuts are appealing to rodents (such as squirrels) who hoard them in the soil in order to avoid starving during the winter, thus giving those seeds that remain uneaten the chance to germinate and grow into a new plant away from their parent. Other fruits are elongated and flattened out naturally and so become thin, like wings or helicopter blades, *e.g.* maple, tuliptree and elm. This is an evolutionary mechanism to increase dispersal distance away from the parent via wind. Other wind-dispersed fruit have tiny *parachutes*, *e.g.* dandelion and salsify. Coconut fruits can float thousands of miles in the ocean to spread seeds. Some other fruits that can disperse via water are nipa palm and screw pine. Some fruits fling seeds substantial distances (up to 100 m in sandbox tree) via explosive dehiscence or other mechanisms, *e.g.* impatiens and squirting cucumber.

### Uses

Many hundreds of fruits, including fleshy fruits like apple, peach, pear, kiwifruit, watermelon and mango are commercially valuable as human food, eaten both fresh and as jams, marmalade and other preserves. Fruits are also in manufactured foods like cookies, muffins, yoghurt, ice cream, cakes, and many more. Many fruits are used to make beverages, such as fruit juices (orange juice, apple juice, grape juice, etc) or alcoholic beverages, such as wine or brandy. Apples are often used to make vinegar. Fruits are also used for gift giving, Fruit Basket and Fruit Bouquet are some common forms of fruit gifts.

Many vegetables are botanical fruits, including tomato, bell pepper, eggplant, okra, squash, pumpkin, green bean, cucumber and zucchini. Olive fruit is pressed for olive oil. Spices like vanilla, paprika, allspice and black pepper are derived from berries.

## Nutritional Value

Fruits are generally high in fibre, water and vitamin C. Fruits also contain various phytochemicals that do not yet have an RDA/RDI listing under most nutritional factsheets, and which research indicates are required for proper long-term cellular health and disease prevention. Regular consumption of fruit is associated with reduced risks of cancer, cardiovascular disease, stroke, Alzheimer disease, cataracts, and some of the functional declines associated with aging.

## Nonfood Uses

Because fruits have been such a major part of the human diet, different cultures have developed many different uses for various fruits that they do not depend on as being edible. Many dry fruits are used as decorations or in dried flower arrangements, such as unicorn plant, lotus, wheat, annual honesty and milkweed. Ornamental trees and shrubs are often cultivated for their colourful fruits, including holly, pyracantha, viburnum, skimmia, beautyberry and cotoneaster.

Fruits of opium poppy are the source of opium which contains the drugs morphine and codeine, as well as the biologically inactive chemical theabaine from which the drug oxycodone is synthysized. Osage orange fruits are used to repel cockroaches. Bayberry fruits provide a wax often used to make candles. Many fruits provide natural dyes, *e.g.* walnut, sumac, cherry and mulberry. Dried gourds are used as decorations, water jugs, bird houses, musical instruments, cups and dishes. Pumpkins are carved into Jack-o'-lanterns for Halloween. The spiny fruit of burdock or cocklebur were the inspiration for the invention of Velcro.

Coir is a fibre from the fruit of coconut that is used for doormats, brushes, mattresses, floortiles, sacking, insulation and as a growing medium for container plants. The shell of the coconut fruit is used to make souvenir heads, cups, bowls, musical instruments and bird houses. Fruit is often also used as a subject of still life paintings.

# TECHNOLOGY OF SPECIAL NON GELIFIED FRUIT MARMALADES

Special non gelified fruit marmalades are products resulting from fruit without stones or seeds, sieved or squashed, concentrated by boiling, without sugar added and non gelified. Their consistency results from a low water content (about 35 per cent) and a high percentage of insoluble substances (5-10 per cent). Sugar the from fruits acts as a preserving agent. Plum special non gelified fruit marmalade is the product representative of this category. Other fruit is used very rarely, because they have a reduced sugar content as compared to plums; though there are some countries producing special non gelified fruit marmalade from pears or sweet apples.

For plums, the finished product in this category must contain minimum 55-60 per cent soluble substances (refractometric extract), rising up to 70 per cent for a high quality product. Basoc Reco[e Settomg for plums is done in relation with sugar content of fruits; when this is higher, the product quality is better and the yield is higher. Thus, if plums with 18 per cent refractometric extract are used, 300-350 kg of fruits are needed for 100 kg finished product. At the same time, concentration of sugar by boiling, also increases by three fold acidity and astringency of this special plum non gelified marmalade; for this reason it is recommended to use only sweet and completely mature plums. Washing is performed in usual conditions. Pre-Boiling of fruits can be carried out in water or vapour, preferably with continuous running and has as its objective the softening of tissues. De-Stoning is performed in a pulper. Boiling of the sieved mass is done in double bottom open kettles with a big evaporation surface or in vacuum evaporators. Boiling in open kettles enables production of a more tasty slightly caramelized, product; boiling in vacuum evaporators has the technological advantages indicated in marmalade production. At the end of boiling and once of necessary concentration is reached, the product is poured directly into receptacles (drums, etc.) and left to cool in order to form a hard surface layer (crust).

Storage of well closed receptacles is carried out as for marmalade. Special non gelified fruit marmalade can also be prepared from chemically preserved semi-processed fruit products, but the quality is lower than that obtained from fresh fruits. Sometimes dried prunes in a mix with preserved semi-processed products can be used for plum special non gelified marmalade preparation. In some countries plum finished products in this category are sweetened by the addition of maximum 30 per cent sugar, calculated in relation to the finished product.

## Technology of Fruit Pastes

These products are obtained in a similar way to marmalades and special non gelified marmalades, but have a lower water content (about 25 per cent).

Reduction of water content can be achieved by continuing the boiling of the product or by natural or artificial drying. A typical example of fruit paste without sugar added is the apricot paste - "pistil", etc. which is a concentrated special non gelified fruit marmalade poured in thin layers and sun dried.

An example of fruit paste with sugar added is quince paste which is a marmalade concentrated by evaporation. Sugar content must be 65 per cent; soluble substances content, 7075 per cent refractometric extract and acidity at least 0.5 per cent expressed as malic acid. Packing is done usually in polyethylene sheets and then in boxes or tins; storage conditions are similar to those for marmalade.

**Technology of Fruit Syrups**

Fruit syrups are products obtained by dissolving sugar in juices obtained from direct pressing of fruits. Sugar dissolving can be done at room temperature or by heating. Syrups have to contain 68 per cent soluble substances (refractometric extract) and minimum 1 g/100 ml malic acid. Up to a maximum 10 per cent of sugar can be replaced by corn syrup. Syrups must be manufactured from the juice of only one fruit species. Juice preparation is carried out at room temperature as described above.

*Sugar dissolving in fruit juice can be performed by one of the following methods:*

- Boiling in open kettles, done by using the basic recipe:
  350-400 kg fruit juice; 650-660 kg sugar; max. 10 kg citric or tartric acid.
  The juice is brought to boiling and the sugar is dissolved; the total time has to be as short as possible in order to avoid flavour loss and a too high sugar inversion degree (optimum inversion degree is 40 per cent). Acid is added preferably towards the end of boiling. During all boiling processes it is necessary to remove foam/scum. In order to avoid caramelization, the syrup has to be cooled rapidly, and this can be carried out in baths with double bottoms through which are circulated water.
  One alternative to this method is to boil syrup in closed vessels to avoid flavour losses.
- Boiling in a vacuum. The basic recipe is the same as above. Sugar and fruit juice are mixed previously in a pre-heating kettle and then transported to vacuum equipment.
  Boiling is performed at 50°C and at the end the temperature is raised slowly up to 65-70°C. The syrup can be cooled directly in vacuum equipment by closing the steam inlet and by increasing the vacuum. In this boiling method it is possible to incorporate a flavour recuperation device.

- A continuous process for syrup preparation can be carried out by dissolving components with heat while passing them through a horizontal cylinder with a screw inside.
  In the methods where sugar is dissolved by heat, it is also possible to use chemically preserved juices.
  In this case it is necessary to first perform the desulphitation of juices preserved with $SO_2$. This can be performed by boiling juice with optional water addition (and before any sugar addition). High quality syrups are obtained however from fresh juices.
- Sugar can also be dissolved at room temperature by using continuous flow percolators. These are similar to those used for salt solution preparation in vegetable canning processes. The juice goes over a sugar layer and is concentrated progressively until saturation (about 65 per cent). The syrup is then passed through a filtration section in the bottom of the percolator.

Syrup filtration is needed in order to clarify crystals; the filtration of syrup is done in warm conditions through cloth. Filling of syrup in bottles is done in aseptic conditions as much as possible in order to avoid syrup infection with osmophile yeasts. Syrup preservation is assured by the high sugar content with respect to a low water activity [unclear]. Storage takes place in well ventilated storage rooms; avoiding sunlight at 10-15°C. The usual product range is: strawberries, cherries, wild berries, citrus fruits.

## TECHNOLOGY OF SEMI-PROCESSED FRUIT PRODUCTS

The semi-processed fruit products are manufactured in order to be delivered to industry processing centres (in the fruit producing country itself or in importing countries) where they will be further manufactured in consumer oriented finished products: jams, jellies, syrups, fruits in syrup, etc.

*In the practice of semi-processed fruit products and for the purpose of this document the following categories are defined:*

- Fruit "pulps": semi-processed products, not refined, obtained by mechanical treatment (or, less often, by thermal treatment) of fruit followed by their preservation. Either whole fruit, halves or big pieces are used which enables easy identification of the species. "Pulps" can be classified in boiled or non boiled (raw).
- Fruit "purées-marks": semi-processed products obtained by thermal and mechanical treatment or, very rare, raw and then refined, operations by which all non-edible parts (cores, peels, etc.) are removed. "Purées-marks" are classified in boiled (the more usual case) and non boiled (raw).
- Semi-processed juices: products obtained by cold pressure or very rare by other treatments (diffusion, extraction, etc.) followed by the preservation.

## Technical Processes for Preservation of Semi-processed Fruit Products

Preservation can be achieved by chemical means, by freezing or by pasteurization. The choice of preservation process for each individual case is a function of the semi-processed product type and the shelf life needed.

### *Chemical Preservation*

In many countries, in practice, this is carried out with sulphur dioxide, sodium benzoate, formic acid and, on a small scale, with sorbic acid and sorbates. Preservation with sulphur dioxide is a widespread process because of its advantages: universal antiseptic action and very economic application. The drawbacks of $SO_2$ are: $SO_2$ turn firms the texture of some fruit species (pomaces), desulphiting is not always complete and recolouring of red fruits is not always complete after desulphitation. Practical preservation dosage levels with $SO_2$ for about 12 months is 0.18-0.20 per cent $SO_2$ (with respect to the product to be preserved). This level could be reduced to 0.09 per cent $SO_2$ for 3 months and to 0.12 per cent $SO_2$ for 6 months preservation. The preservation with sulphur dioxide is in use mainly for "pulps" and for "purées-marks".

Chemical preservation can be performed from a practical point of view by the utilisation of 6 per cent $SO_2$ water solutions or by direct introduction of sulphur dioxide gas in the product (for "purées-marks"). The preparation of 6 per cent $SO_2$ solutions is done by bubbling the gas from cylinders in cold water; from a 50 kg $SO_2$ compressed gas cylinder results 830 l of 6 per cent $SO_2$ solution. These $SO_2$ solutions have to be stored in cool places, in closed receptacles and with periodic concentration control/check by titration or by density measurements approximate results. Preservation with sodium benzoate has the following advantages: it does not firm up the texture and does not modify fruit colour. The disadvantages are: it is not a universal antiseptic, its action needs an acid medium and the removal is partial. Sodium benzoate is in use for "pulps" and for "purées-marks" but less for fruit juices.

Practical dosage level for 12 months preservation is 0.18-0.20 per cent sodium benzoate, depending on the product to be preserved. Sodium benzoate is used as a solution in warm water; the dissolution water level has to be at maximum 10 per cent reported to semi-processed product weight. Formic acid preservation is performed mainly for semi-processed fruit juices at a dosage level of 0.2 per cent pure formic acid (100 per cent). Formic acid is an antiseptic effective against yeasts, does not influence colour of products and is easily removed by boiling.

Formic acid could be diluted with water in order to insure a homogeneous distribution in the product to be preserved; water has to be at maximum 5 per cent of the product weight. Because of a potential effect of pectic substance degradation, formic acid is less in use for "pulps" and "purées-marks" preservation. Sorbic acid used as potassium sorbate (easily water soluble) can be used for preservation of fruit semi-processed products at a dosage level of 0. 1 per cent maximum. Advantages of sorbates are: they are completely harmless and without any influence on the organoleptic properties of semi-processed fruit products.

***Preservation by Pasteurization***

As fruit has a low pH, preservation of semiprocessed fruit products could also be performed by pasteurization (heat treatment step at maximum temperature of 100°C), the length of this step varying with the size of the receptacles. The advantages of this type of treatment are: hygienic process, which assure a long term preservation; the disadvantages are: need for air tight receptacles, and pectic substances could begin to deteriorate if the thermal treatment is too long. Thermal preservation of fruit semi-processed products could also be done by a "selfpasteurization": very hot semi-processed products are filled into receptacles (*e.g.* metal cans) which are sealed and then inverted in order to sterilise the air which goes through the hot fruit mass.

***Preservation by freezing***

This is done on an industrial scale in some countries and can be done with or without sugar addition. The advantages of this process are: absence of added substances; very good preservation of quality of fruit constituents (pectic substances, vitamins, etc.) and good preservation of organoleptic properties (flavour, taste, colour). Freezing is done at about -20 to -30° C and storage at -10 to -18° C. Freezing is applied mainly to semi-processed fruit products aimed at very high quality and high cost finished products. Technological flow-sheet for semi-processed fruit "pulps": chemical preservation. Sorting is needed in order to remove sub-standard fruit (with moulds, with diseases, etc.) and all foreign bodies. Washing is obligatory in order to remove all impurities which cannot be eliminated at the processing step in finished products. Coring and Cutting, mainly for pomace fruits, has as main objective a better utilisation of preservation "space" in receptacles and is not mandatory; this will be defined by customer/supplier agreements/standards. This operation is preferably performed by mechanical means. Preservation is carried out with the 6 per cent $SO_2$ solution which is added to the prepared fruits (placed in bulk in receptacles) in the quantity needed to obtain the preservation dosage level. For a better/homogeneous preservative distribution, the initial 6 per cent $SO_2$ solution could be diluted with water; however, the diluted solution (which will be filled in receptacles) has to be at a dosage level of less than 10 per cent of the semi-processed product weight.

For some soft fruit, especially strawberries, preservation is done with a mix of 6 per cent $SO_2$ solution and calcium bisulphite solution (containing also 6 per cent $SO_2$). Preparation of calcium bisulphite solution is done by the introduction of 30 kg of CaO in 1 $m^3$ $SO_2$ solution and mixing up to clarification. The resulting solution is mixed with the initial 6 per cent $SO_2$ solution, generally in a 1:1 ratio, but the ratio can be adapted to the fresh fruit texture. Firming of soft fruit texture by this treatment is based on the formation of calcium pectate with pectic substances from fruit tissues.

In the case of sodium benzoate, formic acid or potassium sorbate, the dosage levels to be used are as indicated above with the rule that it is not allowed to add more that 10 per cent liquid in receptacles on the prepared fruits. Preservation by pasteurization or "self-pasteurization" will need as additional steps: a) boiling with a minimum water addition (maximum 10 per cent); b) filling of receptacles; c) hermetic closing followed by d) pasteurization or "self-pasteurization".

### Processing of Pineapple-papaya Jam

The fruit should be prepared as per previous instructions. For pineapples, the ends are removed and discarded; the cores and outer parts of the fruits are also removed. The fruit cylinders obtained are pulped through a special extractor (Fitzpatrick communiting machine) equipped with a 0.40-in screen sieve; the pulp thus obtained is used for making jam. The papaya are prepared by hand-peeling the fruit; the fruit is then halved and the seeds removed. It is then pulped in the communiting machine using a 0.40-in screen sieve. When ginger root is used as flavouring, it is peeled and macerated in a Kenwood blender to a very fine consistency. A typical formula for a pineapple-papaya jam (50:50 ratio) with ginger flavouring is given as follows:

| | |
|---|---|
| Pineapple pulp | 25.0 |
| Papaya pulp | 25.0 pounds |
| Cane sugar | 50.0 |
| Apple pectin (150 grade) | 6.0 ounces |
| Citric acid | 6.4 |
| Fresh ground ginger | 7.5 |

Processing is carried out in the following way: The weighed fruit pulp is placed in a stainless steel steam-jacketed kettle and heated to about 110°F under constant stirring. When the product reaches this temperature, the heat is turned off. The pectin (mixed in about ten times its weight with some of the weighed sugar), is then mixed into the fruit pulp, stirring constantly in order to prevent the pectin from clotting.

When the pectin has dissolved, the remainder of the sugar is added and dissolved completely in the mixture. The heat is then turned on and the jam mixture is stirred constantly until it starts boiling vigorously. During the remainder of the cooking, the product is stirred occasionally. Near the finishing point (approximately 221° F), the citric acid and the ginger (if it is used) are also added.

Determination of the finishing point is done by removing samples at intervals, cooling, and reading the soluble solids by means of a refractometre equipped with a Brix scale. After the jam reaches the proper Total Soluble Solids content, the heat is turned off and the surface scum/foam is removed. The jam then is quickly put into receptacles which have been cleaned and sterilised with boiling in water for 30 minutes. The filling operation is done rapidly in order to prevent the temperature of the jam from falling below 190° F. After filling, sterilised lids (boiled for 30 minutes in water) are placed on the receptacles and they are then sealed. After this operation the receptacles are inverted for about 3 minutes to insure that the lids are sterilised. The receptacles are then placed upright. At this stage it is not necessary to do any further processing, therefore the receptacles are cooled in running cold water until they reach a temperature slightly above room temperature. They are then dried in air and labelled.

### *Evaluation of Finished Products*

During production at medium/large scale, it is recommended that quality controls be performed during manufacturing. After ten weeks of storage at room temperature it is recommended that an examination of finished products be performed. The receptacles are opened and contents carefully emptied on to enamel trays without disturbing the formation of the jam. The empty cans (if metal cans were used) are then inspected for signs of corrosion. Factors other than flavour include colour, appearance, syrup separation, firmness and spreading quality. For flavour, jam is tested on pieces of bread. Samples are taken for measurement of pH (with a glass electrode pH metre) and Total Soluble Solids (with a refractometre equipped with a Brix scale). This evaluation enables to have a quality check during product shelf life and to obtain data needed for necessary improvements of future productions. For pineapple-papaya jam, products made with 30 per cent pineapple and 70 per cent papaya with added ginger has the highest score for flavour. The use of plain tin cans causes corrosion problems which is not the case when acid resistant lacquer cans are used.

## FRUIT CULTIVATION IN INDIA

The exhaustive and comprehensive cultivation of vegetables, flowers and fruits is referred to as horticulture. Horticulture in the country has been thoroughly involved with the cultivation of fruits in India, besides also laying simultaneous stress upon vegetables and gardening of rare plants. This very study of fruit and vegetable production is an arena, which provides sizeable subject of enormous scope, if pursued passionately in order to take India to great heights to cultivating fruits in a strategic manner. Cultivation of fruits in India and in the international scenario, involves the consolidation of wide spectrum of disciplines. As the new technologies and developments have gradually become readily available in the country, the cropping and cultivating systems and production practices have also remained witness to significant metamorphoses. Indeed, if dug deep into the fascinating history of fruit cultivation in the country, it can verily be seen that the Silk Route had played a key role in the ushering of various kinds of these delicious sweet and fleshy things called 'fruits' since pre-Christian era, especially from countries like China and Egypt. Fruit cultivation in India is one such major commercial and business sector for exporting merchandise and shipping, from which much of the international revenue is incurred. Since Independence, the country has been trying to come to terms with the dazzling prospects of exporting commercial business and the land being essentially agrarian and rural, possesses ample scope for lands for cultivation and farming. However, the erstwhile tools and implements of the Central Indian administration, is being upgraded every fiscal year, a domain which truly looks towards guaranteed future. Numerous kinds of precepts, rules and practices that were common a few years ago in India, may just not be no longer be current.

India has been perhaps justly renamed as the 'fruit and vegetable basket of the world', a factor that weighs fascinatingly upon the cultivation of fruits in the country. India serving as home to a supreme variety of fruits and vegetables, holds an exclusive position in production figures amongst other countries. More than 90 percent of India's exports in fresh products travels to west Asia and East European markets. India's exports of fresh fruits and vegetables has amplified by 2437.12 crore rupees in the 2007-08 financial year. This amplification list merrily incorporates the products like walnut, fresh mangos, fresh grapes and umpteen other fresh fruits and vegetables. Since prehistoric times, India has served as an agriculturally dependant country. Hundreds of fruits and vegetable types are grown in all parts of India. For such a thriving business to network itself, the Indian industrial strategy has been known to have expanded from grassroots level, onto big markets, dealing globally. The cultivated fresh fruit and vegetable reach small scale fruits and vegetables suppliers, they are then sent to local markets as well as fruits and vegetables exporters, who in turn make baskets of shipping of fruits, dealing with international buyers. The last few decades have witnessed the number of Indian fruit vegetables suppliers and fruits vegetables exporters rising to an all time high.

The total production of fruits and vegetables in the world is approximately 370 metric ton. India proudly ranks first in the world, with an annual output of 32 metric ton, taking the fruit cultivation of Indian graph shooting sky-high. While there are almost 180 families of fruits that are cultivated all over the world, citrus fruits represent approximately 20 percent of world's total fruit production. Major Indian cultivated fruits consist of mango, banana, citrus fruits, apple, guava, papaya, pineapple and grapes. The fruits are further also processed into various products such as fruit juices and concentrates, canned fruit, dehydrated fruit, jams and jellies etc. India with its current production of approximately 32 million metric ton of fruits, accounts for about 8 percent of the world's fruit production. The diverse agro-climatic zones present in the country, makes it possible to practically grow almost all varieties of fresh fruits and vegetables in India. The fruit cultivation and production in India has recorded a growth rate of 3.9 percent, whereas the fruit processing sector has grown at approximately 20 percent per annum. However, the growth rates have been expansively higher for frozen fruits and vegetables (121 per cent) and dehydrated fruits and vegetables (24 percent). There exist more than 4000 fruit processing units in India, with an combined capacity of more than 12 lakh metric ton (less than 4 percent of total fruits produced). It is reckoned that approximately 20 percent of the production of processed fruits is intended for exports, the rest supplies to the defence, institutional sectors and household consumption; fruit cultivation in India however depends a lot upon mango and mango-based products, constituting 50 percent of exports.

The fruits and vegetable cultivation and processing industry in India is highly decentralised in content. A large number of units are in the cottage/home scale and small scale sector, possessing small capacities upto 250 tonnes per annum. But big Indian and multinational companies in the sector possess enormous capacities in the range of 30 tonnes per hour more or less. Since liberalisation and pulling out of excise duty on fruit and vegetable products, there has been substantial rise in the growth rate of the industry. The Indian fruit cultivation, especially mangoes and bananas, are now in huge demand outside the country. India is a producer of tropical fruits like coconuts, jackfruits, cashew nuts, pineapples, bananas and oranges. Of temperate fruits, apples, plums, peaches, almonds, apricots and grapes are grown in abundance. While Jammu and Kashmir and Himachal Pradesh lead in the fruit production of the temperate region, others are grown in various parts of peninsular India and Northern Plains. India brings in pots of foreign exchange by exporting cashew nuts too. Part of the raw cashew nuts is imported and treated here, before they are later re-exported. The north eastern region of India holds mammoth horticulture potential. States like Arunachal Pradesh, Nagaland, Meghalaya and Manipur have favourable soil and climatic conditions and also have great scope for temperate fruit cultivation in these parts of India to augment business management. Presently stone fruits like peach, plum and apricot and pome fruits like apple and pear are grown in small scale but the future is bright and there is great scope for the expansion of temperate fruit culture.

Fruit cultivation in India has received additional impetus in the format of various state governments realising their functionary role in boosting c nationwide commerce. As such, the Haryana government has been lending grants for setting up orchards under the National Horticulture Mission. Under this plan, a special grant of 9,750 rupees per hectare is provided to the farmers for cultivation of guava, 15,750 rupees per hectare for mango, 11,250 rupees for ber and 18,525 rupees for cheeku. This amount is to be disbursed in three years time, according to sources. Ample investment opportunities does exist in expanding the export market for fruit cultivation in India. A modifying acceptance of new products with market development efforts have also been witnessed since long, given the fact that there exists a good international demand for certain fruits and vegetable products. India produces 41 percent of world's mangoes, 23 percent bananas, 24 percent cashew nuts and 36 percent green peas. The total export value of the main exporting fruit crop from India is mango. Exports of mangoes, grapes, mushrooms have started going to the United Kingdom, Middle East, Singapore and Hong Kong.

Mango, termed as the 'king of fruits in India', accounts for 40 percent of the national fruit production almost every year. It occupies 42 percent of the country's 2205.6 thousand hectares land under fruit cultivation in the Indian subcontinent. India had exported 54350.80 metric ton of fresh mangoes with the value of 127.42 crore rupees in 2007-08. The major varieties of mangoes exported include-Dashehri, Alphonso, Kesar, Banganpalli, Kesar, Dusheri, Langra, Chausa, Mallika and. Swarnrekha. The major markets for Indian mangoes comprise U.A.E, Bangladesh, U.K, Saudi Arabia and Nepal. India is verily approximated to account for around 60 percent (9.5 million tonnes) of the world's mango production of 15.7 million tones.

The major cultivation and production areas of this fleshy fruit in Indian subcontinent are in the states of Andhra Pradesh, Uttar Pradesh, Karnataka, Bihar, Gujarat and Maharashtra. India is one of the leading exporters of fresh table grapes to the global market; India's export of grapes has increased from 301.92 crore rupees in 2006-07 to 317.83 crore rupees in 2007-08. The leading destinations for export of Indian comprise Netherlands, U.K, U.A.E, Bangladesh and Belgium. As has already been stated that the country is the largest producer of fruits in the world, acknowledging it as the fruit basket of world, India's fruit cultivation has been orchestrated after several application of masterminds to have thus reached such a position. The major fruits cultivated in India are mangos, grapes, apple, apricots, orange, banana, avocados, guava, lichi, papaya, sapota and water melons. This is due to its potential in different agro climatic zones. India's export of fresh fruits to the major countries consists of U.K, Netherlands, U.A.E, Russia, Bahrain, Qatar, Kuwait, Saudi Arabia, Bangladesh and Nepal.

Fruit cultivation in India further has always remained a recipient of huge respect and admiration, owing to its marvellous produces of deciduous fruits, including pome fruits (apple and pear) and stone fruits (peach, plum, apricot and cherry) in considerable quantities. Deciduous fruits and its cultivation in India mainly takes place in the north Indian states of Jammu and Kashmir, Himachal Pradesh and in the Uttar Pradesh hills. The northeastern hills region, comprising the states of Arunachal Pradesh, Nagaland, Meghalaya, Manipur and Sikkim, also grow some of the deciduous fruits on a restricted scale. Due to initiation and adaptation of low chilling cultivars of crops like peach, plum and pear, they are also presently being grown commercially in certain areas of the north Indian plains too. Out of all the deciduous fruits, apple is the most vital in terms of production and extent.

From amongst the umpteen and healthy and palatable deciduous fruit cultivation in India, apple demands and calls for incredible attention. Indeed, apple was introduced into the country by the British Raj in the Kullu Valley of the Himalayan state of Himachal Pradesh as far back as 1865. On the other hand, the coloured 'delicious' cultivars of apple were introduced to the Shimla hills of the same state in 1917. The apple cultivar Ambri, is measured as indigenous to Kashmir and had been grown long before Western introductions. Pears and other deciduous fruits were domesticated lucratively during the early part of the 20th century, although some of them were accounted to occur under semi-wild conditions much earlier. Apricot was found growing in the drier pockets of northwestern Himalayas and two apricot varieties, locally acknowledged as Halman and Rakchaikarpo, are reported to be indigenous to the Leh-Ladakh region Jammu and Kashmir state. Sweet cherry was introduced from Europe before India's independence in 1947, whereas, commercial cultivars of sour cherry have been fetched primarily from USA in more recent years. The European and Japanese plum kinds are cultivated both in high and low hill areas. A plum variety 'Santa Rosa', reported to be a hybrid between Japanese and American species, predominates (70-80 percent) plantations in the Himalayan hills. Low chilling cultivars of peach and nectarine such as Flordasum, Flordared, and Sunred nectarine are booming introductions to the north-Indian plains. Some local selections of peach, plum and sand pear (Patharnakh) have also taken fruit cultivation in India on a grand commercial scale in subtropical marginal chilling areas of north India. Fruit cultivation in India, quite understandably grows most prolific in the areas of the Himalayas, with abundance speaking out from every angle.

The Uttar Pradesh hills, particularly the Kumaon hills division, possesses exceptional advantage of early harvest of apple, principally due to cultivation of early maturing varieties like Early Shanburry, Fanny and Benoni. The early maturing varieties are harvested 2-3 weeks before the onset of fresh apple from Himachal Pradesh and Jammu and Kashmir and hence, fetch very moneymaking prices. Deciduous fruits, covering pome and stone fruits contribute significantly to the horticulture economy of India. Truly, cultivation of fruits in India, does depend upon the success of plantation, harvest, soils and a perfect ambience of brilliant minds.

## PLANT NUTRITION

Plant nutrition is the study of the chemical elements that are necessary for plant growth. There are several principles that apply to plant nutrition. Some elements are directly involved in plant metabolism. However, this principle does not account for the so-called beneficial elements, whose presence, while not required, has clear positive effects on plant growth. A nutrient that is able to limit plant growth according to Liebig's law of the minimum, is considered an essential plant nutrient if the plant cannot complete its full life cycle without it. There are 17 essential plant nutrients.

*Macronutrients*:

- N = Nitrogen
- P = Phosphorus
- K = Potassium
- Ca = Calcium
- Mg = Magnesium
- S = Sulfur
- Si = Silicon

*Micronutrients include*:

- Cl = Chlorine
- Fe = Iron
- B = Boron
- Mn = Manganese
- Na = Sodium
- Zn = Zinc
- Cu = Copper
- Ni= Nickel
- Mo = Molybdenum

## MACRONUTRIENTS

### Calcium

Calcium regulates transport of other nutrients into the plant and is also involved in the activation of certain plant enzymes. Calcium deficiency results in stunting.

### Nitrogen

Nitrogen is an essential component of all proteins. Nitrogen deficiency most often results in stunted growth.

### Phosphorus

Phosphorus is important in plant bioenergetics. As a component of ATP, phosphorus is needed for the conversion of light energy to chemical energy during photosynthesis. Phosphorus can also be used to modify the activity of various enzymes by phosphorylation, and can be used for cell signaling. Since ATP can be used for the biosynthesis of many plant biomolecules, phosphorus is important for plant growth and flower/seed formation.

### Potassium

Potassium regulates the opening and closing of the stoma by a potassium ion pump. Since stomata are important in water regulation, potassium reduces water loss from the leaves and increases drought tolerance. Potassium deficiency may cause necrosis or interveinal chlorosis.

### Silicon

Silicon is deposited in cell walls and contributes to its mechanical properties including rigidity and elasticity

## MICRONUTRIENTS

### Boron

Boron is important in sugar transport, cell division, and synthesizing certain enzymes. Boron deficiency causes necrosis in young leaves and stunting.

### Copper

Copper is important for photosynthesis. Symptoms for copper deficiency include chlorosis. Involved in many enzyme processes. Necessary for proper photosythesis. Involved in the manufacture of lignin. Involved in grain production.

### Chlorine

Chlorine is necessary for osmosis and ionic balance; it also plays a role in photosynthesis.

### Iron

Iron is necessary for photosynthesis and is present as an enzyme cofactor in plants. Iron deficiency can result in interveinal chlorosis and necrosis.

### Manganese

Manganese is necessary for building the chloroplasts. Manganese deficiency may result in coloration abnormalities, such as discoloured spots on the foliage.

### Molybdenum

Molybdenum is a cofactor to enzymes important in building amino acids.

### Nickel

In higher plants, Nickel is essential for activation of urease, an enzyme involved with nitrogen metabolism that is required to process urea. Without Nickel, toxic levels of urea accumulate, leading to the formation of necrotic lesions. In lower plants, Nickel activates several enzymes involved in a variety of processes, and can substitute for Zinc and Iron as a cofactor in some enzymes.

### Sodium

Sodium is involved in the regeneration of phosphoen-olpyruvate in CAM and C4 plants. It can also substitute for potassium in some circumstances.

### Zinc

Zinc is required in a large number of enzymes and plays an essential role in DNA transcription. A typical symptom of zinc deficiency is the stunted growth of leaves, commonly known as "little leaf" and is caused by the oxidative degradation of the growth hormone auxin.

## PROCESSES

Plants uptake essential elements from the soil through their roots and from the air through their leaves. Nutrient uptake in the soil is achieved by cation exchange, wherein root hairs pump hydrogen ions ($H^+$) into the soil through proton pumps. These hydrogen ions displace cations attached to negatively charged soil particles so that the cations are available for uptake by the root. In the leaves, stomata open to take in carbon dioxide and expel oxygen. The carbon dioxide molecules are used as the carbon source in photosynthesis. Although nitrogen is plentiful in the Earth's atmosphere, relatively few plants engage in nitrogen fixation. Most plants therefore require nitrogen compounds to be present in the soil in which they grow. Plant nutrition is a difficult subject to understand completely, partially because of the variation between different plants and even between different species or individuals of a given clone.

Elements present at low levels may cause deficiency symptoms, and toxicity is possible at levels that are too high. Furthermore, deficiency of one element may present as symptoms of toxicity from another element, and vice-versa. Carbon and oxygen are absorbed from the air, while other nutrients are absorbed from the soil. Green plants obtain their carbohydrate supply from the carbon dioxide in the air by the process of photosynthesis.

## THE ELEMENTS OF COMPLETE PLANT NUTRITION

Complete nutrition results in superior plant growth. Why choose anything less for your plants? There are 20 elements necessary for optimum plant growth. Air and water supply carbon, hydrogen and oxygen. Macronutrients are required by plants in large amounts. Micronutrients are required in smaller amounts. Eliminate any of these elements,and plants will display abnormal growth, deficiencies or may not reproduce. The following is a brief guide to the role played by each of these essential nutrient elements.

**Macronutrients**

- (*N*)-*Nitrogen*: Component of proteins, hormones, chlorophyll, vitamins and enzymes. Promotes stem and leaf growth. The ammoniacal and nitrate forms are used directly by plants for stem and leaf growth. The urea form of nitrogen must be broken down by soil borne microorganisms or urease before it can be utilized by the plant. Urea can cause leaf tip and root burn. Deficiency (Def.): reduced yields, yellowing of leaves, stunted growth. Excess nitrogen can delay fruiting and flowering.
- (*P*)-*Phosphorus*: Seed germination, photosynthesis, protein formation, overall growth and metabolism, flower and fruit formation. Def.: purple stems and leaves, retarded growth and maturity, poor flowering and fruiting. Large amounts without zinc cause zinc deficiency. Low pH (<4) ties up phosphates in organic soils.
- (*K*)-*Potassium*: Formation of sugars, carbohydrates, proteins, cell division. Adjusts water balance; improves stem rigidity and cold hardiness; enhances flavour, colour and oil content of fruits; important for leafy crops. Def.: spotted, curled or burned look to leaves; lower yields.
- (*Ca*)-*Calcium*: Activates enzymes; structural part of cell walls; influences water movement, cell growth and division Required for uptake of nitrogen and other minerals. Leached from soil by watering. Immobile-requires a constant supply for growth. Def.: stunting of new growth in stems, flowers, roots; black spots on leaves and fruit; yellow leaf margins.
- (*Mg*)-*Magnesium*: Critical component of chlorophyll; needed for functioning of enzymes for carbohydrates, sugars and fats; fruit and nut formation; germination of seeds.
- Def.: yellowing between veins of older leaves; chlorosis; leaf droop. Leached by watering. Foliar spray to correct deficiencies.
- (*S*)-*Sulfur*: Component of amino acids, proteins, vitamins, enzymes. Essential for chlorophyll. Imparts flavour to many vegetables. Def.: light green leaves. Water supply may contain sulfur. Leached by watering.

**Micronutrients**

- (*B*)-*Boron*: Affects at least 16 functions: flowering, pollen germination, fruiting, cell division, water relationships, movement of hormones, cell wall formation, membrane integrity, calcium uptake, movement of sugars. Immobile; easily leached. Def.: terminal bud die back causes rosette of thick, curled, brittle leaves or brown, discolored, cracked fruits, tubers and roots.

- (*Cl*)-*Chlorine*: Involved in osmosis (movement of water or solutes in cells), ionic balance necessary to take up mineral elements and photosynthesis. Def.: wilting, stubby roots, yellowing, bronzing. Scents in some plants may be decreased. Leached by watering.
- (*Co*)-*Cobalt*: Required by nitrogen fixing bacteria; formation of B12 vitamin; formation of DNA. Will extend life of cut flowers such as roses. Def.: may result in nitrogen deficiency.
- (*Cu*)-*Copper*: Necessary for nitrogen metabolism; component of enzymes - may be part of enzyme systems that use carbohydrates and proteins. Bound tightly in organic matter. May be deficient in highly organic soils. Not readily lost from soil but may be unavailable. Def.: die back of shoot tips; terminal leaves develop brown spots. Excess is toxic.
- (*Fe*)-*Iron*: Enzyme functions; catalyst for synthesis of chlorophyll; essential for new growth. Def.: pale leaves, yellowing of leaves and veins. Leached by water and held in lower parts of soil. High pH soils may have iron present but unavailable to plants.
- (*Mn*)-*Manganese*: Enzyme activity for photosynthesis, respiration and nitrogen metabolism. Def.: young leaves are pale with green veins similar to iron deficiency; advanced stages-leaves are white and drop; brown, black or gray spots may appear next to veins. Plants in neutral or alkaline soils often show def. Acid soils may increase uptake causing toxicity.
- (*Mo*)-*Molybdenum*: Structural part of enzymes that reduce nitrates to ammonia for amino acid development essential to protein formation; required by nitrogen fixing bacteria. Def.: pale leaves with rolled, cupped margins. Seeds may not form. Nitrogen deficiency may occur if plants are lacking Mo.
- (*Ni*)-*Nickel*: Recently recognized as essential. Required for the urease enzyme to break down urea into usable nitrogen and for iron uptake. Seeds require nickel to germinate.
- (*Na*)-*Sodium*: Improves nitrogen metabolism in many plants, involved in osmotic (water movement) and ionic balance in plants. Def.: yellowing of leaves and leaf tip burn; may inhibit flower formation.
- (*Si*)-*Silicon*: Component of cell walls; enhances resistance to sucking insects and fungi.Foliar sprays reduce populations of aphids on some plants. Enhances leaf presentation; improves heat, drought and cold tolerance; improves photosynthesis; extends bloom life. Def.: wilting, poor fruit and flower set, increased susceptibility to insects and disease. Disease resistance is enhanced by regular foliar feeding.

- (*Zn*)-*Zinc*: Functional part of enzymes including auxin (growth hormone) synthesis, carbohydrate metabolism, protein synthesis, stem growth. Def.: mottled leaves, irregular yellow areas. Zinc deficiency leads to iron deficiency. Occurs in eroded soils; least available at pH of 5.5-7.0. Lower pH can cause availability to the point of toxicity.

## NUTRIENT BALANCE AND NUTRIENT DEFICIENCIES

Plants (like people) need a 'balanced diet'. They need all 13 nutrients to remain healthy. If one is missing, the plant will not grow well. Poor plant nutrition causes plants to grow slowly in the nursery and in the field, and to be more susceptible to diseases. Many people confuse the symptoms of nutrient deficiencies with those of too much or not enough shade or water.

In fact, all three factors, shade, water and nutrients affect plant growth, and interact to produce healthy plants. A plant that grows in full light with abundant moisture and receives all the 13 nutrients will grow fast and have a dark green colour in its leaves. A plant that grows slowly in the shade may also have dark green leaves, but when exposed gradually to the sun, the leaves may turn yellow. This does not mean that plants do not like full sun—it might indicate a nutrient deficiency which did not show up in the shade because the plant did not have enough light to stimulate fast growth. Together, water, shade and nutrients must be monitored and adjusted to produce quality seedlings. It takes practice to learn the signs identifying a missing nutrient or nutrients, but you can learn to do so, and some of the signs are common to many plants. A *good nursery practice* is to carefully monitor the leaves of your plants for signs of nutrient deficiency, and correct them with a better substrate or with fertilizer.

### Inorganic Fertilizers

Inorganic fertilizers are mined from the soil, or produced during complicated chemical reactions. A *good nursery practice* is to read the fertilizer labels. This allows you to apply what the plants need without wasting nursery resources. Fertilizers contain only plant nutrients; they are not used to combat plant diseases or insects. Inorganic fertilizers do not improve the substrate physical properties, whereas organic material such as compost does. Inorganic fertilizers are also expensive and not always available in the stores.

Nursery managers should carefully consider the cost and benefit of buying these products. Granular fertilizers are commonly given names like "17-17-17", or "10-30-10". What do the numbers mean? They represent the percentages of nitrogen (N), phosphorus (P), and potassium (K) in the fertilizer—17% N, 17% P, 17% K. In this case, 51% of the mixture is made up of N-P-K, and the rest is inactive material used to help spread the fertilizer evenly. Urea contains only N, and is labelled as 46-0-0. Urea is very strong and can easily burn the plants if too much is applied.

### Granular Fertilizers

Granular fertilizers can be mixed into the substrate or into the irrigation water, or be applied to older plants on the soil surface. It is better to mix the fertilizer directly into the substrate before planting the seed because the roots can avoid or seek the fertilizer as they need it. Use only small quantities such as 2 or 4 grams (1/2 teaspoon) per 1 kg of soil. It is better to add too little then too much. You need to experiment with different levels. Plants should respond within two weeks. When dissolving fertilizer in warm water, carefully note whether it is thoroughly dissolved. If not, it is probably the phosphorus that remains. It may be better to apply fertilizer in granular form if it does not dissolve thoroughly. Apply liquid fertilizer to the soil, not to the leaves which are easily burnt if fertilizer remains on them. Be extremely cautious when applying fertilizer to young plants.

### Foliar Fertilizers

Foliar fertilizers are used in order to get the nutrients to the plants quickly. They are specially formulated to apply directly on the leaves. Foliar fertilizers are absorbed by the leaves, not by the roots. When plants are acutely deficient in nutrients, foliar fertilizers often help 'green them up'. Frequently, foliar fertilizers only contain the micronutrients, since it is assumed that the macronutrients are available in the substrate.

However, some such as 'GroGreen'® contain both micronutrients and 20-30-10 of N-P-K. Often an adhering agent such as 'Da-Plus' is used to help the fertilizer stay on the leaves so that it is not washed off in the rain. Because foliar fertilizers are expensive, and they may not encourage strong root growth, they should not be used as a long-term solution for plant nutrients.

## COMMON NUTRIENT DEFICIENCY SYMPTOMS OF MACRONUTRIENTS

- *Nitrogen*: This is a mobile nutrient, which means that when nitrogen is deficient, plants move it from the older foliage to the younger, actively growing leaves. The older leaves (the ones lower on the stem of the tree) become yellow first, while the new leaves remain green.
- *Phosphorus*: The entire seedling is stunted, especially during early growth. Depending on the species, the leaves may become dull green, yellow or purpletinged. The purpling of leaves is a classic symptom, but sometimes there are no colour differences in leaves, so visual diagnosis is not always reliable. The purple colour should not be confused with new leaves that often appear purple or red when they first flush out.
- *Potassium*: Symptoms appear in older leaves first. These start to yellow at the edges, and have some green at the base. Later, leaf edges turn brown and may crinkle or curl and small necrotic (dead) spots may

appear. Plants may wilt, even though sufficient water is available in the substrate. When deficiencies are severe, leaves will die. Calcium: This is difficult to detect because signs include slow growth, and dieback of bud or root tips. Seedlings will have stubby little roots with brownish discoloration. The problem is most common in very acidic soils. A well-developed root system with many fine root hairs is important for calcium uptake.

- *Magnesium*: This nutrient is commonly deficient in coarse-structured soils and in acidic soils. Uptake may be blocked if there is too much potassium in the soil. Like nitrogen, magnesium is a mobile nutrient, so deficiency symptoms show up in the older leaves first. These leaves show a very characteristic yellowing between the veins or ribs, and they appear streaked.
- *Sulphur*: Plants will be slightly stunted. This is not a mobile nutrient, so the symptoms show up on younger leaves which are initially light green, but eventually develop scorched and curled margins. Dry areas can form along the margins and then spread inward to the leaf midrib.

## PHOSPHORUS IN PLANTS

Phosphorus is an essential nutrient both as a part of several key plant structure compounds and as a catalysis in the conversion of numerous key biochemical reactions in plants. Phosphorus is noted especially for its role in capturing and converting the sun's energy into useful plant compounds. The two examples that follow illustrate how vital phosphorus nutrition is to normal plant development and production. Phosphorus is a vital component of DNA, the genetic "memory unit" of all living things. It is also a component of RNA, the compound that reads the DNA genetic code to build proteins and other compounds essential for plant structure, seed yield, and genetic transfer. The structures of both DNA and RNA are linked together by phosphorus bonds.

Phosphorus is a vital component of ATP, the "energy unit" of plants. ATP forms during photosynthesis, has phosphorus in its structure, and processes from the beginning of seedling growth through to the formation of grain and maturity. Thus phosphorus is essential for the general health and vigour of all plants. Some specific growth factors that have been associated with phosphorus are: stimulated root development, increased stalk and stem strength, improved flower formation and seed production, more uniform and earlier crop maturity, increased nitrogen N-fixing capacity of legumes, improvements in crop quality, and increased resistance to plant diseases. Phosphorus deficiency is more difficult to diagnose than a deficiency of nitrogen or potassium. Crops usually display no obvious symptoms of phosphorus deficiency, other than a general stunting of the plant during early growth, and by the time a visual deficiency is recognized it may be too late to correct in annual crops. Some crops, such as corn, tend to show an abnormal discoloration when phosphorus is deficient. The plants are usually dark bluish-green in colour with leaves and stem becoming purplish. The degree of purple is influenced by the genetic makeup of the plant, some hybrids showing much greater discoloration than others.

The purplish colour is due to accumulation of sugars which favours the synthesis of anthocyanin (a purplish coloured pigment) that occurs in the leaves of the plant. Phosphorus is highly mobile in plants and when deficient it may be translocated from old plant tissue to young actively growing areas. Consequently, early vegetative responses to phosphorus are often observed. As a plant matures, phosphorus is translocated into the fruiting areas of the plant where high energy requirements are needed for the formation of seeds and fruit. Phosphorus deficiencies late in the growing season affect both seed development and normal crop maturity. The percentage of the total amount of each nutrient taken up is higher for phosphorus late in the growing season than for either nitrogen or potassium.

**Table. Percentage of Each Plant Nutrient Absorbed by a 180 bu/acre Corn Crop During Successive Twenty-five Day Growing Periods**

| | First (Early) | Second (Growth) | Third (Silk) | Fourth (Grain) | Fifth (Mature) |
|---|---|---|---|---|---|
| N | 8 | 35 | 31 | 20 | 6 = 100 |
| $P_2O_5$ | 4 | 27 | 36 | 25 | 8 = 100 |
| $K_2O$ | 9 | 44 | 31 | 14 | 2 = 100 |

## PHOSPHORUS IN SOILS

The total phosphorus content of most surface soils is low, averaging only 0.6% phosphorus. This compares to an average soil content of 0.14% nitrogen and 0.83% potassium. The phosphorus content of soils is quite variable ranging from less than 0.04% $P_2O_5$ in the sandy soils of the Atlantic and Gulf coastal plain to over 0.3% in soils of the northwestern United States. Many factors influence the content of soil phosphorus.

*Among these are*:

- Type of parent material from which the soil is derived;
- Degree of weathering;
- Climatic conditions.

In addition, soil phosphorus levels are affected by erosion, crop removal and phosphorus fertilization. Soil phosphorus is classified in two broad groups, organic and inorganic. Organic phosphorus is found in plant residues, manures, and microbial tissues. Soils low in organic matter may contain only 3% of their total phosphorus in the organic form, but high organic matter soils may contain 50% or more of their total phosphorus content in the organic form. Inorganic forms of soil phosphorus consist of apatite (the original source of all phosphorus), complexes of iron and aluminum phosphates, and phosphorus absorbed on clay particles. The solubility of these phosphorus compounds, as well as organic phosphorus is extremely low and only very small amounts of soil phosphorus are in solution at any one time. Most soils contain less than a pound per acre of soluble phosphorus, with some soils containing considerably less. Through adequate phosphorus fertilization and good crop/soil management, soil solution phosphorus can be replaced rapidly enough for optimum crop production.

## SOIL PHOSPHORUS AVAILABILITY

Soluble phosphorus, either from fertilizer or natural weathering, reacts with clay, iron, and aluminum compounds in the soil and is converted readily to less available forms by the process of phosphorus fixation. Because of these fixation processes, phosphorus moves very little in most soils (less than an inch in most soils), stays close to its place of origin, and crops seldom absorb more than 20 per cent of fertilizer phosphorus during the first cropping season after application. As a result, little soil phosphorus is lost by leaching. This fixed, residual phosphorus remains in the rooting zone and will be slowly available to succeeding crops. Soil erosion and crop removal are the significant ways soil phosphorus is lost.

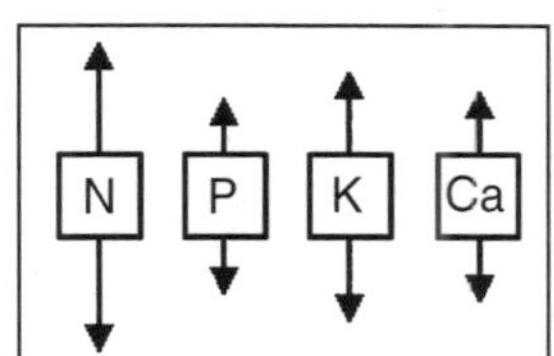


**Fig.** Relative Movement of Nutrients in the Soil

## AVAILABILITY OF PHOSPHORUS

### Soil pH

Precipitation of phosphorus as slightly soluble calcium phosphates occurs in calcareous soils with pH values around 8.0. Under acid conditions, phosphorus is precipitated as Fe or Al phosphates of low solubility.

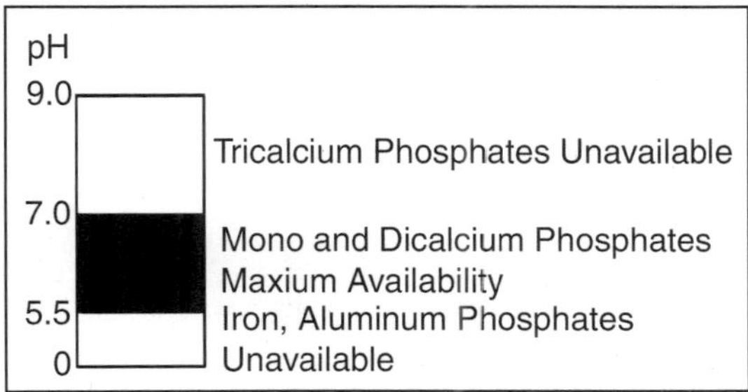


**Fig.** Soil pH Affects Phosphorus Availability

Maximum availability of phosphorus generally occurs in a pH range of 6.0 to 7.0. This is one of the beneficial effects of liming acid soils. Maintaining a soil pH in this range also favours the presence of $H_2PO_4$ - ions which are more readily absorbed by the plant than $HPO_4$ = ions which occur at pH values above 7.0.

**Other Nutrients**

Adequate supplies of other plant nutrients tend to increase the absorption of phosphorus from the soil. Application of ammonium forms of nitrogen with phosphorus increases phosphorus uptake from a fertilizer as compared to applying the phosphorus fertilizer alone or applying the nitrogen and phosphorus fertilizers separately. Applications of sulfur often increase the availability of soil phosphorus on neutral or basic soils where the soil phosphorus is present as calcium phosphates.

**Organic Matter**

Soils high in organic matter contain considerable amounts of organic phosphorus which are mineralized (similar to organic nitrogen), and provide available phosphorus for plant growth. In addition to supplying phosphorus, organic matter also acts as a chelating agent and combines with iron, thereby preventing the formation of insoluble iron phosphates. Heavy applications of organic materials such as manure, plant residues, or green manure crops to soils with high pH values not only supplies phosphorus, but on decomposition provides acidic compounds which increase the availability of mineral forms of phosphorus in the soil.

**Type of Clay**

Clay particles tend to retain or fix phosphorus in soils. Consequently, fine-textured soils such as clay loam soils have a greater phosphorus fixing capacity than sandy coarse-textured soils. Clays of the 1:1 type (kaolinite) have a greater phosphorus fixing capacity than the 2:1 type clays (montmorillonite, Illite, vermiculite). Soils formed under high rainfall and high temperatures contain large amounts of kaolinitic clays and therefore have a much greater fixing capacity for phosphorus than soils containing the 2:1 type clay. High temperatures and high rainfall also increase the amount of iron and aluminum oxides in the soil which contributes greatly to the fixation of phosphorus added to these soils.

**Application Timing**

Fixation of soil phosphorus increases with time of contact between soluble phosphorus and soil particles. Consequently, more efficient utilization of fertilizer phosphorus is generally obtained by applying the fertilizer shortly before planting the crop. This practice is especially effective on soils with high phosphorus-fixing capacities. On coastal plain areas, fertilizers may be applied several months before planting with little or no decrease in availability of the fertilizer phosphorus to the crop. Banding of fertilizer for row crops is also much more likely to increase the efficiency of fertilizer phosphorus on soils of high phosphorus-fixing capacity than on soils of low phosphorus-fixing capacity.

## Soil Temperature/Aeration/Moisture and Compaction

Phosphorus absorption by the plant is decreased by low soil temperature and poor soil aeration. Starter fertilizers containing water-soluble phosphorus are much more likely to increase crop growth during cool weather. Excessive soil moisture or soil compaction reduces the soil oxygen supply and decreases the ability of the plant roots to absorb soil phosphorus. Compaction reduces aeration and pore space in the root zone. This reduces phosphorus uptake and plant growth. Compaction also decreases the soil volume that plant roots penetrate, limiting their total access to soil phosphorus.

## Soil Test Phosphorus Levels

Crop responses to fertilizer phosphorus will be greater and occur more frequently on soils testing low in phosphorus than on high testing soils. However, yields on soils with high P soil test levels usually are higher. The response to phosphorus fertilizer on high testing soils is increasing and it is important to maintain high soil phosphorus levels to support optimum crop production.

# 10

# Healthy and Palatable Deciduous Fruit Cultivation

From amongst the umpteen and healthy and palatable deciduous fruit cultivation in India, apple demands and calls for incredible attention. Indeed, apple was introduced into the country by the British Raj in the Kullu Valley of the Himalayan state of Himachal Pradesh as far back as 1865. On the other hand, the coloured 'delicious' cultivars of apple were introduced to the Shimla hills of the same state in 1917. The apple cultivar Ambri, is measured as indigenous to Kashmir and had been grown long before Western introductions. Pears and other deciduous fruits were domesticated lucratively during the early part of the 20th century, although some of them were accounted to occur under semi-wild conditions much earlier. Apricot was found growing in the drier pockets of northwestern Himalayas and two apricot varieties, locally acknowledged as Halman and Rakchaikarpo, are reported to be indigenous to the Leh-Ladakh region Jammu and Kashmir state.

Sweet cherry was introduced from Europe before India's independence in 1947, whereas, commercial cultivars of sour cherry have been fetched primarily from USA in more recent years. The European and Japanese plum kinds are cultivated both in high and low hill areas. A plum variety 'Santa Rosa', reported to be a hybrid between Japanese and American species, predominates (70-80 percent) plantations in the Himalayan hills. Low chilling cultivars of peach and nectarine such as Flordasum, Flordared, and Sunred nectarine are booming introductions to the north-Indian plains. Some local selections of peach, plum and sand pear (Patharnakh) have also taken fruit cultivation in India on a grand commercial scale in subtropical marginal chilling areas of north India.

Fruit cultivation in India, quite understandably grows most prolific in the areas of the Himalayas, with abundance speaking out from every angle. The Uttar Pradesh hills, particularly the Kumaon hills division, possesses exceptional advantage of early harvest of apple, principally due to cultivation of early maturing varieties like Early Shanburry, Fanny and Benoni. The early maturing varieties are harvested 2-3 weeks before the onset of fresh apple from Himachal Pradesh and Jammu and Kashmir and hence, fetch very moneymaking prices. Deciduous fruits, covering pome and stone fruits contribute significantly to the horticulture economy of India. Truly, cultivation of fruits in India, does depend upon the success of plantation, harvest, soils and a perfect ambience of brilliant minds.

## CONSTRAINTS IN DECIDUOUS FRUIT PRODUCTION

Large number of old orchards (more than 30 years old) are showing decline in terms of growth and fruit yield. Such old trees do not produce adequate extension growth. Large scale replanting is therefore needed. Delicious group of cultivars constitute the major share (about 83% in H.P.) of apple production in the country. These cultivars are self unfruitful and need cross pollination to ensure good fruit set. Interplanting pollinizer cultivars (Golden Delicious, Jonathan, Red Gold, Lord Lambourne etc.) in the proportion of 25 to 33 percent is necessary for good fruit set, and choice of wrong pollinizers and their inadequacy in number often result to low productivity.

In many countries, Delicious group has been replaced or is in the process of replacement with more promising cultivars. The need for injecting new blood into the apple industry through spread of new cultivars (spur types, colour mutants, strains of Gala, Red Fuji; scab resistant cultivars, bud sport selections of Royal Delicious, and some of the promising hybrids) is urgently felt. Some of the spur type and coloured mutants are already popular with farmers and high density planting has also caught the imagination of developmental departments and agencies both in H.P. and J&K. The research system has already identified Early, Mid and Late cultivars for different agro-climatic regions. The low chilling cultivars of stone fruits have also covered large tracts of the subtropical plains of Punjab, U.P and H.P. For the hills, promising cultivars identified need further spread.

Generally, apple is grown in marginal land and fertilizers are not applied according to the requirements of the trees. The water and fertilizer use efficiency is generally poor. Also, spring frost and hailstorms are adverse weather parameters leading to low productivity. Research results have shown that through proper orchard management practices (soil and water conservation and fertilizer application) the fruit yield can be doubled in the existing orchards. The adoption of improved production technology developed by the research system can bring visible and perceptible changes in the temperate fruit industry in India. Technologies like use of clonal rootstocks, introduction of renewal pruning techniques and micro nutrient applications have not been transferred and adopted at a satisfactory level. Apple scab disease has been the major plant protection problem in apple in J&K and H.P, whilst U.P hills are comparatively free from the disease. Apple scab forecasting system developed and the chemical control schedule prescribed have helped in reducing loss due to apple scab to a considerable extent. Apple growers are adopting the prescribed schedule of chemical sprays to control the disease. For checking entries of diseased material in the free areas of U.P. and North-Eastern Hills, strict quarantine and selection of elite disease-free mother plants are very essential.

Often it is not followed strictly. Some of the virus diseases have also been reported in apple for which biological and serological indexing/detection procedures have been developed. Limited quantity of virus-free budwood is also being supplied. Extreme care is now required to be taken to multiply quality planting material (in apple alone approximately 2 million plants/year) for establishing new plantations. Most of the orchardists still sell their crop at flowering to contractors as there is no well organized marketing system. Transportation in the hills itself is problematic. Post-harvest management problems originating from poor harvesting (strip picking) and improper packing system (non CFB boxes) and lack of proper pre-cooling and cold storage facilities result in huge (25-30%) loss of fruits. Capacity of the processing sector is also inadequate. Product diversification, value addition and market infrastructure development would require very substantial investment. The existing processing units are quite old and they require modernization for which substantial investment is required. CA storage trials have shown good promise. Its extension in larger growing areas is needed. Technology for storage of apple is now known, as a result of which apple is now available throughout the year.

## GOVERNMENT POLICIES AND PLANS FOR RESEARCH AND DEVELOPMENT

Research on pome and stone fruits is conducted mainly by three State Agricultural Universities, namely: a) Sher-e-Kashmir University for Agriculture Science & Technology, J&K; b) Y.S. Parmar University for Horticulture and Forestry, H.P.; and c) G.B. Pant University for Agriculture & Technology, U.P. A good number of research stations of these Universities located in major pome and stone fruit growing belts are engaged in temperate fruit research. The Indian Council of Agricultural Research (ICAR) through All India Coordinated Research Projects on Fruits and Post-harvest Technology and another Network project on Apple Scab disease has further strengthened the research activities in deciduous fruits. A few long established temperate fruit research stations namely at Shalimar Bagh in J&K, at Mashoobra in H.P. and at Chaubatia in U.P. hills have made commendable progress in temperate fruits. Both State Governments and the ICAR provide financial support to strengthen fruit research in these Universities. A large number of ad-hoc research projects on pome and stone fruits funded by ICAR are also generating good information on these crops.

During the 8th Government Developmental Plan (1992-1997) the ICAR established a Central Institute for Temperate Horticulture (CITH) with its headquarters at Srinagar in the J&K State, with a regional station at Mukteshwar in the U.P. Hills. Both these research stations will work exclusively on temperate fruit crops.

This new Institute will receive major support during the 9th Plan period (1997-2001). In the North-Eastern Hills region, the ICAR Research Complex for NEH Region with its headquarters in Meghalaya State and regional units in each of the 5 other States are engaged in fruit research.

On the developmental side the State governments are engaged in nursery production of quality planting material. For example, in H.P. alone currently there are 600 nurseries in private and public sector producing and distributing more than 0.8 million plants of apple alone, every year. There are ambitious programmes in all the States to further expand/replant with new improved cultivars. Apple scab disease control and post-harvest processing sectors are getting focused attention in Government developmental plans. The Directorate of Marketing and Inspection of the Government of India has framed grade standards for apple, plum and William pear. The organizations like National Horticultural Board (NHB), National Cooperative Development Corporation (NCDC), Agricultural and Processed Food Products Export Development Authority (APEDA) etc., are providing incentives to traders and exporters to improve their infrastructural facilities like grading and packaging centres, refrigerated transport, setting up of pre-cooling, cold storage, auction platforms etc. The NCDC is undertaking procurement and marketing of apple on a limited scale. The NHB has set up a market information service for the benefit of growers.

## PRESENT SITUATION OF DECIDUOUS FRUIT CROP CULTIVATION

### CROPS GROWN

India grows crops like apple (*Malus pumila Mill.*), pear (*Pyrus communis Berm.f.*), peach/nectarine (*Prunus persica (L). Bats. Ch.*), plum (*Prunus domestica L.*), apricot (*Prunus armeniaca L.*), sweet cherry (*Prunus avium L.*) and sour cherry (*Prunus cerasus L.*) on a commercial scale.

#### Cultivars

The promising cultivars of different temperate fruits in 3 major deciduous fruit growing States of India. Some relevant information on cultivars and cultivar selection is indicated below:

*Apple:* Over 700 accessions of apple, introduced from USA, Russia, U.K., Canada, Germany, Israel, Netherlands, Australia, Switzerland, Italy and Denmark have been tried and tested during the last 50 years. The delicious group of cultivars predominates the apple market. The areas covered under Delicious cultivars are:

83% of the area under apple in H.P., 45% in J&K and 30% in U.P. hills. In more recent times improved spur types and standard colour mutants with 20-50% higher yield potential are favoured. The important selections are:

- Spur types-Red spur, Starkrimson, Golden spur, Red Chief and Oregon spur.
- Colour mutants-Vance Delicious, Top Red, Skyline Supreme.
- Low chilling cultivars-Michal, Schlomit.
- Early cultivars-Benoni, Irish Peach, Early Shanburry, Fanny
- Juice making cultivars-Lord Lambourne, Granny Smith, Allington Pippin.
- Scab resistant cultivars-Co-Op-12, Florina, Firdous, Shirean.
- New Hybrids-Lal Ambri (Red Delicious X Ambri), Sunehari (Ambri X Golden Delicious), Amred (Red Delicious X Ambri), Chaubatia Anupam & Chaubatia Princess (Early Shanberry X Red Delicious) developed in India.

In H.P. monoculture of a few cultivars such as Royal Delicious, Red Delicious and Richared have started showing negative impact on the apple industry. Serious problems of apple scab disease and outbreak of premature leaf fall and infestation of red spider mite are causing great concern. U.P. Hills, particularly the Kumaon hills division, have the unique advantage of early harvest of apple, mainly due to cultivation of early maturing varieties like Early Shanburry, Fanny and Benoni. The early maturing varieties are harvested 2-3 weeks before the arrival of fresh apple from H.P. and J&K, and hence fetch very remunerative prices.

*Pear:* In pear for higher altitude conditions high chilling requirement varieties (like Bartlett) are mainly grown. In more recent years, red-colour strains of pear like Max Red Bartlett, Red Bartlett and Starking are replacing yellow coloured cultivars. In warmer sub-mountainous areas of H.P. and subtropical Punjab oriental pear cultivars like Baghugosha, Kieffer, China and sand pear Patharnakh are cultivated commercially both for table and processing purpose.

*Apricot:* Generally there are two types of apricot, namely, sweet kernel type and bitter kernel type. About 81 exotic accessions and 19 indigenous cultivars were collected and evaluated. The local types Halman and Rakhaikarpo have been popular, whereas exotic introductions namely, Nari, Kaisha, Shakarpa and New Castle are promising. These cultivars are recommended for dry cold areas. The USA variety Nugget is self-fruitful and bears sweet and attractive coloured fruits.

*Peach:* For colder conditions the peach cultivars July Elberta, Elberta, Peshwari, Quetta, Burbank and Stark Earliglo are well adopted. Low-chilling cultivars *viz.* Flordasum, Flordared, Shan-e-Punjab, Sharbati and Sunred (nectarine) have become popular in subtropical belts of U.P. and Punjab States.

*Plum:* A large number of cultivars have been introduced into the country. European plums performed better in the hills, while Japanese plums are more adopted in sub-mountainous lower elevations. Leading cultivar in the hills is Santa Rosa. In the north-Indian plains small fruited cultivars like Titron, Kala Amritsari, Kelsey, and Alubukhara showed better performance. A good number of low-chilling Florida hybrids Sungold, Redgold etc, are under evaluation.

*Cherry:* Many cultivars of sweet cherry have been introduced from Europe, USSR and British Columbia. Promising exotic cultivars like Bigarrean Napoleon, Black Heart, Guigne Noir for J&K and cultivars like Black Tartarian Bing, Napoleon (white) Sam, Sue (White), Shella for H.P. have been identified. For warmer climate, cultivars like Summit, Sunburst, Lapins, Compacat and Stella have been found to be promising.

**Rootstocks**

- Apple has been propagated mainly on seedling stocks of apple (*Malus pumila Mill*). Clonal rootstocks of M and MM series were introduced from East Malling Research Station of U.K. and M-9, M-26, M-4, M-7, MM-106 and MM-111 were identified as promising. Merton 779 was recommended as a commercial rootstock for apple for Kumaon hills of U.P. However, clonal rootstocks have not found favour among the apple growers. *In-vitro* protocols of M-9, MM-106 and MM-111 have been developed. Clonal rootstocks have been preferred only for high density orchards. Four rootstocks, namely, EMLA-111 for drought prone areas, EMLA-7 for slopy land and EMLA-106 for slopy and less clayey soil and EMLA-9 for high density planting on irrigated deep soils have been recommended. In pears, seedling stocks of *Pyrus pashia* (Kainth) is commercially used. Clonal rootstock Quince A with semi-vigorous growth has some promise. Newer clonal rootstocks like Old Home X Farmingdale series and BA 29C are under evaluation.
- For apricot, seedlings of wild apricot have been found to be good giving a smooth union and vigorous scion growth. Peach can also be tried as a rootstock, particularly for dry and light soils. Myrobolan plum is recommended for high moisture soil conditions.
- Peach is normally grafted on wild peach seedlings, plum and even on apricot. Wild peach produced healthy and high-yielding plants. For controlling tree size, apricot rootstocks which impart dwarfing may be used.
- For plum, wild peach and wild apricot are commonly used as rootstocks. Among clonal rootstocks, Myrobolan B. was found to be most promising. Wild apricot seedlings and wild peach were also found to be suitable rootstocks for commercial cultivars like Santa Rosa, Mariposa and Beauty.

- Commercial plantations of sweet cherry were raised mainly on Mazzard and Mahaleb rootstocks. Seedlings are raised from late maturing cherry cultivars and stones of Waterloo or Misri cultivars are often used as rootstocks. Rooting of cherry suckers has been made possible with use of rooting hormones like IBA (6000 ppm) in combination with NAA (2000 ppm).

## PROPAGATION AND ROOTSTOCK INFORMATION

In all the pome and stone fruits vegetative propagation techniques of budding or grafting are followed for multiplication of planting material on standard rootstocks, raised both from seeds or through clonal methods. Cropwise details are as follows:

### Apple

For raising rootstock seedlings, seeds of crab apple or commercial cultivars are stratified during December for 2-3 months at 2-5°C. One year old seedlings are used for budding/grafting. Clonal rootstocks are raised through mound or stool layering. The mother plants are allowed to grow for one season and cut back to 3 cm from the ground level just before the growth begins. When new growth is about 10 cm, the shoots are covered with soil leaving the growing parts exposed. Rooted layers are cut off close to the ground level and planted in nursery beds for grafting/budding.

In H.P. hills, different types of budding/grafting are recommended:

* Chip budding in mid-June and mid-September
* Whip and Tongue and cleft grafting in February-March
* T-budding in May-June.

### Pear

Seedlings of Kainth (*Pyrus pashia Linn.*) or Shiara (*P. serotiana Rehd*) are stratified in moist sand for 35-45 days at 2-5°C. Tongue grafting in February-March or T-budding during June-July is recommended.

### Mango

Mango (Mangifera Indica) is the most ancient among the tropical fruits and believed to have originated in the Indo- Burma region. India is the major mango producing country in the world with an annual production of 8.50 million tonnes from an area of one million hectares. Mango is basically a tropical plant but endures wide range of temperature. It grows well under tropical and sub-tropical conditions. It gives profitable yield in semi-arid conditions, especially with irrigation. It grows where temperature drops as low as 0 deg. Cent. And as high as 46 deg.cent. However, it thrives best at temperature around 27 deg. Cent. It grows well both in low (25 cm) and high (250 cm) rainfall areas. However with annual rainfall around 75 cm, it grows withoug irrigation. The mango flowers and fruits during dry season, which is chracterised by absence of rainfall. Rain or cloudy weather at the time of flowering causes considerable damage to mango as it adversely affects flowering and fruit set and increases incidence of pests and diseases. Mango needs a deep, fertile and perfectly drained soil. Alluvial, lateritic and medium black soils are considered good. Soil depth should be atleast 2.5 m. and soil pH between 5.5-7.5 is good.

*Varieties:* The important commercial mango varieties for different parts of India are:

*North India:* Dashehari, Langra, Chausa, Bombay, Green, Somar Bahist

*Eastern India:* Fazli, Krishna Bhog, Himsagar, Gulab Khas, Jardalu

Western India; Alphonso, Kesar, Pairi, Mankurd, Fernandin.

*South India:* Bangalore (Totapuri), Neelam, Baneshan, Badami, Suverna rekha.

*Newly Evolved Varieties:* Mallika, Amrapali, Arka Aruna (Hy 10), Arka Puneet (Hy 13), Arka Anmol, Ratna, Sindhu, Sai Sugandh.

Orchard establishment and care:

Pits of 1 x 1 x 1 m are dug at a distance of 10 x 10 meter at plant site in summer. Pits are filled with top soil and 5 to 6 baskets of Farm Yard Manure (FYM) and 2.5 Kg. Super Phosphate. To avoid termite attack 100 gm of 10 percent sevin or chlorudane dust is mixed in the pit.

Planting of mango grafts of desired variety is done with commencement of rains while in heavy rainfall regions. Planting is best done after censation of heavy rains. Selected healthy grafts with smooth union and true to type, obtained from reliable source should be planted. At the time of planting, the graft is carefully placed in the pit after removing the Polythene bag and soil is packed firmly around the roots. It is irrigated immediately and stalked with bamboo sticks. Young plants are trained by removing lower branches upto half meter from ground level, only 4 to 5 well spaced branches are allowed to grow in different directions to form main limbs of the Tree. Shoots arising on stem below graft joint are removed regularly.

In young orchards, vegetables and other low growing crops like groundnut, blackgram, green gram, cowpea are grown as intercrops/ covercrops. Cultivation helps to check the weed and improves the growth of the trees besides giving some income to the growers till mango plants start bearing 4-5 years after planting. A young orchard be supplied with 10 kg FYM + 100 Kg N + 50 g $P_2O_5$ + 100 g $K_2O$ per tree in the first year and it should be increased with age. The 10 year old tree should get FYM + 1 Kg N + 500 g $P_2O_5$ + 1000 g $K_2O$. It irrigation is available, additional dose of 500 g N be given in the month of February- March.

Newly planted grafts be irrigated at 3-4 days interval for 6 months, thereafter interval should be 8 - 10 days depending upon climate Irrigation interval of 10-15 days is adequate for 1-5 year old plants. For bearing trees, irrigation should not be given prior to flowering for 2-3 months as it promotes vegetative growth and reduce the crop. When the fruits have developed to the size of a marble, irrigation can be started and continued at weekly or fortnightly intervals. This helps to reduce fruit drop, promotes rapid fruit development, early maturity and improves quality. However, mango is mostly grown as rainfed crop.

*Harvesting:* Mango fruits mature in 3-4 months from flowering, Fruit colour changes from dark green to light green on maturity. Harvesting should be started after few fruits drop, during morning hours. Individual fruits are clipped with 1.5 cm stalk using mango harvester. Harvested fruits are kept on gunny bags under shade. Under sized, bruized and infested fruits are sorted out and healthy fruits are graded into 2-3 grades depending upon colour, shape and size and packed in wooden or corrugated boxes.

*Export Potential:* India is the largest producer of mangoes in the world accounting for over 63 percent of total world production. It is the most important fruit grown in the country, which accounts for over 60 percent of the total value of fruits exported from the country. A substantial quantity of this fruit is also utilised by the fruit processing industry. Maharashtra has clear advantage due to exclusive production of Alphonso variety, the share of which in the export is very high.

*Plant Protection:* Mango hoppers insectpest and powdery mildew disease are the major problems on mango. To control both these dusting with 10 percent Carbaryl dust plus Sulphr 300 mesh in the ratio of 1:1. Three to four times at fortnightly intervals starting from prior to the emergence of first inflorence is recommended. Trunks of the tree should also be dusted thoroughly.

**Ber**

Ber (Zizyphus mauritiana Lamk) is an ancient and indigenous fruit of India, China and Malaysia region. The fruits are very nutritious and are rich in vitamin C, A & B complex. The ber is one of the most common fruit trees of India and is cultivated practically all over the country. Ber fruits can be within the reach of the poor people and hence known as poor man's fruit.

*Importance:* Among the fruit trees, ber cultivation requires perhaps the least inputs and care. It gives good production even without irrigation and can be grown as a rainfed crop in semi-arid and arid regions. The tree can, therefore give assured income even under marginal growing conditions and provides nutritious food at very low cost. The fruit is dried and is used as a dessert fruit. It can also be preserved as a candied fruit.

*Climate:* The ber is a hardy fruit and grows well all over the country under varying climatic conditions and upto an elevation of 1000 meters above sea level. For its successful cultivation, it favours a hot and dry climate. The Ber crop withstands high temperature and aridity by cessation of growth, leaf fall and dormancy phase. In high humidity conditions disease and pest problems increase especially of powdery mildew which is a serious disease.

*Soil:* Ber plant grows on a wide variety of soils ranging from shallow to deep and from gravelly and sandy to clayey. The ber develops a deep tap root system within a short period of its growth, and as such to adverse soil conditions. Ber can also withstand alkalinity and slightly waterlogged conditions. It can, in fact, withstand and often do better than most fruits on poor soils.

*Varieties:* Numerous horticultural varieties of ber are grown all over India. Some of the most popular varieties are Umran, Karaka, Gola, Seb, Chhuhara, Sanaur - 2, Ilaichi and Mehrun. In Maharashtra, cv.Umran Is being grown on commercial scale. It has excellent keeping quality and transportability but flat taste.

*Propagation:* The Ber was commonly propagated by seeds during earlier period. But main disadvantage of this method is of heterozygosity and variability in seedling progeny. Therefore, propagation of superior varieties by patch budding is recommended. For raising a budded plantation, it is considered best to sow ber seeds in the field itself at proper distances and use the seedlings thus raised for budding in situ. For early germination of ber seeds, breaking of endocarp (hard seed coat) gives quick germination.

*Planting Season:* Monsoon season offers best choice, for raising in situ plantation in Maharashtra and arid and semi-arid parts of our country, at a spacing of 6 x 6 m. Protective irrigation during summer months will ensure the survival and good growth of the plants. Deheading these seedlings to ground level in the month of May helps to give new shoots by July. Patch budding these shoots in July helps to convert it into choice variety. In Northern India, planting is done either in February-March or July-September at a spacing of 7-8 months. After the layout, pits of 60x60x60 cm are dug. About 100 g of 10 per cent carbaryl or Aldrex dust is sprinkled on the bottom sides of pits to prevent termites. Pits are filled with top soil mixed with 20 kg farmyard manure and 1 kg super phosphates. Treated seeds or bud grafts are planted in these pits at the onset of Monsoon.

*Interculturing:* Area around the young plant is kept clean by weeding and hoeing. Under irrigated conditions low growing vegetables can be grown as inter crops until the ber plants assume (4-5 years) full grown. Under rainfed conditions legume crops like Moong, Moth (horse gram) and cowpea can be grown as intercrop.

Stirring the soil under the tree canopy after rains provides better aeration around roots and helps in conserving soil moisture and weed control.

**Care of young orchards**

*Training:* If left to itself ber trees would attain bushy, large unmanageable form but the production per unit area is quite low. In order to keep the plant in manageable shape and size, trees be trained properly during first 2-3 years to build a strong framework. By providing support with bamboo stick to the new growth of sprout from either in situ or transplanted seedling, vertical growth is encouraged Once the growth is about 1-1.5 meter, the terminal growth Is pinched, allowing lower buds on main stem to sprout and form main branches. After sufficient growth, these primary branches be pinched to develop secondary branches. Thus plants are trained by pruning and kept in manageable shape, size with well developed framework.

*Pruning:* Pruning is an essential operation in ber production as fruits are borne in the axil of leaves on the young shoots of current season. Pruning is therefore, done every year to induce maximum number of new healthy shoots which bear good quality fruits. The best time for pruning was observed to be 15th April to 15th May under prevailing climatic condition of Maharashtra.

*Special Horticultural Practices:* Attempts to increase fruit set, fruit size and early/delayed maturity through application of growth regulators proved to effective. The growth regulators like GA, NAA, CCC and Ethephon were used. GA (10 ppm) and NAA (10 ppm) during fruit development and Ethephon one month before harvesting were sprayed. Application of NAA increased the yield of fruits with cost benefit ratio of 1:3:30.

*Irrigation:* Ber is mostly grown as a rainfed crop. Irrigation as a rule not advisable. But if available could increase yield. However, increase in yield also diminishes the quality of fruit.

*Nutrition:* Like other fruit trees, ber also requires regular application of manures and fertilisers for good yields. Application of 30-50 kg Farm yard manure, 250 g N per tree (split in 2 equal doses) and 250 g $P_2O_5$ per tree + 50 g $K_2O$ per tree (single dose) per year for full grown tree (5 years and thereafter) is recommend for better yield and quality of fruits.

*Plant Protection:* Ber fruitfly is one of the important pests of ber, which is widely distributed throughout India. The infested fruits turn brown, rot and smell offensively. The pest can be controlled by spraying 3 to 4 times with Carbaryl 50 WP 0.2 per cent or Dimethoate 30 EC 0.03 per cent commencing from the attainment of fruits of the pea size. Among the diseases powdery mildew is very common on ber fruits. Small whitish spots appear on young fruits, which later enlarge and cover the entire fruit. The affected fruits either drop off or become corky, mis-shapen and under-developed. Disease can be controlled by dusting with sulphur @ 150 to 200 g /tree and subsequent 3 dustings at an interval of 15-20 days. Alternaria leaf spot and Cercospora leaf spot also appears in the form of grey spots on leaves. Both these diseases are effectively controlled by spraying dithane M-45 (0.25%) or Foltart (0.1%) as soon as the disease appear. Subsequent 2 to 3 sprays be given at an interval of 15 to 20 days depending upon the intensity of the disease.

Harvesting and yield: In Northern India, peak period of harvesting falls between February and April. While in Maharashtra harvesting extends from

November to January. The fruits are harvested in 4 or 5 pickings since all the fruits on the tree do not mature at one time. The fruit picking is done by hand using a ladder. The fruits should be harvested at proper stage of maturity. The best index of the correct picking stage is the characteristic maturity colour and softness of particular cultivar after the fruit has attained the full size. The fruit requires about 120 days to reach maturity.

Under dryland (rainfall) conditions, on an average 60-80 kg fruits per tree per year can be harvested. Under irrigated conditions yield will be 3-4 times higher. Post harvest handling and marketing: The underripe, overripe and damaged fruits are sorted out. The remaining sound fruits are graded in two grades -large and small according to size. Fruits are packed in gunny bags, wooden boxes, cardboard boxes or nylon knot bags. The growers generally auction their crop to the contractors. The contracts are fixed either on the share of accruing incoming basis or on the basis of a lump sum to be paid in installments to the grower.

**Grapes**

*Introduction :* The grape is the most important crop grown in the world. Mostly it grown for making wines and preparation of raisin and then as a table fresh fruit. While in India, it is mainly grown for table use. Grape cultivation is believed to have originated near Caspian Sea, however, Indians know grapes since Roman times. Total area under grapes in India is about 40,000 ha, distributed mainly in Maharashtra, Karnataka, Andhra Pradesh and Tamil Nadu.

*Economic Importance:* At present, grape is the most important fruit crop grown commercially with the objectives.

- For table purpose
- For export purpose
- For making wines and
- For making raisins.

Fresh grapes are a fairly good source of minerals like calcium, phosphorous, iron and vitamins like B. Famous champagne and other desert wines are prepared from grapes.

*Climatic Requirements:* The ideal climate for grape growing is the Mediterranean climate. In its natural habitat, the vines grow and produce during the hot and dry period. Under South Indian conditions - vines produce vegetative growth during the period from April to September and then fruiting period from October to March. Temperatures above 100C to 400C influence the yield and quality. High humidity and cloudy weather invite many fungal diseases, besides lowering the T.S.S. : Acid ratio.

*Soil Preference:* The grape is widely adopted to various soil conditions, but the yield and quality reach to the highest on good fertile soils have pH 6.5 to 8.5, organic carbon above 1.0%, free of lime and having a medium water holding capacity. Early but medium yields with high T.S.S. are harvested on medium type of soils.

*Varieties:* Table purpose varieties:

- *Seeded Varieties:* Cardinal, concord Emperor, Italia, Anab-e-shahi, Cheema sahebi, Kalisahebi, Rao Sahebi,
- *Seedless Varieties:* Thompson seedless, flame seedless, kishmish chorni, perlette, Arkavati.

*Raisin Purpose Varieties:* Thompson seedless, manik chaman, sonaka, Black corinth, Black monukka, Arkavati, Dattier

*Wine Varieties:* Chardonnay, Cabernet Saurignnon, Bangalore Blue, Muscat, Blanc, Pinot Noir, Pinot Blane, White Riesling, and Merlot.

*Propagation:* Grapevine is most commonly propagated by hard-wood cuttings, though propagation by seed soft wood cutting, layering, grafting and budding is specific to certain situations. Occasionally, unrooted cuttings are also planted directly in the field in the pre-determined position for a vine. For hardwood cuttings, IBA, 1000 ppm treatment is useful for early, better and uniform rooting of cutting. For grafting Dog ridge, Ramsey, 1616, 1613,1103P, So4, etc. are used. Sometimes the rootstocks are planted in the field and there they are grafted with suitable varieties.

*Planting and Season:*Usually planting is done from October onwards till January. Rarely planting is also done during June-July where the monsoon is late. Monsoon planting is avoided mainly for avoiding diseases on young growth. For planting N-S direction the trenches are opened. The size of trench may be 60 to 75 cm. Deep wide. Then these trenches are filled with FYM, organic manures, 5:10:5 organic mixtures, single super phosphates, biofertilisers, neem cakes, etc. Spacing for planting is maintained depending on soil type, variety and method of training. The distance between two rows may be 2 to 3 m while distance between vines within a row will be half of that, accommodating vines from 2000 to 5000 per hectare.

*Interculturing:* The following aspects are important:

- *Gap Filling:* To be done preferably during one month after planting.
- *Recut:* Basal cut keeping 2/3 buds is taken one month after planting with an objective to get uniform new growth.
- *Supporting:* The bamboo supports are fixed for vine support and young growing points are trained on them
- *Weeding:* The vine rows are weeded out, twice/thrice depending on the intensity of weeds.
- *Irrigations:* regular depending on soil and season are given.
- Fertilizers are applied with cow during slurry to hasten the growth.
- Suitable plant protection measures are following depending on incidence of pests and diseases.

Grapes vines takes about 1.5 to 2 years after planting to bear the first crop. During this period the care of young vines is taken as under:-

- *Training:* The vines are trained first on bamboo and then on support—trellis. A suitable method of training is adopted.
- *Pruning:* Initial pruning is done only for training *i.e.* for developing trunk, arm, fruiting, canes, etc.
- *The Fertilizer Doses:* Including organic, inorganic and bio-fertilizers are applied twice in a year.
- Plant protection schedule is prepared and followed for the total initial period of growing.

Special Horticultural practices:

- *Pruning and Training:* The vines are trained on a suitable trellis *i.e.* 'T', 'Y', 'H' or bower and regularly pruned twice in a year. First annual pruning is done during the month of April to get the new vegetative growth while second pruning to get the crop is done during the month of October. While doing April pruning 0 to 2 buds on arm are kept while doing October pruning 5 to 10 buds on fruiting cane are kept. Use of HCN is done to have early, uniform and higher sprouting particularly after winter pruning is made.
- *Girdling:* Vines are trunk girdled at bloom period to increase the fruit set, to increase the weight and T.S.S. and also to enhance maturity.
- *Use of Hormones:* The following plant hormones at various stages and concentrations are usually used to increase the yield and to improve the quality of bunches.

| | | | |
|---|---|---|---|
| A.At bloom | GA3 | 20 to 30 ppm; CCC | 500 ppm |
| B.At setting | GA3 | 30 to 40 ppm; 6BA | 5-10 ppm |
| C.At 4-6 mm size | GA3 | 40 to 50 ppm; Brassinos | 100 ppm |
| D.At 6-8 mm size | GA3 | 30 to 50 ppm; Cppu | 2-3 ppm |

*Irrigation:* Grape is strictly irrigated perennial crop and regularly irrigated. For flood irrigation, 5-7 days during summer 8-10 days during winter and 15-20 days during rainy season- interval is maintained while for drip irrigation, 40-50 L; 30-40, 20-30 L of water per day per vine, water is applied.

*Nutrition:* Balanced nutrition and use of chemical, organic and bio-fertilisers is essential to get a good crop of good quality every year. About 700 to 900 N, 400 to 600 P and 750 to 1000 K Kgs/ ha/year are applied to get about 30 to 35 tonnes produced yearly.

The use of vermiphos, biomeal, mixtures of 5:10:5 ormichem, micronutrient mixtures have proved useful in grape production. Fertilizers are applied mainly twice in a year at the time of pruning, besides occasional foliar sprays are also practiced. Now-a days, Fertigation techniques is being popular in grape growers.

*Plant Protection:* Grape shoots, leaves, blossoms and berries are attacked by many fungi and insect pests, besides some nematodes are also cause damage to roots.

*Major Fungal Diseases:* Anthracnose, Powdery mildew, Downy mildew, Dead arm, Botraytis and Botrodiplodia.

*Major Bacterial Diseases:* Xyantomonas, blight.

*Viral Diseases:* Fan leaf disease.

*Major Insect Pests:* Flea beetle, Mealy bug, Red mite, Thrips, caterpillars.

*Soil Born Pests:* Nematodes, phyloxere, white ants, and white grub.

*Control Measures:* A number of systemic and contact fungicides and pesticides are available and are to be used as per following local schedule.

Besides pests and diseases the crop is to be protected against weeds, Cyprus, doob grass, Parthenium olerace are some of the common and important weeds found in vineyards. They are controlled by frequent weeding fruit ching/growing cover crops or by using chemical weedicides as Gramaxone, Basaline, Roundup, Glycel, etc. Grape bunches are also to be protected against hot sun, Cold wave, dry air spell, Dew and Storm. Some chemical some physiological and some mechanical means methods are adopted.

*Harvesting and Yields:* Normal grape harvest season starts in February and continuous up to end of April. Well matured bunches having at least 180 Brix are harvested

Av. yields - For seedless varieties - 20 to 30 t/ha/y

For seeded varieties - 40 to 50 t/ha/y

*Post Harvest Handling:* Harvested grapes are packed in 2 to 4 kg-corrugated boxes. Grape guards, pouches are kept inside the boxes for distant markets. Pre-cooling and use of grape guards are the musts for cold storage and export markets.

Mumbai, Delhi, Calcutta, Ahmedabad, Ludhiana, Patna, Jamshedpur, Bangalore, Hyderabad, are the main market places in the country.

Some additional features:

- *Export:* Day by day increased quantities are exported to Europe, Middle East, Dubai, etc.
- *Winemaking:* Wine making and champagne making are profitable and started at few places in Maharashtra and Karnataka.
- *Raisin Making:* A good quality black and golden raisins are prepared from the varieties. The raisins are increasing demand from Indian as well as from outside countries.

## Strawberry

Strawberry (Fragaria sp.) is a native of temperate regions, but varieties are available which can be cultivated in sub. Tropical climate. In India it is generally cultivated in the hills. Its main centres of cultivation are Nainital (district) and Dehradun in Uttar Pradesh, Mahabaleshwar (Maharashtra), Kashmir Valley, Bangalore and Kalimpong (West Bengal). In recent years, strawberry is being cultivated successfully in plains of Maharashtra around

Pune, Nashik and Sangali towns. The strawberry is the most widely adapted of the small fruits. Strawberries are grown throughout Europe, in every state of the United States, as well as in Canada and South America. The wide variation in climates within these regions and the wide adaptation of the strawberry plant permit harvesting and marketing, the fruit during greater part of the year.

*Importance:* Strawberry is a delicious fruit taken fresh in several ways. It also makes excellent ice cream and Jam on account of its rich aroma, and is also a good source of vitamin C. It is a soft and a highly perishable fruit, often shipped in frozen condition in Western countries.

*Climate:* Strawberry thrives best in temperate climate. It is a short day plant which requires exposure to about 10 days of less than 8 hours sunshine for initiation of flowering. In winter, the plants do not make any growth and remain dormant. The exposure to low temperature during this period helps in breaking dormancy of the plant. In spring when the days become longer and the temperature rises. The plants resume growth and begin flowering. The varieties grown in milder sub. Tropical climate do not require chilling and continue to make some growth during winter. From the standpoint of response to length of the light period, strawberries are placed in two groups: (1) varieties that develop flower buds during both long and short light periods, the everbearing varieties and (2) varieties that develop flower buds during the short light periods only, most commercial varieties.

*Soil Requirement:* Strawberry requires a well-drained medium loam soil, rich in organic matter. The soil should be slightly acidic with pH from 5.7 to 6.5. At higher pH root formation is poor. The presence of excessive calcium in the soil causes yellowing of the leaves. In light soils and in those rich in organic matter, runner formation is better. Strawberry should not be cultivated in the same land for a number of years. It is preferable to plant it in green manured field. Alkaline soils and soils infected with nematodes should be avoided.

*Varieties:* A large number of varieties are available. For the hilly areas, varieties Royal Sovereign, Srinagar and Dilpasand are suitable. Some of the introductions from California, such as Torrey, Toiga and Solana may prove even more successful. The variety found successful in Bangalore has been named Bangalore and which has performed well at Mahabaleshwar also. For the north Indian plains, Pusa Early Dwarf which has dwarf plants, large firm wedge-shaped fruits, has been recommended. Another variety with rich aroma but softer fruits is Katrain Sweet. Some of the varieties found successful in warmer parts of the U.S.A. are: Premier Florida-90, Missionary, Blackmore,Klonmore & Klondike. Some of these may prove successful for cultivation in Indian plains.

*Propagation:* Propagation is done by means of runners that are formed after the blooming season. The plants may be allowed to set as many runners as possible but not allowed to set any fruits. All the plants with good root system should be utilised to set a new plantation. Given the best attention and care, a single plant usually produces 12 to 18 runners.

*Planting:* The land for strawberry planting should be thoroughly prepared by deep ploughing followed by harrowing. Liberal quantities of organic manure should be incorporated in the soil before plating. Strawberry can be planted on flat beds, in the form of hill rows or matted rows, or it can be planted on raised beds. In irrigated areas, plantings on ridges is advised. In Mahabaleshwar, the usual practice is to plant on raised beds 4 x 3 meters or 4 x 4 meters.

The planting distance should be 45 cm from plant to plant and 60 to 75 cm. from row to row. In the hills, Transplanting is done in March-April, September-October, but in the plains, the months of January-February may be utilised for this purpose. At Mahabaleshwar normally strawberry is planted during November-December. The plants should be set in the soil with their roots going straight-down. The soil around the plant should be firmly packed to exclude air. The growing point of the plant should be just above the soil surface. During planting, the plants should not be allowed to dry out and should be irrigated immediately after planting.

*Care of Young Plantation:* The roots of strawberry plants spread out close to the surface. Therefore, the soil should be well-supplied with moisture, and hoeing should be done lightly and young plantation be kept weed free.

*Special Horticultural Practices:* In cold climate the soil is covered with a mulch in winter to protect the roots from cold injury. The mulch keeps the fruits free from soil, reduces decay of fruits, conserves soil moisture, lowers soil temperature in hot weather, protects flowers from frost in mild climates and protects plants from freezing injury in cold climates. Several kinds of mulches are used, but the commonest one is straw mulch. The name strawberry has been derived from this fact. Black alkathine mulch is also used to cover the soil. It saves irrigation water, prevents the growth of weeds and keep the soil temperature high.

*Irrigation:* Since strawberry is relatively shallow-rooted, it is susceptible to conditions of drought. Planting early in autumn allows the plants to make good vegetative growth before the onset of winter. However, in this case it is necessary to ensure that newly planted runners are irrigated frequently after planting, otherwise the mortality of the plants becomes high. During September and October, irrigation should be given twice a week if there is no rain. It may be reduced to weekly intervals during November. In December and January, irrigation may be given once every fortnight. When fruiting starts, the irrigation frequency may should again be increased. At this stage frequent irrigation gives larger fruits.

*Nutrition:* Strawberry requires moderate amounts of nitrogen. Addition of organic matter to the soil, in the form of 50 tons of Farm Yard manure per hectare is highly desirable. It improves the water holding capacity of the soil and also gives better runner formation. Farm yard manure may be supplemented by chemical fertilizers to make up the total quantity of nitrogen from 84 to 112 kg per hectare, Phosphorus 56 to 84 kg per hectare, and Potash 56 to 112 kg per hectare.

The Phosphatic fertilizer should be incorporated into the soil before plantings. The nitrogenous fertilizer be applied in Two doses (Three weeks after planting and again at the time of flowering) and potash at the time of flowering only. Application of adequate amounts of nitrogen gives higher yield of early berries.

*Plant Protection:* Red spider mites and cutworms are important pests of strawberry. The mites can be controlled with 0.05 per cent Monocrotophos + 0.25 per cent wetable sulphur. The cut worms can be controlled by dusting the soil before planting with 5 per cent chloradane or Heptachlor dust at the rate of 50 kg per hectare and mixing it thoroughly in the soil by cultivator.

The two commonest diseases of strawberry are red stele, caused by the fungus Phytophthora fragariae and black root rot. The remedy for the former lies by growing resistant varieties like stelemaster and for the latter to maintain the vigour of the plants and rotate strawberry with other crops like legume vegetables (beans, peas etc.).

Strawberry also suffers from virus diseases known as yellow edge, crinkle and dwarf. Raising of strawberry nursery in the hills helps to check these. Strawberry also throws some chlorotic plants, which result from genetic segregation. These should not be confused with virus affected plants and should be rogued out.

*Harvesting and Yield:* The fruit ripens during late February to April in the plains and during May and June at high elevations like Mahabaleshwar, Nainital and Kashmir. For local market the fruit should be harvested when fully ripe, but for transport to distant markets, it should be harvested when still firm and before colour has developed fully all over the fruit. Harvesting should be done preferably daily.

Since fruit is highly perishable, it is packed in flat shallow containers of various types (card board, bamboo, paper trays etc.) with one or two layers of fruits. Harvesting should be done early in the morning in dry conditions. Washing the fruit bruises it and spoils its lustre. The yield varies according to season and locality. A yield of 20 to 25 tons per hectare is excellent, though yields upto 50 tons per hectare have been reported under ideal conditions.

*Post Harvest Handling and Marketing:* Strawberries are highly perishable and hence a great deal of care in harvesting and handling as well as its marketing also requires to be organised carefully. Usually the fruit is picked in the early morning and sent to the market in the afternoon of the same day or is picked in the late afternoon, stored overnight in a cool place, and sent to market the following morning.

### Others

Plum and apricot are generally grafted on wild apricot seedling rootstocks. Peach is also used as a rootstock. Peach is budded/grafted on wild peach and peach-almond hybrid rootstocks. Hard wood cuttings (treated with 500 ppm IBA) can also be employed for raising rootstocks. For apricot, peach, plum and cherry shield budding during July-August has been recommended for the hills of J&K. Micro-propagation through tissue culture technique has not yet been commercialized for temperate fruits, although tissue culture protocols have been developed at research Institutes/Universities. Colt, a rootstock for cherry can be multiplied in large numbers by following the tissue culture technique. Shoot buds are cultured in MS medium containing 1-2 mg/litter benylaminopurine. Individual shoots are rooted on MS medium containing 1 mg/litter Indole Butyric Acid (IBA). The rooted plants are transferred to sterilized sand + soil (1:1) mixture and covered for 7-10 days for hardening.

## FRUIT THINNING

In apple, for obtaining better grade fruits of optimum size a post-bloom spray of 100-200 ppm ethephon 2-3 weeks after full bloom has been recommended both in H.P. and J&K.

### Pear

Harvesting time of pear is determined by slight change in fruit colour and easy separation of fruits. Fruits should not be allowed to ripen to soft condition on the tree and should be harvested when they are still hard. Depending on cultivars fruit may be ready for harvest from 70 days after full bloom to 135 days after full bloom. The Bartlett pear normally takes 122 ± 3 days in valley areas of H.P.

### Cherry

Sweet cherry should be harvested when fruit stalk separates easily from the spur and taste develops typical to the cultivar. From full bloom it takes 45 days in cultivars like Guigne Noir and Gross Lucenta, whereas in cultivars like Bigarrean Napoleon the number of days varies from 65-75 days.

### Peach

For harvesting peach at the right time, the proper colour development in fruits and pit browning are considered as reliable guides. In yellow fleshed cultivars, deep orange colour development on fruits is associated with proper maturity. At least 5% of the pit area also should be brown if the fruit is to develop good flavour at ripening. In the peach variety July Elberta, it takes about 90-95 days after full bloom, whereas on cv. Elberta it normally takes

100-105 days. In low chilling cultivars like Flordasun, fruits should be harvested at 50% colour development stage.

**Plum**

In free-stone cultivars of plum, pit browning up to 5% of the pit area is a good index of maturity. For proper fruit maturity, early cultivars like Sharp Early and Formosa take 95 to 105 days, mid season variety Santa Rosa requires 115-120 days and late cultivars like Grand Duke take 130-135 days after full bloom.

**Apricot**

Apricot fruits of a single tree ripen for a period of about 3 weeks. The fruits develop detectable flavour when allowed to ripen on the tree. The early cultivar Charmagz matures within 75 days from full bloom, while the late cultivar Quetta requires 125-130 days for proper harvest maturity.

**Yields**

There is a wide range of variation in fruit yield in different States and in different cultivars. The average yield of apple and pear have been indicate. Although average productivity of apple in H.P. has been shown as 9.4 t/ha, some well managed orchards of Delicious apple in Kotgarh area of the same State yield about 50 t/ha. Similarly, yield record of 35 t/ha has been achieved in 12 year old plantations of coloured mutant apple cultivars on MM 106 rootstocks under high density planting system (2222 plants/ha) in a research station in Shimla hills of H.P. Nutrition takes from Fruit Trees are:

## MANGO MALFORMATION

It is the most important malady of mango and was first reported by Burn (1910). It causes the losses from 50-80 per cent. It is also reported to be due to other disorders like physiological and acarological. Indications of its virus nature has also been postulated. The maximum incidence of the disease has been recorded in the north, north-west and north-east parts of the country. The incidence is, however, less in western and southern India. There are two types of symptoms, namely, floral malformation and witches broom or bunchy top or vegetative malformation with a proliferation of infected tissue. The flowering panicles turn into a compact mass of flowers.

This compact mass is very hard and not soft like normal panicle. Individual flower is greatly enlarged and has a large disc. The percentage of bisexual flowers in malformed panicles is very low. In bunchy top, compact leaves are formed in a bunch at the apex of shoot or in the leaf axil. A similar bunch consisting of small rudiments crowded together on short shootlets is seen in vegetative malformation in which the growth of shootlet is arrested. Vegetative malformation is more pronounced in young seedling and seedling trees. The malformed heads dry up in black masses and persist on the trees for a long time. The malformed inflorescences contain more of endogenous cytokinins than healthy ones (Nicholson and van Staden 1988).

The variety Zebda has been reported to be resistant in Egypt (El Ghandour *et al.* 1979). Considerable incidence reduction has been reported by spraying 100-200 ppm NAA during October. Use of disease free planting material and prophylactic spray of insecticides and fungicides can keep the orchards healthy. In areas with less than 5-10 per cent infection, pruning of diseased plants should be made compulsory; pruning of diseased parts along the basal 15-20 cm apparently healthy portions. This is followed by the spraying of Bavistin (0.1%) or Captaf (0.2%). Partial control of malformation has been achieved by spraying mangiferin-Zn2 and mangiferin-Cu2+chelates.

**Stem End Rot (Botryodiplodia theobromae Pat.)**

It is an important disease of ripe mango. This fungus also causes die-back symptoms in orchards. It is known to occur in India, Myanmar, Sri Lanka, the Philippines, Mauritius and the U.S.A. In India, it is recorded from all the mango growing areas. The disease is characterized as the dark epicarp around the base of the pedicel in the initial stage. In the next few hours, the affected area enlarges to form a circular, black patch which in humid atmosphere extends rapidly and turns the whole fruit completely black within two or three days. The pulp of the diseased fruits become brown and somewhat softer. The diseased fruits showed increased phosphorus and reduced ascorbic acid contents. Pectolytic enzyme activity was also lowered in such fruits. The disease is carried from the dead twigs and bark of the trees harbouring the pathogen and with the onset of rains the orchards get contaminated.

Dipping of mangoes in 6 per cent borax solution at 43°C for 3 minutes reduces the incidence. Fruit dipping in Aureofungin and wrapping was also found effective. Pathak (1980) suggested to harvest mangoes on clear dry day. Hot water treatment at 53°C for 10 minutes and at 51 to 55°C for 15 minutes have been recommended elsewhere. Care should be taken to prevent snapping-off of the pedicels. Injury should be avoided to fruits at all stages of handling. Spalding (1982) in his experiments found thiobendazole and thiophenate methyl to be the most effective ones. Recently, Rawal and Ullasa (1988) have reported that Bavistin (0.1%) or Topsin-M (0.1%) or chlorothalonil (0.2%) spraying in the field before harvesting give very effective control.

**Die-back (Botryodiplodia thiobromae Pat)**

It is a descructive disease of mango and is known to occur in India and other mango growing countries. The disease is manifested by discolouration and darkening of the bark some distance from the tip of the twigs. The dark area advances and young green leaves start withering first at the base and then extending outwards along the vein. The affected leaves turn brown and the margins roll upward. At this stage, the twig or branch dies, leaves shrivel and fall, and this may be accompanied by exudation of gum. The infected twigs show internal discolouration. Brown streaks of vascular tissues is seen on splitting the twigs lengthwise along the long axis. In early stages, epidermal and sub-epidermal cells of twigs appear slightly shrivelled. The areas of cambium and phloem show brown discolouration and yellow gum like exudate flows out of the cells. The disease is known to be enhanced by a beetle, Xyleborus affinis (Batista 1947). Relative humidity above 80% and temperature of 25 to 31.5°C, and rains enhance disease development. The die-back is also caused due to Colletotrichum gloeosporioides. Both these pathogens also attack fruits which get spoiled during ripening.

For effective control of the disease, pruning and destruction of infected twigs is the foremost practice. Spraying the trees periodically with copper oxychloride sulphate is recommended by Alvarez and Lopez (1971). Pasting of trees with a mixture of oil and 5% phenol was found effective. Sprays of carbendazim (0.1%) or methyl thiophenate (0.1%) or chlorothalonil (0.2%) at fortnightly interval during rainy season is important. Sometimes short hole borers also predispose the trees to infection and hence proper insecticides are also to be sprayed. Healthy twigs should be selected for grafting of seedling during propagation.

**Leaf Blight (Macrophomina mangiferae Hingorani & Sharma)**

It was first reported from Brazil to be caused by Macrophomina sp. by Lacca (1922). In India, it was reported by Patel *et al.* (1949). Later, it has been reported by various workers from Delhi, U.P. and Gujarat. The other pathogens causing the disease are grey blight (Pestalotiopsis mangiferae), phoma blight (Phoma glomerate), and twig blight and fruit rot (Phoma sp.). The disease usually appears as yellowish pin head like spots on leaves and twigs of the affected plants. These gradually enlarge discolouring the surrounding tissues which first become brown and then dark brown with slightly raised and brown purplish margins and later ash coloured due to the appearance of pycnidia.

Spots are round to start with but later become oval or irregular in size depending upon environmental conditions such as humidity and temperature. Infection is mainly observed on leaves and rarely on stems. On stem, the lesions are elliptic which later girdle the stem at the point of infection. On fruits, water soaked circular lesion are produced which enlarge rapidly and cause rotting. The pathogen can survive for more than a year on the leaves of mango. The varieties, namely, Khandeshi, Borasio and Asal Damido are reported to be resistant. Field sanitation is by collecting and destruction of diseased parts. Spraying of Burgandy mixture, Perenox, lime sulphur and dithane have been recommended by Hingorani *et al.* (1960).

**Red Rust (Cephaleuros mycoides Karst)**

This or red spot is a common algal disease on the mango in the tarai and in the other humid regions of India. Its occurrence has been reported from Bihar, Karnataka and Uttar Pradesh. The alga attacks the foliage, bark and twigs of the host plant and is both parasitic and epiphytic in nature. In serious infections, the bark becomes thickened, the twigs get enlarged and remain stunted and the foliage becomes sparse and finally dries up. Initially the spots are greenish-grey in colour and velvety in texture, but later on the surface bears reddish-brown appearance. Algal spot is circular to irregular in shape, slightly elevated and with usually 2 mm in diameter, though in some cases it may be as much as 1 cm. Bordeaux mixture (6:6:100), Supramar, Fytolan, Blitox-50 or lime sulphur are recommended.

**Black Spot (Xanthomonas compestris pv mangiferae indicae)**

It is one of the serious diseases and is reported from many countries. In India, it is reported from Maharashtra, Delhi, Uttar Pradesh, Tamil Nadu and Karnataka. It affects leaves, petioles, fruits and tender stems. Numerous small angular water soaked lesions appear in groups towards the tip of the leaf blade. These are first light yellowish but later turn dark brown to black and get surrounded by a distinct halo. Several lesions are often found to coalesce forming large necrotic spots. On young fruits, water-soaked lesions develop which also turn dark brown to black. Infected fruits may show skin cracking and badly affected ones dry prematurely. Infection by wounding produces diseased spots on tender stems, petioles and stalk bearing fruits. These spots turn into deep longitudinal scars accompanied by appreciable amount of bacterial gum-like ooze. The incubation period on leaves varies from 2-14 days.

In India, the disease remains dormant during November to March due to low temperature (11.8-22°C) and the leaf infection is considerably reduced by the fall of the infected leaves. Kent mangoes showed close relationship of rainy season with incidence of bacterial leaf spot. One day's rainfall did not affect the disease appearance, whereas four days of rainfall led to the 36.9% disease incidence. Twig canker persists and initiates fruit infection. The disease spread is rapid during the rains and becomes severe in July-August. Five application of Bordeaux mixture (4:4:50) with spreader was recommended by Wager (1937), whereas four sprays of 6:6:50 Bordeaux mixture was found to be effective by Marloth (1947). Sprays with fungicides such as copper containing materials and agrimycin have also been advocated. Shekhawat and Patel (1975) recommended orchard sanitation by way of removal of infected materials and seedling treatment as preventive measures. Agrimycin-100 proved best in arresting the disease development. Plantomycin (200 ppm) followed by agrimycin + bavistin (1000 ppm) was reported to be the best against this disease. Streptomycin sulphte followed by aureofungin have been recommended by Prakash and Raooff (1985) to control bacterial disease of mango.

**Black Mould Rot (Aspergillus niger Van Teigh)**

The affected fruits show yellowing of base and development of irregular, hazy, greyish spots and coalesce into dark brown or black lesions. The mesocarp of the rotted areas becomes depressed and soft. The diseased fruits show rapid decrease in ascorbic acid content. The pathogen infects injured fruits. Totapuri variety is highly susceptible. A fruit dip treatment with benlate at 1500 ppm can control the rot.

**Soft Rot (Rhizopus arrhizus Fisc.)**

This pathogen alone is reported to be responsible for the decay of 6.3 per cent fruits in Delhi markets. The pathogen incites a typical soft rot. Development of disease is slow and less intensive at low temperature (7-8°C). The disease develops rapidly between 20 and 40°C. Thakur and Chenulu (1970) found that 2-aminothiozole with 2-aminopyridine each at 7 per cent concentration protected mango fruits for 20 days.

Some of the other post-harvest diseases and pathogens are Alternaria rot (Alternaria tenuissima (Neer. Ex. fr.) Wiltshire), Sclerotium rot (Sclerotium rolfsii Sacc.), dry rot (Boothiella tetraspora Lodhi and Mirza), brown rot (Pestalotiopsis mangiferae P. Henn. and P. glandiola (Cast) Stey), fruit rot (Phomopsis mangiferae Ahmed) and black spot, Actinodichium jenkinsii.

**Black Tip**

This disease is peculiar to India and does not occur in any other country. It has been reported to occur in regions of West Bengal, Uttar Pradesh, Bihar and Punjab in orchards located in close proximity of a brick kiln. South India is free from this disease. The symptoms manifest as small etiolated area at the distal end of the fruit, which gradually spreads, turns nearly black and covers the tip completely. Before the etiolation is complete, isolated greyish spots appear which become dark brown, enlarge and coalesce into a continuous necrotic area. The brown discolouration spreads to the neighbouring parenchyma and the deposits also appear in the ducts. The browning and deposits gradually spread throughout the mesocarp and the affected cells distintegrate and coalesce into a dead tissue. In severe cases, the necrosis extends to the endocarp. The practical control measure lies in keeping the brick kilns away from the mango orchards. The use of telescoping chimney 12-15 m high has also been recommended. Spraying of borax (0.6%) at 10-14 day intervals has been recommended.

**Soft Nose**

The problem was first observed as physiological breakdown in cvs. Kent and Haden by orchardists in 1950 in Florida. The soft nose is characterized by hallowing of the green skin in the areas between the apex and the stigmal point. A lack of firmness in diseased fruits can also be felt with experience. Upon cutting the flesh on the ventral side towards the apex of some soft nose fruits appear to be over ripe, while on shoulder and dorsal side it is unripe. The over ripe flesh surrounds a mass of yellowish to brown tissue which is firmer than the surrounding affected tissue and is bitter in taste.

Young (1957) concluded that the disorder is not due to the microorganisms. However, Young *et al.* (1963) reported the loss of only few percent in fruits harvested from trees grown on calcareous rock soil, but on acidic soil, sandy soil it is common for 15-20 per cent incidence. According to Young *et al.* (1963), soft nose disorder increased with N fertilization but was reduced to greater extent by increasing Ca level of the soil. Nitrogen fertilizer in sandy acidic soil increased the incidence whereas in calcareous soil there was no effect. To overcome this disorder, the high N level of soil should not be allowed in the sandy or acidic soil and proper Ca level should also be maintained in soil. Early picking is also suggested. The other disorders are little leaf caused by zinc deficiency, tip burn due to moisture stress and salt accumulation in soil and leaf scorch due to accumulation of chloride ion in leaf.

**Phanerogamic Parasites (Dendrophthoe (Loranthus) viscum)**

The mango trees are severely affected by this parasite particularly in neglected orchards. They are commonly present on the trunk or branches of the trees. The foliage of the infected tree is sparse, reduced in size and its bearing capacity and quality of fruits is considerably reduced. The fruits of this parasite are baccate with viscous mesocarp and attract the birds by its attractive colour. The seeds of this parasite are spread through bird and animal droppings. Such seeds, germinate and produce haustoria. Sometime tumour like swellings are observed on the affected tissue. There are many other species of parasites which also attack mango trees. All epiphytic plants can be removed easily in early stages before their aerial roots penetrate the ground and surround the host trunk.

Some more diseases reported by various workers attacking the mango crop in the field are leaf spots due to Rhizoctonia bataticola (Taub) Butt., Cercospora mangiferae indica Munjal Lal & Chona, Alternaria alternata Fr. & Keissler, Phoma sorghicola (Sacc.) Boerema. Dorem and Venkast; Scab (Elsinoe mangiferae Bit & Jen; black band (Rhinocladium corticolum Massee); pink disease (Pellicularia solmonicolor (Berk & Br.) Dastur); root rot and damping-off (Rhizoctonia solani Kuhn) and wilt due to Verticillium albo-atrum Reinke and Berth and Fusarium solani (Mart.) Sacc. There are several more fungi causing different diseases but the above are the most important among them.

## THE PAPAYA

The papaya (carica papaya) is one of the important delicious fruit crops grown in the tropical and subtropical parts of the world. It is originated in Mexico and spread to almost all the corners of the tropical world.

The papaya is highly productive, problematic and interesting crop. It is easy to grow and short duration crop. As a raw fruit, it is used in cooking and some preparations. Its latex is used as a papain in food and medicine industry. Ripe fruit is very delicious containing vitamin A and Carbohydrates.

*Climate:* The papaya being tropical crop, favour high temperature and high humidity. It is very susceptible to frost and hail storm. The long days are favourable for good quality and flavour. During flowering, high rains are injurious and cause heavy damage.

*Soil:* The papaya grows under wide types of soils. However, very shallow and very deep black soils are not suitable medium, fertile, well drained and lime free soils are preferred for papaya cultivation.

*Varieties:* Important varieties under cultivation

*Pusa varieties:*

- Pusa delicious
- Pusa majesty.
- Pusa jiant
- Pusa dwarf
- Pusa nanha

*Coimbatore varieties:*

- $Co_2$
- $Co_3$
- $Co_5$
- $Co_4$
- $Co_5$
- $Co_6$

*Bengalore varieties:*

- Coorg honey
- Pink flesh

*Pantnagar varieties:*

- Pant 1
- Other varieties: Ranchi, Washington, Honey dew, Taiwan varieties like, golden queen, No 815 and others.

*Propagation:* Commercially the papaya is propagated by seeds. The tissue culture technique is limited to research laboratories only. The seeds loose viability in a short period and therefore the seeds should not be stored for more than a season. The seedlings in polybags are prepared. Due care is taken to avoid damping off of the newly germinated and young seedlings. The seedlings become ready for transplanting within 6-8 weeks.

Planting and season: Planting is done during the flowering seasons:

- Spring season ( February - March)
- Monsoon season (June-July)
- Autumn season (October-November).

Heavy rains, hot air, frost, etc. are considered while selecting the season for planting in a particular area. The pits of 30x30 x 30 are prepared in already selected and prepared field at the distance of 2.5 to 3 meters distance. The pits are filled with well decomposed FYM and NPK mixtures. A care is taken not to disturb the roots while transplanting the seedlings.

*Intercultivation:* Interculturing is mainly done to remove the weeds during the early period of growth, weeding and hoeing in between rows also favour better aeration to root zone. Some times pre-emerging weed killers like Basalin are used. Secondly, roughing is done to remove extra plants, weak plants and affected plants. After ensuring one plant per pit, earthing up is done 30 cm in radius around the plants.

*Special Horticultural Practices:* Planting of 3 to 4 seedling at one pit and then removing extra plants and keeping one plant per pit while doing so 10 per cent male plants are kept in female plant population to have pollination and to improve the fruit set.

*Irrigation:* For better growth, production and quality, the optimum soil moisture is maintained by irrigating the crop judiciously. Irrigation interval well depends on season, crop growth and soil type. In no case, water should be allowed to stagnate causing root and stem rot. Drip system of irrigation is beneficial and the actual quantity of water to be given per plant per day should be worked out critically.

*Nutrition:* The papaya is a very feeder and requires the application of chemical, organic and biofertilisers. The dose of NPK @ 500 kg each per ha, along with 20-25 tonnes of FYM, 50 to 100 kg of ormichemi mirconutrients and ultrazyme sea weed extract granules 25 kg is found for a crop of about 50 tonnes within 18-20 months. Additional 60% of this dose is again applied for the second flush.

*Plant Protection:* The papaya is susceptible for virus disease, which are spread through insect vector. The rouging of is strictly followed to avoid further spread of the disease, in addition the insecticidal sprays are taken against aphids, white flies and other sucking pests. Powdery mildews, downy mildew and stem rot.

Providing wind breaks, well drained conditions to soil avoiding planting papaya after papaya and following a suitable crop rotation keeping the field clean and weedfree are the some of the measures which will help to keep the papaya crop in good healthy condition.

*Harvesting, Packing, Post Harvest Handling:* Usually the fruits are harvested when they are full size, light green with a tinge of yellow at epical end. When the latex ceases to be milky and become watery the fruits are considered suitable for harvesting. First picking may start at 14/15 months after planting. Three to five picking for one season are often taken fetching about 30-35 tonnes per hectare. Suitable grading must be done before packing. Since the fruits are easily perishable, care should be taken to wrap the individual in paper and finally packed in crates.

*Special Features—Papain Production:* Crude papain in flakes or powder crude papain white brown. Spray dried crude papain are the three types of papain extracted from raw-mature papaya fruits. About 8-10 g of final produce can be obtained from a well developed fruit weighing about 2 kg. Tapping may be done 2/3 times preferably during cool hours. The fruits after latexing can be used as a table fruit or for preparing fruity or jams etc.

## CITRUS

*Pollen Collection :* The species handled for pollen collection include Citrus limon, Citrus aurantifolia, Citrus sinensis and Poncirus trifoliata.

*Inventory:* Petri dishes, butter paper, forceps, muslin cloth fixed to a 10 cm. cylindrical ring with the help of a firm rubber band, clean razor blade, etc.

*Procedure:* Pollen collecting are usually made on a bright sunny day between 8 - 10 AM, staminate flowers are harvested at peak anthesis and the dehiscing anthers are gently caressed over a muslin cloth sieve. Anthers are carefully removed from pistillate flowers and pollen is extracted over a muslin cloth sieve. Thus, pure pollen is extracted in clean petri plates or butter paper.

*Precautions:* The following points should be noted:

- Quality of pollen collected depends on the correct stage of anthesis/ anther dehiscence.
- Pollen should be free of anther debris.
- Do not collect pollen from infected or insect damaged flowers.
- Decide upon bagging of inflorescence depending on insect activity.
- Do not venture to collect pollen on a rainy day or if it had rained overnight.
- Do not force out pollen from anthers.

*Viability Testing:* Pollen collections are subjected to viability indexing by germination in vitro by the hanging drop technique (Shivanna and Rangaswamy 1993). This process involves suspension of pollen grains in a drop of liquid medium consisting of a carbohydrate source, inorganic and organic compounds.

*Inventory:* Cavity slides, cover glasses (22 mm square), microslides, petroleum jelly, large petri plates, filter paper discs, incubator, double distilled water, needle, forceps, "Quick fix" (epoxy resin).

*Medium:* Prepared in deionised, double distilled water. 20% sucrose, 100 ppm $H_3BO_3$, 300 ppm $Ca(NO_3)24\ H_2O$, 200 ppm Mg $SO_4$ $7H_2O$, 100 ppm $KNO_3$, pH 7.3.

*Procedure:* Apply petroleum jelly to the edges of a new cover glass. Place this cover glass on the work table so that the edge on which petroleum jelly is applied faces the roof. Place a 25 to 50 l drop of nutrient medium on the centre of the cover glass assuming the shape of a small bubble. Add pollen grains with the help of a needle carefully filling the drop with optimum quantity of pollen. Invert a cavity slide over the cover glass with the cavity over the drop.

The petroleum jelly helps to adhere to the cover glass around the cavity, sealing the cavity from all sides. Reverse the cavity slide and place in a petri plate fixed with moistened filter paper disc cut to its exact size. Place the cavity slide over a microslide in such a way that the cover glass with the hanging drop over the cavity is well clear of its surrounding area, intact and hanging. Cover petri plate with the top lid (also fixed with moistened filter paper) and keep in an incubator at 25 ± 2°C, dark, for 18 to 20 hours.

After the prescribed duration of incubation, the hanging drop preparation is separated carefully from the cavity slide. After removing the residual petroleum jelly from the slides of the cover glass, a drop of Alexander's differential stain is added over the stained drop and mixed slowly. Place a clean microslide over the stained drop. The drop is flattened with the cover glass sticking to the underneath of the microslide. Remove excess stain and seal the edge of the cover glass microslide interface with a thin layer of quick acting adhesive. The slide with germinated pollen is ready for observations under the microscope for qualitative and quantitative viability estimates.

*Precautions:* Noteworthy points are:

- Prepare medium fresh.
- The medium on the cover glass should be just optimal for it to hang on the cover glass when inverted.
- Add just enough stain to the drop containing germinated pollen. If excessive stain is added, the germinated pollen drifts away outside the cover glass when the drop is flattened, leading to errors in quantitative estimation.

**Storage**

*Inventory:* Butter paper, glass vials, silica gel, gelatin capsules, polyethylene aluminium laminated pouches, head sealer, domestic refrigerator, freezer, liquid nitrogen storage, cryoflask filled with liquid nitrogen.

*Procedure:* Pollen samples are packed in glass vials/gelatin capsules/butter paper packets depending on the method of storage. Lemon pollen contained in glass vials are best stored at 5 or -18°C for short duration of 8 weeks (Ganeshan and Sulladmath 1983) and medium term duration of 6 months with a progressive decline in viability index by germination in vitro at weekly or monthly intervals. Cryogenic temperatures (-196°C) afford much more prolonged duration storage, beyond two years as demonstrated by Japanese workers (Kobayashi *et al.* 1978). Ganeshan and Alexander (1991) reported cryogenic preservation of lemon pollen for 3.5 years.

*Storage of 5°C and -18°C:* Storage in liquid nitrogen: Pollen samples are packed either in gelatin capsules or butter paper packets, sealed air tight in polyethylene aluminium laminated pouches and lowered into a canister of a cryoflask. The canister is capped with a perforated lid and plunged slowly in liquid nitrogen contained in the cryoflask. Frequent refilling of the cryoflask with liquid nitrogen at least once every 10 to 15 days, ensures a constant cryogenic temperature. For proper ease of operation, samples to be cryopreserved for a long duration are located at the bottom of the canister and those that are to be retrieved after a short duration, at the top of the canister.

*Retrieval and Post-storage Viability Assessment:* There are not many critical steps to be followed for retrieval of pollen samples stored at low or sub-zero temperatures. Pollen is removed from the glass vial and directly tested for viability as described above. Pollen samples retrieved from cryogenic temperatures have to be carefully pulled out of the canister with the help of a one foot long blunt forceps, held at ambient temperature for 10-15 minutes prior to a viability test or field pollination.

Pollen samples are viable if:

- Germination profiles in vitro are as good as fresh pollen.
- Germination time taken is not more than that of fresh pollen (pollen vigour).
- Estimated germination is in the range of 60 to 80 per cent of fresh pollen.

***Field Pollinations with Stored Pollen for Fertility Assessment***

*Inventory:* Breeder's crossing kit.

*Procedure:* Stored pollen are required to be tested for fertility under field conditions through controlled pollination. For this, pollen samples retrieved from storage are transported to field locations in an ice box and used in crosses with compatible seed parents in order to obtain an estimate of fertility, fruit and seed set.

Stored pollen is applied to pre-bagged emasculated perfect flowers at peak stigma receptivity with the help of a painting brush, followed by bagging to avoid contamination with stray pollen. For comparative fertility estimates, crosses with fresh pollen must be carried out as far as possible at least initially. If the pollen parent in question is not in flowering, crosses may be repeated with the fresh pollen over the same seed parent within a short period.

Observations to be recorded in field are:

- Number of crosses made with fresh and stored pollen.
- Number of fruits set to maturity.
- Number of seeds per fruit.
- Number of aborted seeds.

Seeds set through stored pollen are germinated and observation of any marker characters transmitted by pollen parent (for example, trifoliate leaf character in F1 population of crosses involving pollen of Poncirus trifoliata), would confirm that the stored pollen was fertile and has transmitted the character conserved to the next generation.

## KOKAM

Kokam (Garcinia indica chaisy) is an evergreen tree, native of India mostly found in Western Ghats along the seacoast in Konkan region of Maharashtra state.

*Importance:* The fleshy ripe purplish red fruit possesses sweetish to acidic pulp. Apart from traditional use for preparing "Amsul" or "Kokam Sol" and "Amrit Kokam" squash, it is gaining importance for its taste. The Kokam seeds contain 23-26 per cent fat which remains solid at room temperature. Hence, it is being used as a substitute for Coca butter. It is also being used in cosmetics. Due to this, demand for kokam both in internal market as well as in export market is increasing. It is, imperative to extend the area under Kokam plantation on a large scale using cultivable waste land in Western Ghats.

*Climate:* Kokam is a Tropical fruit,grows and fruits well in high rainfall areas of the sea coast of about 1000 to 2000 mm having warm and humid climate.

*Soil:* Lateritic, alluvial or medium deep well drained soils are suitable. It is mostly grown in hill slopes mixed with other fruit trees like jamun, mango, cashew etc. in Kokan region of Maharashtra.

*Varieties:* There are no standard varieties. Kokam is a slow growing plant and exhibits a large variation in sex forms and bearing capacity. The Kokan Krishi Vidyapeeth has spotted a high yielding (35 kg seed/ tree) under natural conditions. Such selected trees could be clonally propagated by soft wood grafting.

*Planting Season:* Pits of 60X 60 X 60 cm be dug during summer at 6 X 6 m distance and filled with well decomposed farm yard manure and top soil at a ratio of 1:3 and 1 kg superphosphate be added at the bottom of the pit. The planting of sapling be done at the onset of monsoon in June. At the time of planting 100g carbaryl dust (10 per cent) be mixed in each pit to avoid termite attack.

*Interculturing:* Field be kept weed free by regular weeding and hoeing. Short duration, low growing, inter crops like cowpea, vegetable can be grown during first four five years.

*Care of Young Orchards:* Young plants be given support of bamboo sticks. Shoots arising from the main stem just before the bud /graft joint be removed regularly. Wherever possible young saplings be irrigated during dry period and summer months for initial 2/3 years.

*Special Horticultural Practices:* Since it is grown as a rainfed crop no special treatment is required. However, if the orchard is established by planting seedlings about 50 per cent plants will be male which will not bear any fruits. This could be identified 7-8 years after planting when they start flowering. This could be converted by grafting with selected scion stick taken from female (fruit bearing) plants. However, it is necessary to maintain 10 per cent male plants in an orchard for pollination.

*Irrigation:* Kokam is grown as rainfed fruit crop and irrigation is not given except during summer or dry spells in the initial 2/3 years.

*Nutrition:* One year plant be given 2 kg FYM + 50g nitrogen + 25g $P_2O_5$ + 25 g $K_2O$ during August- September. The quantity be increased by 2 kg FYM + 50 g nitrogen + 25g each of $P_2O_5$ and $K_2O$ every year. From 10 years onwards each tree be supplied with 20 kg FYM + 500g N and 250g each of $P_2O_5$and $K_2O$ during August September.

*Plant Protection:* No serious pests/ diseases are reported on Kokam and hence it is normally does not require any plant protection measures.

*Harvesting and Yielding:* Seedling trees starts fruiting after 7-8 years while grafted/ budded plants bear fruits after 4-5 years. Flowering starts in October November and continues upto Feb-March. Fruits are ready for harvest during April-May. Red ripe fruits be harvested with the help of 'Atul' harvestor specially developed for this purpose by Konkan Krishi Vidyapeeth, Dapoli, Most of the existing plantations are by seed. Hence the yield varies from place to place depending upon orchard management practices. On an average individual trees yield 30-50 kg ripe fruits per year.

*Post Harvest Handling and Marketing:* Kokam fruits are juicy and highly perishable and hence need careful handling and after harvesting. Fruits are collected in bamboo baskets lined with rice straw and stored under shade. Harvested fruits are sorted removing under signed,damaged fruits and graded as per the size and colour into two grades. Mostly fruits are processed into 'Amsul' or 'Amri Kokam' squash because of poor transportability and highly perishable nature.

## LIMES AND LEMONS

*Lime or Acid lime (Citrus aurantifolia swing):* Is more popular in India than lemon (Citrus limon). Acid lime is cultivated largely in Andhra Pradesh, Maharashtra, Tamil Nadu, Gujarat, Rajasthan, Bihar and to a limited extent in other states.

Lemons are grown only to a limited extent in homestead gardens. Limes are available throughout the year in some part or the other in our country. They are used as fresh fruit or for the preparation of pickles and beverages. They are rich in vitamin C, minerals and salts.

*Climate:* Areas with dry climate and low rainfall are best suited for growing limes. Lemons can be grown in heavy rainfall humid regions.

*Soil:* Lime and Lemon may be planted in medium black, loamy or alluvial soils having perfect drainage and devoid of calcium carbonate layer.

*Varieties:* Commonly grown acid lime cultivar is Kargzi lime, but recently identified four high yielding cultivars Vikram, Prumalini, PKM, and Sai Sharbati are now gaining popularity. Commonly grown lemon varieties are Assam lemon, Italian lemon, Pant lemon, Galgal and Eureka lemon, Sevilla and Malta lemon varieties are popular in South India.

*Propagation:* Acid lime can be propagated by seed, budding or air layer. Since it is highly polyembryonic, seedlings are true to type and resistant to diseases. Propagation by seedling is recommended. Budded plants on Jambheri or Gajanimma are used for planting lemons.

*Planting:* Pits of box 60 x 60 cm size are dug 5 to 6 meters apart each way and filled with 30 to 40 kg. Farm yard Manure or compost and Top soil along with 2 kg superphosphate. One year old, healthy seedlings of recommend acid lime variety of bud grafts of choice lemon variety is used for planting. Most ideal season for planting is the onset of monsoon (June - July) for both time and lemon. However, in heavy rainfall area September-October would be appropriate planting season.

*Interculture:* Low growing vegetables, leguminous or green manure crops may be grown as inter crops between young trees upto 4 to 5 years of their age. Area around plants may be kept weed free by weeding and hoeing regularly.

*Care of young Orchids:* Soon after planting plants be supported with bamboo stick and irrigated thoroughly. Main stem of the young sapling be topped at a height of about one meter from ground level within 25-30 cm from cut end 3-4 well spaced branches be allowed to grow in different directions to form main limbs of the tree. Plants be protected against pests and diseases by spraying with pesticides and fungicides regularly.

*Special Horticultural Practices:* Acid lime flowers and produces fruits almost all the year round, but there are three major flushes in June-July, September-October and January- February. These are known as Mrig Bahar, Hast Bahar and Ambia Bahar respectively.

Fruits of September-October (Hast Bahar) flowering harvested during summer months are most remunerative. To achieve this, irrigation be suspended for 1 to 11/2 months before actual flowering. During this period field may be cultivated and applied with manure and fertiliser towards the end and irrigation be resumed. Plants be sprayed with 1000 ppm cycocil twice during September-October which has been found to increase the fruit yield two times than the normal yield.

*Nutrition:* For each completed year of growth till 5 years, 20 kg Farm Yard Manure, 100 g urea and 1 kg Superphosphate be applied per plant in two equal split doses in June and December every year. Fifth year onwards application of 40 kg FYM, 400 g N + 200 g P + 400 g K per plant in two split doses in June-July and September-October be applied. In addition, 150 g N be applied per plant 15 days after fruit set. Manures and fertilizers are applied in a circular trench 20-30 cm deep about a meter wide around the periphery of the tree. Manures are thoroughly mixed with soil, and the tree irrigated profusely.

*Irrigation:* Young plantation needs regular watering during post-monsoon period at an interval of 10-15 days and at 8-10 days during summer months. Water is applied around the periphery of the tree in a ring and basin system so that tree trunk should not come in direct contact with water. The irrigation is suspended for 1 to 11/2 months during Bahar treatment. It is resumed during flowering. Bearing trees should get regular watering at 10-15 days during summer and 15-20 days interval watering at 10-15 days during summer and 15-20 days interval during winter.

*Plant Protection:* The major pests affecting the Acid lime and lemon plants include Lemon butterfly, leaf eating caterpillars, whitefly, leaf miner and mealy bugs. These pests can be controlled by spraying with Monocrophos, Endosulphon or Phosphamidon 1.5 to 2 ml.per litre of water at an interval of 10-15 days. Instead of repeating the same pesticide different pesticide be used at each spray.

Among the diseases Gummosis and canker are very devastating. For control of Gummosis, use ring and basin system of irrigation to avoid contact of water with main stem so that the disease causing soil borne fungus will not attack the stem. If disease is seen remove the affected portion of bark and 1 cm of surrounding healthy bark with knife and apply Bordeaux paste (one part copper sulphate plus one part quick lime and sufficient water to make paste) thoroughly on the wound. Canker is caused due to bacterial (xanthomonas sp.) infection, which develops considerably during rainy season. It appears as a minute yellowish-brown spots on stem, leaves and fruits which turn into corky brown specks. Affected leaves turn yellow and drop down. Remove all the affected parts and fallen leaves and burn them. Spray the trees with Bordeaux mixture 3:3:50 by using 1.5 kg copper sulphate plus 1.5 kg quick lime in 225 litres of water or 1250 gm of copper oxychloride in 500 litres of water also can be sprayed. Repeat, if necessary at an interval of 10-15 days.

*Harvesting and Yield:* Maturity of citrus fruits depends upon the climatic conditions, nutrition, heat units and moisture availability. The period of maturity is shorter in acid lime (5-6 months) and longest in mandarin and sweet orange. The lime and lemon fruits are harvested when they just start changing. They are harvested in installments as they ripen. Major harvesting period in July-September and November-January.

*Post Harvest Handling and Marketing:* Harvested fruits are graded according to size and colour and packed in bamboo baskets or wooden crates lined with neem foliage. Sometimes fruits are packed in gunny bags and transported by rail or road to distant markets.

## BANANA

Banana is the oldest and commonest fruit known to the mankind. It is one of the important fruits, and constitutes second largest fruit industry in India. It is nutritious palatable and easily digestible fruit. It is available throughout the year. Banana is rich in carbohydrates, minerals such as calcium, potassium, Mg, Na and phosphorous. Other than fresh fruits, it can be consumed as processed in various forms like chips, powder, flakes, etc. Banana pseudostem is chopped and used as cattle feed. Also, the leaves are used as plate. The botanical names of banana are Musa cavendish and Musa paradisiaca, which belongs to the family Musaceae.

*Climate:* Being a tropical crop, banana requires warm, humid and rainy climate. The optimum temperature range is 10 to 400C and the relative humidity is 90% or above. It is highly susceptible to frost and cannot tolerate arid conditions. Strong desiccating winds cause considerable reduction in the growth of the plant and yield and quality of fruits.

*Soil:* Banana is a heavy feeder crop/. Therefore, fertility of soil is very important. Rich, well drained, fertile, free working, soils with plenty of organic matter are best suited for cultivation. The optimum range of pH of soil should be 6 to 8.

*Propagation:* Commercial edible bananas do not produce viable seeds. So, the banana is commonly propagated by suckers and sword suckers with narrow leaves. Rhizomes whole or in bits from fruited and non-fruited plants with atleast one sound bud can be successfully used as propagating material.

*Planting:* Planting of banana is done by two methods *viz.* Pit method and furrow method. Planting is done from February to May whereas in North India, it is done during July-August. In South-India, it can be done any time of year except summer. Tall varieties should be planted at 3x3 m whereas dwarf ones at 2x2 m apart.

*Manuaring:* Banana is a heavy feeder and responds well to manuring. Banana is a quick growing and short-lived plant. Therefore, it is more beneficial, if quick growing fertilizers are applied. In Tamil Nadu, application of 100 g of N, 30 g of $P_2O_5$ and 300 g of $K_2O$ per plant to be supplied in three doses during the second, third and fifth months is recommended. In Maharashtra, 100 g N, 40 g $P_2O_5$ and 100 g $K_2O$ per plant is recommended. Out of these, $P_2O_5$ and $K_2O$ is applied at the time of planting and N is given is three split doses: third, fourth and fifth month after planting.

*After Cares:*

- *Desuckering:* Desuckering is an important operation in banana cultivation. In this, the unwanted suckers, which develop near the base of the plant, are removed. Cutting the suckers from ground level and then pouring the kerosene (2-3 drops) to kill the growing point of pseudostem does it.

- *Propping:* It is an essential cultural practice to give proper support to the plants with bamboos. It is done to avoid falling down of plants due to high winds.
- *Wrapping:* To protect the fruits from sunburn, hot wind and dust the bunch is covered. Wrapping is also done to improve the colour of the fruit.

*Harvesting and Yield:* Harvesting of banana is done 12 to 15 months after planting in dwarf and 15 to 18 months after planting in tall varieties. Signs of maturity of banana fruits are, fruit becomes plumpy and angles are filled in completely, when tapped gives metallic sound, drying off of top leaves and change in colour of fruits from deep green to light green. Tall varieties like Poovan yield 15-25 tonnes/ha, while Dwarf Cavenshish yield 25-50 tonnes/ ha. It can be stored at temperature slightly above 550F and relative humidity of about 85-95% for about three weeks.

*Varieties:* Following are the major banana varieties grown in India.

- *Poovan:* It is the most important commercial variety in Tamil Nadu, Andhra Pradesh and West Bengal. It is also known as Lal velchi in Maharashtra. It is resistant to Panama wilt,
- *Dwarf Cavendish or Basarai:* It is a dwarf variety. It is resistant to Panama disease. It is a high yielding variety with fruits large and of good quality.
- *Robusta or Harisal:* Fruit colour remains green when ripe. It is best variety for the export purpose.
- *Rasthali or Mutheli:* It is a good variety but susceptible to Panama disease
- Rajeli or Nendran
- Sonakela
- Safed Velchi.

*Plant Protection:*

*Disease:* Important diseases of banana are:

*Panama Wilt:* It is caused by the soil bourne fungi Fusarium oxysporum.

*Symptoms:* Leaves become yellow, Pseudostem splits and breaks.

*Control Measures:* Eradication of the infected plants. Growing resistant varieties like Dwarf Cavendish. Soil drenching with Vapam @ 0.85% and mercuric chloride @ 3000 PPM in nursery.

*Bunchy Top:* It is caused by the virus, which is transmitted by an aphid Pentalonia nigroneruosa. Symptoms: Infected plants have short, narrow, erect leaves with short petiole. Plants remain slanted.

*Control Measures:* Infected plants should be uprooted and burnt. Spraying of 2-4, D to control weeds.

*Insect Pests:* Important insect pests are:

*Root Stock Weevil:* Damaging stage is adult and grub

*Symptoms of Damage:* Grub bores into stem which is affected by fungi or bacteria.

*Control Measures:* Use healthy suckers and rhizomes. Before planting treat the pit with 0.65 %. Lindane 60-80 g/pit. Soak the sucker in 0.1% Lindane solution, spray 0.05% endosulfan.

## BANANA AND PLANTAIN PROCESSING TECHNOLOGIES

### Traditional Processing

Most of the world's bananas are eaten either raw, in the ripe state, or as a cooked vegetable, and only a very small proportion are processed in order to obtain a storable product. This is true both at a traditional village level with both dessert and cooking bananas and when considering the international trade in dessert bananas. In general, preserved products do not contribute significantly to the diet; however, in some localised areas the products are important in periods when food are scarce. Probably the most widespread and important product is flour preparation from unripe banana and plantains by sun-drying. In Uganda, dried slices known as "mutere" are prepared for storage from green bananas, the dried slices being either used directly for cooking or after grinding into a flour. "Mutere" is used chiefly as a famine reserve and does not feature largely in the diet under normal conditions.

In Gabon, plantains are sometimes made into dried slices which can be stored and used on long journeys, and plantains are used in Cameroon to prepare dried pieces which are stored and ground as needed into flour for use in cooking a paste known as "fufu". Dried green banana slices are also used in parts of South and Central America and West Indies for preparing flour. The other nutritionally important product is beer which is a major product in Uganda, Rwanda and Burundi where green banana utilisation is particularly high.

### Preservation Methods and Processes

#### *Drying*

Both ripe and unripe bananas and plantains are normally peeled and sliced before drying, although banana figs are sometimes prepared from whole ripe fruit. Sun drying is the most widespread technique where the climate is suitable but drying in ovens or over fires is also practiced. In west Africa, plantains are often soaked and sometimes parboiled before drying. The slices of unripe fruit are normally spread out on bamboo frameworks; or a cemented area; or on a mat; or on a swept-bare patch of earth; or on a roof; or sometimes on stones outcrops or sheets of corrugated iron. Oven-drying of ripe bananas is practiced in Polynesia as a mean of preserving the fruits, which are then wrapped in leaves and bound tightly to store until needed. In East Africa a method has been reported that involves drying the peeled bananas on a frame placed over a fire for 24 hr before drying in the sun, to accelerate the process.

***Product Stability and Storage Problems***

There is little experimental data on the storage life of the traditionally made banana and plantain products.

***Potential for Scaling up of Traditional Processes to Industrial Level***

Many banana products are now produced on an industrial scale, including the traditional banana figs and flour, and the processing techniques. One of the main problems encountered has been the susceptibility of banana products to flavour deterioration and discolouration and in the past many products reaching the market have been of poor quality. A great deal of research has been directed to overcoming these problems, although however good the resultant products are they cannot compare in flavour and other characteristics with the fresh banana fruit. Indeed, an important constraint on the large-scale development of banana processing is the lack of demand for banana products since the fresh fruit is available throughout the year in most parts of the tropical world. The production of beer from banana and plantains has not been scaled up to an industrial level, and while an important product in localised areas of tropical Africa, the market is rapidly declining in favour of European-type brews produced locally.

**Industrial Processing**

***Products and Uses***

The main commercial products made from bananas are canned or frozen purée, dried figs, banana powder, flour, flakes, chips (crisps), canned slices and jams. Banana products can be divided roughly into two types - those for direct consumption, such as figs, and those for use in food manufacturing industry, for example purées and powder. Banana figs, or fingers as they are sometimes known, are usually whole, peeled fruit carefully dried so as to retain their shape, although sometimes the fruit is sliced or halved to facilitate drying. Banana and plantain chips (crisps) are thinly sliced pieces of fruit fried in oil and eaten as a snack like potato chips (crisps). The main use of canned slices is in tropical fruit salads. Banana flakes are used as a flavouring or in breakfast cereals. Banana purée find use mainly in the production of baby foods. Banana flour is said to be highly digestible and is used in baby and invalid foods, but can also be used in the preparation of bread and beverages. Banana powder is used chiefly in the baking industry for the preparation and fillings for cakes and biscuits and is also used for invalid and baby foods.

***Processing Technology***

In general, to obtain a good-quality product from ripe-bananas the fruit is harvested green and ripened artificially under controlled conditions at the processing factory. After ripening, the banana hands are washed to remove dirt and any spray residues, and peeled. Peeling is almost always done by hand using stainless steel knives, although a mechanical peeler for ripe bananas has been developed, capable of peeling 450 Kg of fruit per hour. The peeling of unripe bananas and plantains is facilitated by immersing the fruit in hot water. For example, immersion in water at 70-75 ° C for 5 min. has been suggested as an aid for peeling green bananas for flour production, while the peeling of green bananas for freezing has been facilitated by immersion in water at 93° C for 30 min.

### *Banana Figs*

Fully ripe fruits with a sugar content of about 19.5 per cent are used and are treated with sulphurous acid after peeling, then dried as soon as possible afterwards. Various drying systems have been described using temperatures between 50 and 82° C for 10 to 24 hr to give a moisture content ranging from 8 to 18 per cent and a yield of dried figs of 12 to 17 per cent of the fresh banana on the stem. One factory in Australia uses a solar heat collector on the roof to augment the heat used for drying bananas. Bananas can also be dried by osmotic dehydration, using a technique which involves drawing water from 1/4-in. thick banana by placing them in a sugar solution of 67 to 70 deg. Brix for 8 to 10 hr. followed by vacuum-drying at 65 to 70° C, at a vacuum of 10 mm Hg for 5 hr. The moisture content of the final products is 2.5 per cent or less, much lower than that achieved by other methods.

### *Banana Puree*

Banana purée is obtained by pulping peeled, ripe bananas and then preserving the pulp by one of three methods: canning aseptically, acidification followed by normal canning, or quick-freezing. The bulk of the world's purée is processed by the aseptic canning technique. Peeled, ripe fruits are conveyed to a pump which forces them through a plate with 1/4-in. holes, then onto a homogeniser, followed by a centrifugal de-aerator, and into a receiving tank with 29in. vacuum, where the removal of air helps prevent discolouration by oxidation. The purée is then passed through a series of scraped surface heat exchangers where it is sterilised by steam, partially cooled, and finally brought to filling temperature. The sterilised purée is then packed aseptically into steam-sterilised cans which are closed in a steam atmosphere.

### *Banana Slices*

Several methods for canning of banana slices in syrup are used. Best-quality slices are obtained from fruit at an early stage of ripeness. The slices are processed in a syrup of 25 deg. Brix with pH about 4.2, and in some processes calcium chloride (0.2 per cent) or calcium lactate (0.5 per cent) are added as firming agents. A method for producing an intermediate-moisture banana product for sale in flexible laminate pouches has been developed. Banana slices are blanched and equilibrated in a solution containing glycerol (42.5 per cent), sucrose (14.85 per cent), potassium sorbate (0.45 per cent), and potassium metabisulphite (0.2 per cent) at 90 deg. C for 3 min. to give a moisture content of 30.2 per cent.

### *Banana Powder*

In the manufacture of banana powder, fully ripe banana pulp is converted into a paste by passing through a chopper followed by a colloid mill. A 1 or 2 per cent sodium metabisulphite solution is added to improve the colour of the final product. Spray- or drum-drying may be used, the latter being favoured as all the solids are recovered. A typical spray dryer can produce 70 kg powder per hour to give yields of 8 to 11 per cent of the fresh fruit, while drum-drying gives a final yield of about 13 per cent of the fresh fruit. In the latter method the moisture content is reduced to 8 to 12 per cent and then further decreased to 2 per cent by drying in a tunnel or cabinet dryer at 60° C.

### *Banana Flour*

Production of flour has been carried out by peeling and slicing green fruit, exposure to sulphur dioxide gas, then drying in a counter-current tunnel dryer for 7 to 8 hr. with an inlet temperature of 75° C and outlet temperature of 45° C, to a moisture content of 8 per cent, and finally milling.

### *Banana Chips (Crisps)*

Typically, unripe peeled bananas are thinly sliced, immersed in a sodium or potassium metabisulphite solution, fried in hydrogenated oil at 180 to 200° C, and dusted with salt and an antioxidant. Alternatively, slices may be dried before frying and the antioxidant and salt added with the oil. Similar processes for producing plantain chips have been developed.

### *Banana Beverages*

In a typical process, peeled ripe fruit is cut into pieces, blanched for 2 min. in steam, pulped and pectolytic enzyme added at a concentration of 2 g enzyme per 1 kg pulp, then held at 60 to 65° C and 2.7 to 5.5 pH for 30 min. In a simpler method, lime is used to eliminate the pectin. Calcium oxide (0.5 per cent) is added to the pulp and after standing for 15 min. this is neutralised giving a yield of up to 88 per cent of a clear, attractive juice. In another process banana pulp is acidified, and steam-blanched in a 28-in Hg vacuum which ensures disintegration and enzyme inactivation. The pulp is then conveyed to a screw press, the resulting purée diluted in the ratio 1:3 with water, and the pH adjusted by further addition of citric acid to 4.2 to 4.3, which yields an attractive drink when this is centrifuged and sweetened.

## PINEAPPLE

*Pineapple is Tropical Fruit:* It is a good source of Vitamin A and B and rich in Vitamin C and calcium. It contains phosphorus and iron. The fruit contains a special enzyme called 'Bromelin' which digests protein. Fruit is used for table purpose, preparations of juices, slices, tit bits, squash, jam, mixed jam, candy etc. Dried waste after extraction (pomace) is valuable cattle feed. Other byproducts are alcohol, calcium citrate, citric acid and vinegar. Leaves of pineapple are used for fibre preparation. Fruit of pineapple is botanically called sorosis. Botanical name of pineapple is Ananas comosus and family is Bromeliaceae.

*Climate:* Pineapple is tropical fruit. It requires moderate temperature and high humidity. The optimum temperate range is 210C to 230C. At low temperature, no fruit bud differentiation takes place. A well-distributed rainfall of 150 c. m. per year is adequate.

*Soil:* Crop is grown on light to medium soils. Soils should be well drained. Sandy loams, laterites and slightly acidic soils with pH 5.5 to 6.00 are suitable for crop cultivation.

*Propagation:* Pineapple is propagated by vegetative method, suckers, slips, crown and disc are used for its multiplication. Slips are best for planting.

*Planting:* For planting rainy season is the best, planting is done at beginning of monsoon in dry region and at the end of the monsoon in high rainfall area. Planting is done either in flat beds or in shallow trenches. The planting distance in the single row system should be 25 x 60 x 90 cm and in double row system 25 x 30 x 90cm.

*Manuring:* Pineapple is heavy feeder. It requires 600 kg N, 400 kg P and 400kg K per ha. Nitrogen is given in two split doses, first at begining of monsoon and second in the month of February.

*Harvesting and Yield:* It matures in both 15 to 20 months after planting. Usually flowering takes place from February to April and fruits are ready from July to September. The fruits are harvested when they just become yellow. An average yield is 10 to15 tonnes per hectare.

*Varieties:*

- *Gaint Kew or Kew:* It is one of the most important cultivars of Pineapple. It is ideal for canning industry. Flesh of the fruit is yellow, fibreless and juicy. The fruit is cylindrical in shape with average weight 1.9 kg. It is a late maturing variety.
- *Queen* : It is the second important cultivar which is grown in our country. It is an early maturing variety, used as table variety. It is mostly grown in Bengal region. The fruits are small cylindrical, with average weight 1.2 kg. Fruit colour is golden yellow while the flesh is deep golden yellow
- *Mauritius* : It is a mid season variety. Fruits of this variety are medium sized with yellow and red skin.

- Typhone No.1 and No.3
- Red spanish

## POTENTIAL FOR DECIDUOUS FRUITS

### AREA EXPANSION

Both North-West and North-Eastern regions of India offer large areas ideally suitable for cultivation of pome and stone fruits. In the North-West India, stone fruits like peach, plum and apricot come up well at elevations between 900 to 1500m with an annual rainfall of 90-100cm. Apple, cherry and pear are commercially successful at elevations between 1500-2700 m above msl. The cold arid regions between 1550-3650m with annual rainfall of 25-40 cm are again highly suitable for dry nuts and drying type of apricots. There are vast tracts of land still available for further expansion of these crops. In the North-Eastern Hills, excepting the rain shadow belts of Arunachal Pradesh and high altitude Lachung area of Sikkim, apple may not be successful commercially. Other crops like pear, peach, plum and apricot offer good scope for further expansion.

### Widening the Cultivar Base

The low productivity and poor quality of apple are often linked with monoculture of a few old cultivars and their degeneration over the years. The U.P hills, particularly Kumaon division, has the unique advantage of harvesting apple fruits for early market. Similarly, the rainshadow belts in North-Eastern Hills can offer good quality apple for the eastern Indian markets, thus reducing the cost of long distance transportation from North-Western Hills. Part of the markets of Bangladesh can be captured by the fruits of the North-East. There is good scope for introduction of new promising cultivars, replacing the Delicious group. Similarly, use of clonal rootstocks of Malling and Malling-Merton series and even their indexed material 'EMLA' selections will greatly change the productivity and quality of fruits. High density planting with spur type cultivars offer good scope. The identification of low chilling peach, plum and pear cultivars offer good possibilities for their cultivation in the low hills and in sub-tropical plains. Some of the new hybrids, including scab resistant apple cultivars need verification trials on a commercial scale.

### Management Practices

Scientific water management and practicing proper training and pruning of trees including introduction of renewal pruning techniques, will make significant impact on increased production even in the existing orchards. Drip irrigation, in-situ water harvesting and correcting macro and micro-nutrient deficiencies will go a long way in bringing notable improvement in productivity as well as fruit quality. By adopting IPM strategy and organic farming practices, selected export markets can be targeted well.

## Processed Products

There is immense scope for increasing various processed products of pome and stone fruits, for which technologies are available. The existing capacity of the most organized processing unit of HPMC is only marginal as compared to the volume of fruit available for processing. The HPMC utilized only 1 to 1.5 percent of total cull fruit available. The present combined capacity of two units of HPMC is 30,000 tons annually, which can be easily raised to 50 to 75 thousand tons. The expansion of grading and packing stations, their further modernization with mechanical grading equipment, use of CFB boxes and more number of pre-cooling and cold storage units will improve the marketing system and enhance marketability of the produce. Apple has been identified as one of 6 most promising fruits for fresh fruit exports.

## FRUIT CULTURE

*Introduction:* The Cashewnut (Anacardium occidentale L.) is one of the important nut tree grown in the tropical world. This fruit tree was introducted into Malbar coast of India in 16th century by Portuguese. It is the native of Brazil. It is grown mostly on West Coast area of Maharashtra, Goa, Karnataka, Kerala and Tamil Nadu.

*Importance:* The dry nuts - Kernels are of high nutritive value and also rich in protein, carbohydrates, minerals like calcium, phosphorus and iron. This has a great export value and is considered as one of the important horticultural crop in India.

*Climate:* The Cashew requires a minimum rainfall of 60 cm per annum, but can stand extremes of rainfall from 20 cm to 400 cm. If there is sufficient water supply, it can withstand a long period of dry spell and low humidity. The Cashew is a sun loving tree and does not tolerate excessive shade, also it does not favour very high temperatures above 45 deg cent. During the fruit set and development. Heavy rains and cloudy weather during flowering adversely affect the yield.

*Soil:* The Cashew is cultivated on a wide variety of soils. It is considered a crop of marginal land and is recommended for slopy and light soils. The best soils for better production are deep, friable, well drained and without a hard pan upto 2/3 m in depth. The Cashew is mainly grown on Laterite, red and coastal sands in India.

*Varieties:* The following are the important varieties of Cashew grown in different parts of the country.

- Vengurla 1
- Vengurla 2
- Vengurla 3
- Vengurla 4
- Vengurla 5
- Vengurla 6

- Bhubaneshwar
- Kanaka
- Dhana

x. Selection

*Propagation:* The cashew is grown by the following methods:

- *Seed Propagation:* It is the oldest and cheapest method of propagation. It is also used to raise the plants for the purpose of grafting.
- Layering this method is more successful in coastal and more humid areas.
- *Soft Wood Grafting:* This is followed both in the nursery as well as in the field of in-situ. For this purpose, veener grafting is done.
- *Top-working:* This method is used not for propagation but for converting inferior free into desirable one.

*Planting and Season:* Normally cashew grafts are planted at the spacing of 7.0 x 7.0 or 7.5 x 7.5 or 8.0 x 8.0 m on plain land or on desirable contours on slopy lands. After marking, the pits of 60x60x60 cm are dug and filled with a compost,top soil, single super phosphate and karnaj cake. This is to be well before the monsoon starts. Planting should be done during June-July *i.e.* in the beginning of monsoon.

*Interculturing:*

- Removal of weeds is done once twice
- Intercrops are planted in interspace
- Some perenial forest plants and/or some seasonal crops can be taken as per the local need and situation.

*Care of young orchard:*

- Gap filling for the missing plants
- Removal of outgrowth on stocks in case of grafted plants
- Staking with bamboos.
- Cover cropping
- Providing protective irrigation during first few summers. These are some of the points for young plantations.

*Special Horticultural practices:*

- Pruning of dead and dried shoots alongwith crisscross branches and water shoots
- Pruning of leader shoots in June followed by 2% $KNO_3$ spray
- Spraying with 10 ppm NAA twice during flowering for increasing fruit set and minimizing flower and fruit drop.

*Irrigation:* The cashew is mostly grown as the rainfed crop and requires no irrigation in the high rainfall areas. However, if the rainfall is low, the cashew responses well to irrigation water at the time of fruit set upto full development stage of nuts. Irrigation should not be given before or at the time of flowering, as it is likely to promote vegetative growth.

*Nutrition:* To improve the growth and yield the cashew trees should be manured and fertilized with bio-fertilizers like biomeal and NpK 500,200, 200 gm per tree per year. Spraying with ultrazyme and 8:12:24:4 NPKmg twice, during fruit development is useful to increase size and yield of nuts.

*Plant Protection:* The crop particularly blossom and fruits should be protected against Tea Mosquito, Stem and root borer is also an important pest. Leaf minor, Leaf and blossom webber, Flower thrips are the common pests in neglected orchards. Inflorescence blight, pink disease, and anthrac nose are some diseases found on cashew. Clean cultivation, pruning, better aeration and preventive measures should be adopted besides occassional sprays of selected pesticides and fungicides.

*Harvesting and Yield:* The Cashew comes to its full bearing stage at the age of 10th year and continuous to give yield further for 30-40 years. Usually the nuts are collected after they fall off from the tree. They are separated from apples and dried for 2/3 days to reduce the moisture. Regarding yield there is a wide variation and differences of opinions. However, the desirable variety at good management yields 50 kg of raw nuts.

*Post Harvest Handling:* The processing of raw nuts involves roasting, shelling, drying, peeling, grading and packing `A' grade nuts are always in demand from the outside markets and fetch a good price. A Fenny like drinks are prepared from the apples.

# Index

## M



## N



## P



## R



## S



## T



## V



## W